Engineering Statics with MATLAB®

This text makes use of symbolic algebra and vector-matrix algebra to demonstrate a new approach to learning statics. Symbolic solutions are obtained, together with the types of solutions covered in other texts, so that students can see the advantages of this new approach.

This innovative text is an extension of second-generation *vector Statics* courses to a new, third-generation *matrix-vector Statics* course, a course that addresses deformable as well as rigid bodies and employs MATLAB®.

MATLAB® is used as a "calculator" whose built-in functions are used to solve statics problems. This text uses vectors and matrices to solve both statically determinate rigid body problems and statically indeterminate problems for deformable bodies.

The inclusion of statically indeterminate problems is unique to this text. It is made possible by using symbolic algebra and a new, simplified vector-matrix formulation that combines the equations of equilibrium, the homogeneous solutions to those equations, and a description of the flexibilities found in the deformable elements of a structure to solve directly for the unknown forces/moments.

Lester W. Schmerr Jr. holds a PhD in Mechanics from the Illinois Institute of Technology and a BS in Aeronautics and Astronautics from MIT. He is a Professor Emeritus at Iowa State University, where he taught and conducted research for four decades.

Advances in Applied Mathematics

Series Editor: Daniel Zwillinger

Experimental Statistics and Data Analysis for Mechanical and Aerospace Engineers
James Middleton

Advanced Engineering Mathematics with MATLAB®, Fifth Edition
Dean G. Duffy

Handbook of Fractional Calculus for Engineering and Science
Harendra Singh, H. M. Srivastava, Juan J. Nieto

Advanced Engineering Mathematics
A Second Course with MATLAB®
Dean G. Duffy

Quantum Computation
Helmut Bez and Tony Croft

Computational Mathematics
An Introduction to Numerical Analysis and Scientific Computing with Python
Dimitrios Mitsotakis

Delay Ordinary and Partial Differential Equations
Andrei D. Polyanin, Vsevolod G. Sorkin, Alexi I. Zhurov

Clean Numerical Simulation
Shijun Liao

Multiplicative Partial Differential Equations
Svetlin Georgiev and Khaled Zennir

Engineering Statics with MATLAB®
Lester W. Schmerr Jr.

https://www.routledge.com/Advances-in-Applied-Mathematics/book-series/CRCADVAPPMTH?pd=published,forthcoming&pg=1&pp=12&so=pub&view=list

Engineering Statics with MATLAB®

Lester W. Schmerr Jr.

CRC Press
Taylor & Francis Group
Boca Raton London New York

CRC Press is an imprint of the
Taylor & Francis Group, an **informa** business

A CHAPMAN & HALL BOOK

First edition published 2024
by CRC Press
2385 Executive Center Drive, Suite 320, Boca Raton, FL 33431

and by CRC Press
4 Park Square, Milton Park, Abingdon, Oxon, OX14 4RN

CRC Press is an imprint of Taylor & Francis Group, LLC

ISBN: 978-1-032-43700-2 (hbk)
ISBN: 978-1-032-44526-7 (pbk)
ISBN: 978-1-003-37259-2 (ebk)

DOI: 10.1201/9781003372592

Typeset in Palatino
by MPS Limited, Dehradun

Access the Instructor Resources: www.routledge.com/9781032437002

Contents

Preface

This book is based on a course in Engineering Statics I have taught and developed for over 40 years at Iowa State University. Since statics is an entry-level course in engineering that has a highly developed role and place in an engineering curriculum, the topics covered are many of the same found in a wide variety of universities and contained in many textbooks. However, the approach here has some new and unique aspects that I would like to highlight. Statics texts have changed little over the years since the 1960s when I was first learning the subject. Certainly, changes in homework problems have reflected the emergence of hand-held calculators and personal computers in engineering. A stronger emphasis on student participation has reflected a new understanding of the role of active learning in the material being presented. But the basic structure and content of the course have remained relatively static (pun intended) for many decades – a situation that I have sought to address. Let me explain.

In the late 1950s and early 1960s, in response to Sputnik and the space race, engineering curricula developed a new, more heavily math- and science-based approach. In statics, that push was reflected in the introduction of vectors and vector algebra into "second generation" statics texts. Since that time, there have been many changes to how students work and learn. The introduction of the hand-held calculator and the personal computer has been a major part of those changes. But the development of computer software environments such as MATLAB® has also had a major impact. In MATLAB, for example, students and practicing engineers now have access to a wide range of tools for solving the statics problems of the past as well as new problems. MATLAB is also ideally suited for solving problems described in vector-matrix and symbolic terms. In this book, we will closely incorporate MATLAB into the statics teaching and learning environment. This will let us broaden the horizons of statics without sacrificing the "traditional" content of statics courses that educators (but not the students, perhaps) cherish. This new approach will include:

Using symbolic algebra to define and solve problems, compute integrals, etc.

Dealing with matrices as well as vectors.

Examining systems and structures from a complete standpoint rather than emphasizing problems where, for example, a student is asked to find only a particular force or moment within an entire structure.

Dealing with statically indeterminate problems as a precursor to later courses such as Strength of Materials.

The symbolic algebra capabilities of MATLAB warrant special mention. Statics problems have traditionally been done with paper and pencil, and consequently, much attention is often paid to finding "efficient" solutions for unknowns that avoid having to solve simultaneous equations. With MATLAB, it is possible to write the equations of equilibrium in a form that is almost identical to what is placed on paper and then solve them in that symbolic form, regardless of the number of equations. The answers can either be numerical or in terms of some symbolic variables of the problem, for which specific values can then be substituted. This makes statics less about finding "clever" solution paths through a problem to more about accurately representing the problem from free body diagrams and then letting MATLAB handle the solution process. If MATLAB's symbolic math toolbox capabilities are not available, then one can use the numerical matrix-vector algebra capabilities of MATLAB in much the same fashion. For testing purposes, of course, one may design problems that do not need symbolic or matrix-vector capabilities, but having those capabilities offers many new possibilities for the examples one covers in class as well as the homework that is assigned. Just as in the past there came a time to replace the graphical methods taught in statics with vector algebra, it is time to move from vector algebra to matrix-vector algebra and symbolic methods, a move that can be effectively done with MATLAB.

The inclusion of statically indeterminate problems in statics is unusual, so let me explain how it is possible. Traditionally, statics has only dealt with rigid bodies where the equations of equilibrium are all that are needed to solve problems that are statically determinate. Those equations can be written in matrix-vector form as $[E]\{F\} = \{P\}$, where $[E]$ is the equilibrium matrix, $\{F\}$ is a vector of the unknown forces/moments, and $\{P\}$ is a vector of known forces/moments. Statically indeterminate problems have not been considered because of the necessity of also describing deformations and their relationship to displacements and forces/moments, requiring a more detailed look at topics that are thought better suited to courses such as Strength of Materials. However, a recent and deeper understanding of statically indeterminate problems has shown that, besides the equilibrium matrix, the only thing we really need to solve for all the unknown forces is a flexibility matrix that relates the forces to the deformations [1]. If one stays with simple problems such as systems with springs, 2-D and 3-D trusses, and wire-supported structures (problems that are commonly used examples in statics anyway), the flexibility matrix is a diagonal matrix that is simple to obtain and whose physical meaning is also easy to understand at the level of a statics course. To solve statically indeterminate problems, it is well known that one must augment the equilibrium equations with compatibility equations for the deformations. Normally, compatibility equations are obtained by looking at the conditions the geometry places on the deformations. *Compatibility equations, however, can also be obtained directly from homogeneous solutions of the underdetermined equilibrium equations found in statically indeterminate problems!* Relating the deformations to the forces through the flexibility matrix gives the additional equations needed to turn the underdetermined equilibrium equations into a determinate set. Solving the homogeneous underdetermined equilibrium equations is no more difficult than solving the equilibrium equations for statically determinate problems (which we do all the time in statics), and MATLAB has built-in methods one can use in both cases. In addition, the equilibrium matrix and the flexibility matrix are all that are needed to determine the displacements present in a deformable structure. In problems such as statically indeterminate trusses, for example, one can then illustrate the geometry changes of the truss. Thus, with a modest knowledge of matrix-vector algebra and an understanding of the roles that equilibrium, compatibility, and flexibilities play in deformable bodies, statically indeterminate problems can and should be included in statics. This provides engineering students with a new and practical understanding of the statics of both rigid and deformable bodies and better paves the way for later courses such as Strength of Materials. This extension of a traditional statics course can only be found in this text.

Some may ask if such changes to a statics course can be made to an already full course. The answer is yes. Example problems are solved using the traditional approaches found in other texts alongside solutions obtained with MATLAB. In this way, students can see the advantages that MATLAB provides. All of this can be done without sacrificing the traditional learning objectives and content of the course. It should be noted that *the approach of this text does not require students to learn MATLAB programming*. MATLAB is instead used here as a powerful "calculator" where students need only learn how to manipulate the built-in matrix-vector and symbolic functions of that "calculator", just as they learn how to manipulate the built-in functions of a hand-held calculator.

The text includes the topics most instructors cover in a statics course, but there are additional topics that one might consider. These have been included in an addendum to the text that is available for students and instructors on the website www.eng-statics.org. There are four additional chapters in that addendum on (1) fluid pressure forces and forces in cables, (2) the principle of virtual work for rigid bodies, (3) work-energy principles for deformable bodies and stability, and (4) an introduction to the finite element method. Virtual work principles for rigid bodies, fluid pressure forces, cables, and stability are topics that can often be found in other texts but are likely covered lightly, if at all, in a statics course. With the addendum, one or more of these topics can be readily included. The work-energy principles for deformable bodies and a very gentle introduction to finite elements are present to introduce statics students to topics that are foundations of some of the most powerful tools used in engineering analysis and practice.

There are also differences between the homework problems here and those found in other texts. There are three types of homework problems. First, there are multiple-choice problems. These are generally simple problems that can be worked out in class. In my classes, I have used "clickers" or learning software (such as Top Hat [2]) and cell phones to get and tally the students' answers to these types of problems. However, these multiple-choice problems also can be used by students to validate their basic understanding of how to solve problems. The second type of homework problem generally asks for more results than found in the multiple-choice problems. They often include asking for specific forces or moments, as found in other texts, but they also may include determining complete solutions involving all the unknowns, which generally is best done with MATLAB. An example of the latter case would be obtaining a solution for all the member forces in a truss and the reactions using the method of pins (joints). Like the examples given in the text, many homework problems are solved both by "hand" (i.e., with a calculator) and with MATLAB. This approach again shows the value that MATLAB has for solving many of the problems we deal with in statics. The third type of homework problem I have called review problems – two- or three-step problems that are like what might be placed on a test and do not require MATLAB. Answers to all the homework problems except the multiple-choice problems and those problems having graphical answers are given at the end of the text and at the end of the addendum.

Another unique feature for instructors is that all the homework problems that have numerical answers also include a MATLAB script that solves the problem with either the parameters that are given in the text or for other values that can easily be specified before running the script. This allows instructors to "refresh" almost all the homework problems without waiting for a new edition of the text. The homework solutions are available to instructors in a PDF solutions manual as well as individual PowerPoint slides. All instructor materials are available from the publisher. Lecture slides (with commentary notes) that present the content of the text and an addendum are also available for both students and instructors on the author's website, www.eng-statics.org, as is a tutorial on solving statics problems with MATLAB and other course materials.

[1] Schmerr, L.W., *Advanced Mechanics of Solids – Analytical and Numerical Solutions with MATLAB®*, Cambridge University Press, Cambridge, U.K., 2021.
[2] https://tophat.com

1

Introduction to Statics and the Concept of Force

OBJECTIVES

- To provide an overview of the basic quantities and numerical calculations used in statics.
- To define forces as vectors and the use of the parallelogram law of addition.
- To express a vector in terms of its Cartesian components or in terms of its magnitude and a unit vector that defines its direction.
- To introduce the notations used for scalars, vectors, and matrices.
- To define the dot product and its relationship to direction cosines.
- To describe the use of MATLAB® in numerical and symbolic solutions.
- To introduce position vectors.

1.1 About This Book

Archimedes (287–212 BCE) is often considered to be the founder of statics, making statics one of the oldest engineering subjects. He described the principle of the lever, the concept of the center of gravity, and the principle of buoyancy for floating objects, all of which are part of the foundations of statics. Since much of the subject matter in statics is based on centuries of experience, statics texts are often highly uniform in the topics that they cover, but the way in which problems are solved has changed significantly over time. Early statics texts, for example, solved many problems with the

DOI: 10.1201/9781003372592-1

use of geometrical methods that could be executed by hand with paper and pencil graphics and with drawing instruments. From the late 1950s to early 1960s, in response to Sputnik and the space race, engineering curricula developed a new, more heavily math and science-based approach. In statics, those changes were reflected in the introduction of vectors and vector algebra. The emergence of the hand-held calculator and personal computers has also been a major part of those changes. More recently, the development of computer software environments has also had a significant impact. In MATLAB, for example, students and practicing engineers now have access to a wide range of numerical and symbolic tools for solving the statics problems of the past as well as new problems. In this course, we will incorporate MATLAB into both how subjects are presented and how problems are solved. These changes will be reflected in using matrices as well as vectors, so it is important that you are familiar with the properties of matrices and how to obtain matrix-vector solutions. Appendix A in this book can give you much of the background that is needed. The matrix-vector approach to statics using MATLAB found in this text is the latest example of the evolution of statics over many, many years.

Traditionally, statics has only dealt with rigid bodies (bodies whose size and shape are fixed) and examined the conditions under which a rigid body remains at rest. Dynamics of rigid bodies, in contrast, studies their motion. While the main emphasis in this book is on the statics of rigid bodies, we will also examine the equilibrium of bodies that can deform. Normally, this topic is left to later courses such as strength of materials, but we will show you how we can use the conditions of equilibrium and the "flexibility" of a deformable body to solve statics problems that are not able to be solved by equilibrium conditions alone. We will see how even very small deformations that are present in a structure play a crucial role in the ability of an engineer to analyze the statics of those structures. Thus, in this book, we will give you a glimpse of how statics is applied in engineering practice to the many complex structures and systems found in the world.

1.2 Quantities and Units

Statics is concerned with the conditions under which structures are at rest (in "equilibrium"). The basic quantities we will deal with in statics are:

Scalar Quantities
lengths, angles

Vector Quantities
forces, moments and couples, positions, displacements

Matrix Quantities
area moments, mass moments of inertia, equilibrium matrices, stiffness matrices, flexibility matrices, compatibility matrices

The fundamental units of the quantities found in statics are normally given either in the U.S. Customary System of Units or the International System of Units (SI):

Length
U.S. Customary: foot (ft), inch (in.)
Metric (SI): meter (m), centimeter (cm), millimeter (mm)
where 1 in. = 25.4 mm, 1 m = 39.37 in. = 3.281 ft

Angle
Both systems: radian, degree
where 1 radian = $360/2\pi$ degrees = 57.3 degrees

Force
U.S. Customary: pound (lb)
Metric (SI): Newton (N)
where 1 lb = 4.448 N, 1 N = 0.2248 lb

Moment
U.S Customary: foot-pounds (ft-lb) or inch-pounds (in-lb)
Metric (SI): Newton-meters (N-m)

SI System: In the SI system, the fundamental unit of mass is the kilogram (kg). The force, F, measured in newtons (N) is obtained from the mass through Newton's law $F = ma$, where a is an acceleration measured in m/sec^2 as follows:

$$\text{Force} = \text{mass} \times \text{acceleration}$$

$$1 \text{ newton} = (1 \text{ kg})(1 \text{ m/sec}^2)$$

The weight, W, of a body is a force given by $W = mg$, where m is the mass in kg and $g = 9.80665$ m/sec^2 is the acceleration of gravity on the surface of the earth (usually used in engineering calculations as $g = 9.807$ or 9.81)

U.S. Customary System: In the U.S. Customary System, the fundamental unit of force is the pound (lb). The mass in this system is called the slug, where

$$\text{Force} = \text{mass} \times \text{acceleration}$$

$$1 \text{ lb} = (1 \text{ slug}) (1 \text{ ft/sec}^2)$$

On the surface of the earth, the acceleration of gravity $g = 32.1740$ ft/sec^2 (usually used as $g = 32.2$ in engineering calculations).

Example 1.1

Suppose I have 454 grams of sugar. What is its weight in Newtons? In pounds?

The weight is $W = (0.454$ kg$) (9.81$ m/s$^2) = 4.454$ Newtons. In terms of pounds, $W = (4.454$ N$)(0.2248$ lb/N$) = 1.00$ lb. [Note that is common to say that one pound of sugar "weighs" 454 grams but grams is a measure of mass and pounds is a measure of force, so we are really mixing two different units.]

Example 1.2

Suppose I have a 100 lb sack of flour. What is its mass in slugs? In kg?

We have $(100$ lb$) = (m$ slugs$)(32.2$ ft/sec$^2)$, so $m = 100/32.2 = 3.11$ slugs. Since $(100$ lb$) (4.448$ N/lb$) = 444.8$ N, the mass is $m = 444.8/9.807 = 45.4$ kg.

1.3 Numerical Calculations

Numerical engineering calculations in statics are now routinely done with pocket calculators and computers. This technology has made it possible to evaluate results to many decimal places. In MATLAB, for example, numerical calculations are done in floating point double precision, which gives 15–17 decimal digits of precision. The accuracy of most engineering calculations, however, is normally much less than this machine-based precision and is reflected in the number of *significant digits* in the result. In *scientific notation* (where a number is expressed in the form $a \times 10^b$, with $1 \le a < 10$), the number of significant digits (or *significant figures*) is the number of digits needed to express the value of a within a specified level of accuracy. For example, for a value $(5.123 \pm 0.0004) \times 10^3$, four significant figures would be sufficient. In this book, we will normally give answers to problems done "by hand" (i.e., with the aid of a pocket calculator) to 3–4 significant figures, with more significant figures present in any interme-diate calculations (to prevent roundoff errors from accumulating). In solutions done with MATLAB, numerical answers are typically displayed in a "short format" where four decimal places are present. We will leave such answers in that format with the understanding that those MATLAB answers may not reflect the actual significant figures present.

1.4 Forces and Vectors

A force is a measure of a "push" or "pull" action exerted on a structure or system. It is a vector quantity having both a magnitude and a direction and satisfies the parallelogram law of addition (that we will define shortly). In two dimensions, we can characterize a vector as a directed arrow, having a length of a given dimension (called the magnitude), such as the 10 lb magnitude of the force shown in Figure 1.1, and whose direction can be specified by the two angles (θ_x, θ_y) that the arrow makes with respect to a set of x- and y-rectangular (Cartesian) axes. These angles are not independent since

$$\cos^2 \theta_x + \cos^2 \theta_y = 1 \tag{1.1}$$

In three dimensions, we can also characterize a vector by a directed arrow having a length (magnitude) and whose direction is described by the three angles $(\theta_x, \theta_y, \theta_z)$ with respect to the x-, y-, and z-rectangular (Cartesian) axes, such as the force of 10 lb magnitude shown in Figure 1.2. These three angles are also not all independent since they must satisfy

$$\cos^2 \theta_x + \cos^2 \theta_y + \cos^2 \theta_z = 1 \tag{1.2}$$

The cosines of the angles appearing in Eq. (1.1) and Eq. (1.2) are called *direction cosines*. We will discuss them more shortly.

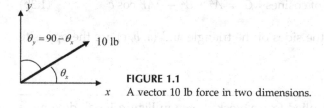

FIGURE 1.1
A vector 10 lb force in two dimensions.

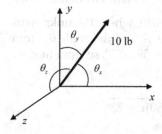

FIGURE 1.2
A vector 10 lb force in three dimensions.

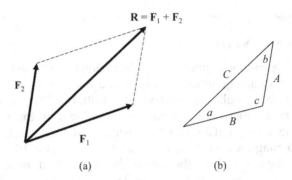

(a)   (b)

FIGURE 1.3
(a) The parallelogram law of addition of vectors. (b) An oblique triangle.

Forces obey the *parallelogram law of addition*. The sum of two or more forces is called the *resultant force*. Figure 1.3(a) shows the resultant, **R**, of two forces, (**F**$_1$, **F**$_2$), which lies in the plane of those two forces. The resultant **R** = **F**$_1$ + **F**$_2$ forms the diagonal of the parallelogram shown in Figure 1.3(a), which is why this type of addition is called the parallelogram law. Because the resultant and its constituent forces form an oblique triangle (Figure 1.3(b)), we can use the *law of sines* and the *law of cosines* to effectively solve force addition problems. For the oblique triangle of Figure 1.3(b), we have

$$\text{Law of sines} \quad \frac{\sin a}{A} = \frac{\sin b}{B} = \frac{\sin c}{C} \quad (1.3a)$$

$$\text{Law of cosines} \quad C^2 = A^2 + B^2 - 2AB\cos c \quad (1.3b)$$

where (A, B, C) are the sides of the triangle and (a, b, c) are the opposite angles.

Example 1.3

For the two forces applied to the block shown in Figure 1.4(a), determine the magnitude of the resultant force and the angle it makes from the x-axis.

The oblique force triangle is shown in Figure 1.4(b), where the unknowns we seek are the magnitude of the resultant force, R, and the angle θ. From the geometry, the angle a is given as $a = 180 - 48 - 59 = 73°$, so from the law of cosines

$$R = \sqrt{780^2 + 650^2 - 2(780)(650)\cos 73°} \quad (1.4)$$
$$= 857 \text{ N}$$

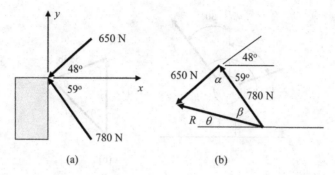

FIGURE 1.4
(a) Forces acting on a block. (b) The geometry for finding the resultant, R.

From the law of sines

$$\frac{\sin \beta}{650} = \frac{\sin 73°}{857} \quad \to \quad \beta = 46.5° \tag{1.5}$$

and from the geometry

$$\theta + \beta = 59° \quad \to \quad \theta = 12.5° \tag{1.6}$$

_____

This last example was straightforward, but we should realize that in a force addition problem, there may be more than one possible solution, as the next example shows.

Example 1.4

Determine the tension T and the angle θ for the forces acting on a support as shown in Figure 1.5(a) so that the resultant force is 100 lb acting vertically.

From the force triangle of Figure 1.5(b) and the law of sines

$$\frac{\sin \beta}{100} = \frac{\sin 30°}{55} \to \beta = 65.4°$$
$$\theta = 180 - \beta - 30 = 84.6° \tag{1.7}$$

and from the law of cosines

$$T = \sqrt{100^2 + 55^2 - 2(100)(55)\cos(84.6°)}$$
$$= 109.5 \text{ lb} \tag{1.8}$$

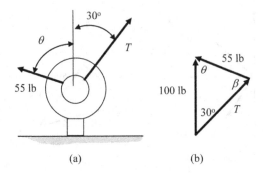

FIGURE 1.5
(a) Forces acting on a support, generating a vertical 100 lb force. (b) The oblique force triangle.

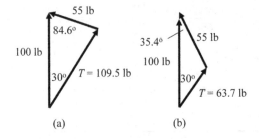

FIGURE 1.6
(a) One solution to the force addition problem of Figure 1.5. (b) A second solution.

which gives the solution shown in Figure 1.6(a). However, since $\sin(180 - \beta) = \sin\beta$, $\beta = 114.6°$ is also a solution and in this case $\theta = 180 - \beta - 30 = 35.4°$. From the law of cosines

$$T = \sqrt{100^2 + 55^2 - 2(100)(55)\cos(35.4°)}$$
$$= 63.7 \text{ lb}$$

(1.9)

and this solution is shown in Figure 1.6(b). Although either solution is acceptable, this second solution is likely better since it requires a smaller tension to produce the 100 lb force.

Even when we have multiple solutions to a force addition problem, we may not be able to accept all the possible solutions, as this next example shows.

Example 1.5

Determine the magnitude of the force, **F**, and the angle, θ, in Figure 1.7(a) so that the sum of the two forces shown has a magnitude of 20 lb and is directed horizontally to the left.

The oblique force triangle is shown in Figure 1.7(b). From the law of cosines

$$F = \sqrt{20^2 + 100^2 - 2(20)(100)\cos 70^\circ}$$
$$\rightarrow \quad F = 95 \text{ lb} \tag{1.10}$$

If we use the law of sines

$$\frac{\sin \theta}{100} = \frac{\sin 70^\circ}{95}$$
$$\rightarrow \quad \theta = 81.41^\circ \text{ or } \theta = 98.59^\circ \tag{1.11}$$

but also from the law of sines

$$\frac{\sin \beta}{20} = \frac{\sin 70^\circ}{95}$$
$$\rightarrow \quad \beta = 11.41^\circ \text{ or } \beta = 168.59^\circ \tag{1.12}$$

The angle $\beta = 168.59^\circ$ is clearly not possible since the sum of the interior angles of the triangle would be greater than 180°, so we must have $\beta = 11.41^\circ$. This in turn requires that $\theta = 98.59^\circ$ and not $\theta = 81.41^\circ$ so there is only one acceptable answer. The interesting thing about this problem is that if you use a calculator or any software program (including MATLAB) to calculate the angle θ from the law

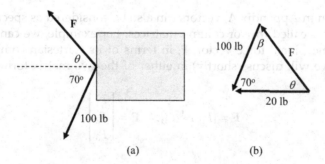

(a) (b)

FIGURE 1.7
(a) Two forces applied to a block, generating a 20 lb force acting to the left. (b) The force triangle.

of sines, as done above, you will always get the wrong answer of
$\theta = 81.41°$! This is because the solution to the inverse sine function
is typically given as a "principal angle", which lies between $-90°$ and
$+90°$, and we must choose a value outside that range to solve this problem.

1.5 Scalars, Vectors, and Matrices

Before proceeding with our discussion of forces, in this section, we want
to say a few words about notation for the quantities appearing in this
text. Most statics texts describe the magnitude of a vector by an un-
bolded lower-case or upper-case symbol such as a or A, while the vector
itself is given in boldface letters such as **a** or **A**. Occasionally, vectors are
also written as a symbol (or symbols) with an arrow overhead, such as
$\overrightarrow{AB}$, for example. The magnitude of a vector is described by a numerical
value (usually with some dimension) and is called a scalar quantity while
a vector quantity, as already discussed, has both a magnitude and a
direction. In the following chapters, we will also use matrices and matrix
algebra to formulate and solve statics problems. Appendix A gives an
overview of the properties of matrices that should provide sufficient
background for the topics found in this book. The components of
matrices will be expressed as a square or rectangular array of numbers
(or symbols) arranged in a series of rows and columns such as

$$\begin{bmatrix} 3 & 6 & 4 \\ 6 & 2 & -1 \\ 4 & 2 & 5 \end{bmatrix}, \quad \begin{bmatrix} 1 & 3 & 5 \\ 4 & 4 & 2 \end{bmatrix}, \quad \begin{bmatrix} 1 & 3 & 6 \\ 9 & 2 & 2 \\ 4 & 3 & 5 \\ 3 & 5 & 7 \end{bmatrix}$$

As shown in Appendix A, vectors can also be considered as special cases
of matrices called row or column matrices. For example, we can write a
three-dimensional force vector, **F**, in terms of its Cartesian components
(which we will discuss shortly) in either of the two matrix forms:

$$\mathbf{F} = [F_1 \ F_2 \ F_3], \quad \mathbf{F} = \begin{bmatrix} F_1 \\ F_2 \\ F_3 \end{bmatrix}$$

where the first form is a row matrix representation of the vector and the
second form is a column matrix representation of the same vector and

these two forms are related through the matrix transpose (T) operation discussed in Appendix A, i.e.,

$$\begin{bmatrix} F_1 \\ F_2 \\ F_3 \end{bmatrix} = [F_1 \ F_2 \ F_3]^T, \quad [F_1 \ F_2 \ F_3] = \begin{bmatrix} F_1 \\ F_2 \\ F_3 \end{bmatrix}^T$$

In MATLAB, these vectors can be directly entered (and echoed) in either of these forms as, for example,

```
F = [2 1 4]            F = [2; 1; 4]
F =          or        F = 2
    2 1 4                  1
                           4
```

where a semi-colon is used to start a new row. The transpose can always be used to convert one form to the other with the MATLAB symbol " .' ". For vectors or matrices composed of real numbers or real symbolic variables, the symbol " ' " can be used instead (i.e., without the "dot") but for matrices or vectors composed of complex numbers or symbolic variables, the use of " ' " gives the complex conjugate as well as the transpose, as described in Appendix A. In statics, we do not deal with complex numerical values but we will deal with symbolic values and symbolic vectors and matrices, so we either need to declare any symbolic variables as real variables (see Appendix A) or always use the " .' " for the transpose. We will see examples that use these choices shortly.

If **F** is the row vector just given, for example, the transpose gives us the same vector as a column vector:

```
F.'
F = 2
    1
    4
```

To explicitly indicate a vector quantity is in a column format (without showing the individual components), we will use a "curly" bracket notation and write a column vector as, for example, {F}, and its corresponding row vector form as {F}^{T}. Matrices will almost always be written in this book as upper-case symbols in square brackets such as [A], but they are also commonly written in other texts as A or **A**, depending on the context. These notations and basic matrix algebra can be found in Appendix A. MATLAB is a software package that uses matrices as its fundamental data type so it is especially useful, as we will see, for defining and manipulating matrices and vectors and it will be used throughout this book. Appendix A also describes many of the matrix operations we will use in MATLAB. You are strongly encouraged to read Appendix A to become familiar with matrix notation and matrix algebra.

1.6 Unit Vectors and Cartesian (Rectangular) Components

A *unit vector* is simply a dimensionless vector with a magnitude of one. Given any vector **A** whose magnitude is A, a unit vector, $\mathbf{e}_A$, in the direction of **A** is given by

$$\mathbf{e}_A = \mathbf{A}/A \tag{1.13}$$

(see Figure 1.8). Three unit vectors of particular significance are the unit vectors (**i**, **j**, **k**) acting along the x-, y-, and z-axes of a Cartesian (rectangular) coordinate system, as shown in Figure 1.9. Statics texts will sometimes write these unit vectors as either ($\mathbf{e}_x$, $\mathbf{e}_y$, $\mathbf{e}_z$) or ($\hat{\mathbf{i}}$, $\hat{\mathbf{j}}$, $\hat{\mathbf{k}}$), where the caret is used to explicitly denote that these are unit vectors. These coordinate unit vectors are fundamental in the representation of a vector in terms of its Cartesian coordinates. Consider, for example, the vector **F** shown in Figure 1.10. The origin of a Cartesian coordinate

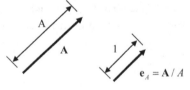

FIGURE 1.8
A vector, **A**, having a magnitude A and a unit vector, $\mathbf{e}_A$, in the direction of **A**.

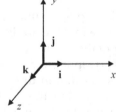

FIGURE 1.9
A Cartesian coordinate system with unit vectors (**i**, **j**, **k**) acting in the x-, y-, and z-directions.

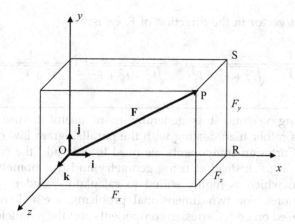

FIGURE 1.10

A vector F represented in terms of its components (F_x, F_y, F_z) in a Cartesian coordinate system.

system is placed at the start of this vector and a box is drawn with the vector as the main diagonal of that box. By the vector law of addition, the vector can then be represented as the sum of the directed line segments (OR, RS, SP) as

$$\mathbf{F} = \overrightarrow{OP} = \overrightarrow{OR} + \overrightarrow{RS} + \overrightarrow{SP} \tag{1.14}$$

If the lengths of the sides of the box are labeled as (F_x, F_y, F_z), then combining these with the unit vectors $(\mathbf{i}, \mathbf{j}, \mathbf{k})$, Eq. (1.14) can be written as

$$\begin{aligned}
\mathbf{F} &= F_x\mathbf{i} + F_y\mathbf{j} + F_z\mathbf{k} \\
&= \mathbf{F}_x + \mathbf{F}_y + \mathbf{F}_z
\end{aligned} \tag{1.15}$$

where (F_x, F_y, F_z) are called the Cartesian (rectangular) scalar components of F and $(\mathbf{F}_x, \mathbf{F}_y, \mathbf{F}_z)$ are called the vector Cartesian (rectangular) components of F. It follows that the magnitude of F, $F \equiv |\mathbf{F}|$, satisfies:

$$\begin{aligned}
F^2 &= |\mathbf{F}|^2 = |\overrightarrow{OP}|^2 = |\overrightarrow{OR}|^2 + |\overrightarrow{RS}|^2 + |\overrightarrow{SP}|^2 \\
&= F_x^2 + F_y^2 + F_z^2
\end{aligned} \tag{1.16}$$

so that

$$F = |\mathbf{F}| = \sqrt{F_x^2 + F_y^2 + F_z^2} \tag{1.17}$$

and the unit vector in the direction of **F**, $\mathbf{e}_F$, is:

$$\mathbf{e}_F = \frac{\mathbf{F}}{F} = \frac{F_x}{\sqrt{F_x^2 + F_y^2 + F_z^2}}\mathbf{i} + \frac{F_y}{\sqrt{F_x^2 + F_y^2 + F_z^2}}\mathbf{j} + \frac{F_z}{\sqrt{F_x^2 + F_y^2 + F_z^2}}\mathbf{k} \quad (1.18)$$

When adding vectors, it is generally more useful to use Cartesian components rather than dealing with the parallelogram law of addition since with Cartesian components we need to only add the components themselves. Thus, instead of using geometry and trigonometry, we can use algebra, which is highly suited to calculator-based or computer-based solutions. For two-dimensional problems, we will often show solutions based on both Cartesian components and the parallelogram law of addition so that you can see both types of approaches. Three-dimensional problems will be solved with Cartesian components.

As an example of using Cartesian components, consider the addition of two forces ($\mathbf{F}_1$, $\mathbf{F}_2$) in a two-dimensional problem, as shown in Figure 1.11, to find the resultant **R**. If we identify the Cartesian components of all these vectors, then it follows from Figure 1.11 that

$$\begin{aligned} R_x &= F_{1x} + F_{2x} \\ R_y &= F_{1y} + F_{2y} \end{aligned} \quad (1.19a)$$

and the same result is true in the more general three-dimensional case as well, where we would find:

$$\begin{aligned} R_x &= F_{1x} + F_{2x} \\ R_y &= F_{1y} + F_{2y} \\ R_z &= F_{1z} + F_{2z} \end{aligned} \quad (1.19b)$$

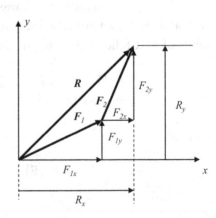

FIGURE 1.11
Vector addition by Cartesian components.

In MATLAB, where vectors can be written as row or column matrices, vector addition is done exactly in this manner with the Cartesian components written as $[F_x \; F_y \; F_z]$ or $[F_x: F_y; F_z]$. For example, consider two three-dimensional force row vectors F1 = [3 4 5], F2 = [1 1 1]. Then, adding these forces in MATLAB gives:

```
F1 = [3 4 5];
F2 = [1 1 1];
R = F1 + F2
R = 4 5 6
```

or, if we want to use column vectors:

```
F1 = [3; 4; 5];
F2 = [1; 1; 1];
R = F1 + F2
R = 4
    5
    6
```

However, the use of MATLAB also gives us a way to do vector addition with the use of the magnitudes and unit vectors. For example, suppose we have two forces $(\mathbf{F}_1, \mathbf{F}_2)$ whose magnitudes are (F_1, F_2) in the directions defined by the unit vectors $(\mathbf{e}_1, \mathbf{e}_2)$, respectively. Then, we can add those two vectors in MATLAB in this form as well. Here is an illustration.

Consider two unit vectors in MATLAB:

```
e1 = [1 1 1]/sqrt(3);
e2 = [3 4 12]/13;
```

You can verify these are unit vectors with the norm function, which computes the magnitude of a vector. Consider, for example, e2:

```
norm(e2)
ans = 1
```

Now, suppose the magnitude of a vector force, F1, in the direction of e1 is 100 lb and the magnitude of a vector force F2 in the direction of e2 is

200 lb. Then, we can sum these forces to obtain the resultant vector force, R, by simply adding these two vectors in this form and showing the result:

```
F1 = 100*e1;
F2 = 200*2;
R = F1 + F2
R = 103.8889   119.2735   242.3504
```

In statics, we often deal with forces whose direction is given but whose magnitude is unknown. In MATLAB, we can compute the resultant of these forces by writing them in symbolic form. Consider adding two forces of magnitudes (T_1, T_2) in the directions defined by the same unit vectors (e_1, e_2) just defined. Then, in MATLAB, the vector resultant $\mathbf{R} = T_1 \mathbf{e}_1 + T_2 \mathbf{e}_2$ can be computed symbolically as:

```
syms T1 T2 real      % define T1 and T2 as real symbolic variables
Rs = T1*e1 + T2*e2;  % the symbolic resultant vector
```

It is more convenient to display this resultant vector as a column vector rather than as a row vector, which we can do by taking the transpose with " ' " since the unit vectors are real numerical vectors and we have declared the symbolic variables as real also. For symbolic vectors or matrices that are defined without declaring them as real, we must use the form " .' " to obtain the transpose, as discussed earlier. Here we can use " ' ":

```
Rs'
ans = (3*T2)/13 + (3^(1/2)*T1)/3
      (4*T2)/13 + (3^(1/2)*T1)/3
      (12*T2)/13 + (3^(1/2)*T1)/3
```

which gives us the three rectangular components symbolically. We can substitute numerical values for T1 and T2 to get a specific value for the resultant force. Let us use the previous values of T1 = 100 lb and T2 = 200 lb. We can substitute these values with the MATLAB subs function where we place the symbolic variables for which we want to make replacements within curly brackets (this is called a cell vector) and the values we want to substitute in a row vector (see Appendix A for more details)

Rn = subs(Rs, {T1 T2}, [100 200])
Rn = (100*3^(1/2))/3 + 600/13
        (100*3^(1/2))/3 + 800/13
        (100*3^(1/2))/3 + 2400/13

The vector Rn, however, is still a vector in symbolic form (you can verify this by examining the properties given for this vector in the MATLAB workspace). We can convert it to a numerical vector with the double function:

double(Rn)
ans = 103.8889
        119.2735
        242.3504

which gives us the same resultant as before, but now written as a column vector.

Now, consider a vector addition problem with Cartesian components.

Example 1.6

Determine the resultant of the two forces shown in Figure 1.12(a) by giving the magnitude of the resultant and the angle it makes with respect to the positive *x*-axis.

First, we must use the geometry to determine the Cartesian components of the forces:

$$F_{1x} = \frac{12}{13}(210) = 193.85 \text{ N}$$

$$F_{1y} = \frac{-5}{13}(210) = -80.77 \text{ N}$$

$$F_{2x} = \frac{-2}{\sqrt{29}}(170) = -63.14\text{N}$$

$$F_{2y} = \frac{5}{\sqrt{29}}(170) = 157.84 \text{ N}$$

(1.20)

Then, we must add those components and compute the magnitude of the resultant

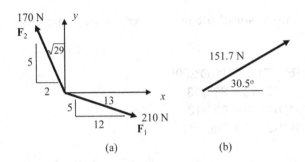

FIGURE 1.12
(a) Two forces, where (b) the resultant force is shown.

$$R_x = F_{1x} + F_{2x} = 130.71$$
$$R_y = F_{1y} + F_{2y} = 77.07$$
$$R = \sqrt{R_x^2 + R_y^2} = 151.7 \text{ N} \qquad (1.21)$$

and finally, we obtain the angle:

$$\theta = \tan^{-1}\left(\frac{77.07}{130.71}\right) = 30.5° \qquad (1.22)$$

The result is shown in Figure 1.12(b). Doing this in MATLAB instead:

```
e1 = [12 -5]/13;           % define unit vectors in the direction of the forces
e2 = [-2 5]/sqrt(29);
F1 = 210*e1;               % form up forces from their unit vectors and magnitudes
F2 = 170*e2;
R = F1 + F2;               % compute the resultant force
magR = norm(R)             % and its magnitude, in Newtons
magR = 151.7402
angR = atand(R(2)/R(1))    % determine the angle in degrees
angR = 30.5253
```

1.7 Oblique Components

In contrast to obtaining the Cartesian scalar components (R_x, R_y) of a resultant **R**, as shown in Figure 1.13(a), if we ask to find the magnitude of the two scalar components, A and B, along the oblique lines a-a and b-b, as shown in Figure 1.13(b), these are *oblique components*. There are several

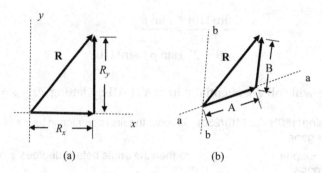

FIGURE 1.13
(a) The Cartesian components of a force **R**. (b) The oblique components along lines a-a and b-b.

ways to solve the problem, either using the parallelogram law or Cartesian components. MATLAB can still be used effectively in all these ways, as we will now show with an example.

Example 1.7

A 250 N force is to be resolved into two oblique components whose magnitudes are A and B, acting along lines a-a and b-b, respectively (Figure 1.14(a)). If A is 160 N, determine the scalar component B along b-b and the angle β.

From the law of sines (Figure 1.14(b)), we can develop two relations

$$\frac{\sin(110)}{250} = \frac{\sin(180 - 110 - \beta)}{160}$$
$$\rightarrow \sin(70 - \beta) = 160 \sin(110)/250 \qquad (1.23)$$

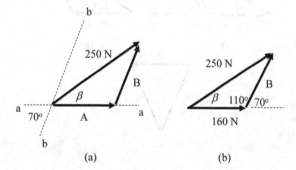

FIGURE 1.14
(a) Resolving a 250 N force into oblique components along a-a and b-b. (b) The force triangle.

$$\frac{\sin(110)}{250} = \frac{\sin \beta}{B}$$

$$\rightarrow B = 250 \sin \beta / \sin(110)$$ (1.24)

which we will solve sequentially in MATLAB by letting $70 - \beta = \alpha$. Then

```
alpha = asind(160*sind(110)/250)   % solve the first equation in terms of alpha
alpha = 36.9705
beta = 70 - alpha                  % then the angle beta in degrees is obtained
beta = 33.0295
B = 250*sind(beta)/sind(110)       % solve the second equation for B in Newtons
B = 145.0132
```

Even though we used MATLAB, we could have solved this problem just as easily by hand with a calculator.

Now, let us examine this problem using Cartesian components. From Figure 1.15(a), we could equate the x- and y-components of the resultant 250 N force to the sum of the x- and y-components of the components of **A** and **B**. We find:

$$R_x = A_x + B_x \rightarrow 250 \cos \beta = 160 + B \cos(70^\circ)$$
$$R_y = A_y + B_y \rightarrow 250 \sin \beta = B \sin(70^\circ)$$ (1.25)

It is possible to solve these equations by squaring them and using $\sin^2 \beta + \cos^2 \beta = 1$ to eliminate β so we can solve for B. Once B is found

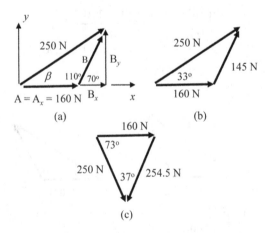

FIGURE 1.15
(a) Obtaining oblique components of a force using rectangular components. (b) One possible solution. (c) A second solution.

then we can find β. This is more complicated than using Eq. (1.23) and Eq. (1.24) directly from the parallelogram law of addition. Thus, most statics texts will only describe our first solution procedure. MATLAB, however, has powerful symbolic equation-solving abilities that we can employ that make the solution of Eq. (1.25) relatively simple. First, we define our unknowns B and β as symbolic MATLAB variables B and beta:

```
syms B beta
```

Next, we define the two equations in Eq. (1.25) as two symbolic expressions in a symbolic vector Eq which has two components:

```
Eq(1) = 250*sind(beta) − B*sind(70) == 0;
Eq(2) = 160 + B*cosd(70) − 250*cosd(beta) == 0;
```

A solution of these equations can be obtained with the MATLAB function solve as:

```
S = solve (Eq)
S = struct with fields:
    B: [2×1 sym]
    beta: [2×1 sym]
```

S here is a MATLAB data structure that contains our answers for B and beta in the fields S.B and S.beta. MATLAB tells us that there are two answers for both B and beta which are placed in (2×1) column vectors (two rows and one column). The values for B and beta, however, are not displayed. We can access the answers for B, for example, by typing S.B but the answer will be in a symbolic form that is difficult to read. We can change it to a numerical double precision form by using the MATLAB function double. For example:

```
double(S.B)
ans = 145.0132
      −254.4596
```

Look at MATLAB documentation, which is easily found on the web, if you want to learn more about MATLAB structures. We will use the MATLAB function solve in many later problems. Both answers in S.B and S.beta are in a symbolic form, so let us now change all of them to a numerical form through the MATLAB function double. The argument of the function double here is a row vector that contains the values in the fields of the structure S we want to convert to numerical values:

```
double([S.B S.beta])
ans = 145.0132    33.0295
      -254.4596   -73.0295
```

We see in the first row of results the previous values of B and β that we obtained (Figure 1.15(b)). However, we also see another solution in the second row. This is also a legitimate solution, as shown in Figure 1.15(c).

It is not necessary to place the symbolic equals zero "== 0" in these equations to use the MATLAB function solve. We need only place all the non-zero terms on the left side of a set of equations we want to be equal to zero in a symbolic vector Eq and then solve them with solve (Eq):

```
Eq(1) = 250*sind(beta) - B*sind(70);
Eq(2) = 160 + B*cosd(70) - 250*cosd(beta);
S = solve(Eq);
double([S.B S.beta])
ans = 145.0132    33.0295
      -254.4596   -73.0295
```

Finally, there is a closely related way to solve for oblique components with the symbolic capabilities of MATLAB that is a different way of looking at the solutions we just obtained. In oblique component problems in two dimensions, we are asking to decompose a resultant vector **R**, as **R** = $F_u\mathbf{e}_u + F_v\mathbf{e}_v$, where ($F_u$, F_v) are the magnitudes of the oblique components along the directions defined by the unit vectors ($\mathbf{e}_u$, $\mathbf{e}_v$). We can obtain three-dimensional oblique components **R** = $F_u\mathbf{e}_u + F_v\mathbf{e}_v + F_w\mathbf{e}_w$ in the same fashion. Consider the problem of Figure 1.14 again. Unit vectors along a-a and b-b are easy to write down, as is the resultant:

```
ea = [1, 0];
eb = [cosd(70) sind(70)];
R = [250*cosd(beta) 250*sind(beta)];
```

Then, we can pose the resultant problem as determining when the resultant vector minus the oblique component vectors equals zero. Thus, we find the set of equations, which we place in the vector Eq2:

```
Eq2 = R - 160*ea - B*eb;
```

which we then solve for B and beta and convert to numerical form

```
S = solve(Eq2);
double([S.B  S.beta])
ans = 145.0132    33.0295
     -254.4596   -73.0295
```

which gives the same solution as before. This solution was performed in purely vector form and we never had to explicitly write down the Cartesian components of Eq2.

A MATLAB symbolic solution is a powerful method that we will use often in this book. By simply writing a system of equations in symbolic form (almost identical to how we write them on paper in the first place) or forming those equations vectorially, as shown in the last example, we can then solve those equations symbolically. In most cases our equations will be systems of linear equations which can also be solved numerically in MATLAB, so we will often show both the numerical and symbolic solutions, but you will likely find symbolic solutions to be easier to set up. In the problem just examined, the equations are not linear but can be solved algebraically. More complex non-linear problems can be solved numerically, if not symbolically, in MATLAB.

To further illustrate the value of this use of MATLAB, let us return to Example 1.5 and find the solution to that problem in vector form with the function solve. We will take the x-direction as positive to the left and the y-direction as positive upward.

```
syms F theta                      % the unknown force F magnitude and angle theta
eR = [1 0];                       % unit vector in the 20 lb resultant direction
eP = [cosd(70) − sind(70)];       % unit vector in the 100 lb force direction
eF = [cosd(theta) sind(theta)];   % unit vector for the force of magnitude F
Eq = F*eF + 100*eP − 20*eR;       % form up the vectors which sum to zero
S = solve(Eq);                    % solve for F and the angle theta
double([S.F S.theta])             % convert to numerical

ans = −95.0364   −81.4057
       95.0364    98.5943
```

In the second row of the answer, we see the solution we obtained earlier. The first row also is a legitimate answer if we allow F and theta to be negative. When we used the parallelogram law, we drew F and theta as if

they were positive, so we never saw this second solution. The MATLAB
function solve obtained this "hidden" solution.

1.8 The Dot Product and Direction Cosines

1.8.1 The Dot Product

Determining the Cartesian components of a vector is closely related to a
way to multiply two vectors together called the scalar or dot product. As
its name implies, the result of this multiplication is a scalar quantity. The
dot product is defined as follows:

Let **A** and **B** be two vectors and let θ be the smallest positive angle
between them (Figure 1.16). Then, the scalar (dot) product of **A** and **B** is
written as **A · B**, where

$$\mathbf{A} \cdot \mathbf{B} = AB \cos \theta \tag{1.25}$$

and where A and B are the magnitudes of **A** and **B**.

There are a few important properties of dot products that we will use
often:

1. The order in which we take the product is immaterial, i.e.

$$\mathbf{A} \cdot \mathbf{B} = \mathbf{B} \cdot \mathbf{A} \tag{1.26}$$

2. The dot vectors of the unit-vectors along a set of Cartesian axes
   are given as

$$\begin{aligned} \mathbf{i} \cdot \mathbf{i} = \mathbf{j} \cdot \mathbf{j} = \mathbf{k} \cdot \mathbf{k} = 1 \\ \mathbf{i} \cdot \mathbf{j} = \mathbf{i} \cdot \mathbf{k} = \mathbf{j} \cdot \mathbf{k} = 0 \end{aligned} \tag{1.27}$$

3. If the dot product **A · B** = 0, then either **A** = 0, or **B** = 0 (or both),
   or else **A⊥B** (**A** is perpendicular to **B**)

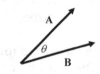

FIGURE 1.16
Two vectors separated by the angle θ. Their dot product is $AB\cos\theta$.

4. The dot product of any two unit vectors is the cosine of the angle, θ, between them:

$$\mathbf{e}_A \cdot \mathbf{e}_B = \cos\theta \qquad (1.28)$$

which is obviously true just by placing $A = B = 1$ in Eq. (1.25)

5. The dot product can be written simply as the sum of the products of the Cartesian components of $\mathbf{A}$ and $\mathbf{B}$, i.e.

$$\mathbf{A} \cdot \mathbf{B} = A_x B_x + A_y B_y + A_z B_z \qquad (1.29)$$

This very important result simply follows from Eq. (1.28) and Eq. (1.27) and the definition of the dot product since recall:

$$\mathbf{e}_A = \frac{A_x}{A}\mathbf{i} + \frac{A_y}{A}\mathbf{j} + \frac{A_z}{A}\mathbf{k}, \quad \mathbf{e}_B = \frac{B_x}{B}\mathbf{i} + \frac{B_y}{B}\mathbf{j} + \frac{B_z}{B}\mathbf{k}$$

so, carrying out the dot products of the unit vectors along the Cartesian axes

$$\mathbf{e}_A \cdot \mathbf{e}_B = \frac{A_x B_x}{AB} + \frac{A_y B_y}{AB} + \frac{A_z B_z}{AB} = \cos\theta$$

and we have

$$A_x B_x + A_y B_y + A_z B_z = AB\cos\theta = \mathbf{A} \cdot \mathbf{B}$$

6. If we have a vector $\mathbf{A}$ and we want to find the scalar Cartesian component of $\mathbf{A}$ along a line L whose direction is defined by a unit vector $\mathbf{e}_L$, then that component, A_L, is

$$A_L = \mathbf{A} \cdot \mathbf{e}_L \qquad (1.30a)$$

The vector Cartesian component is

$$\mathbf{A}_L = (\mathbf{A} \cdot \mathbf{e}_L)\mathbf{e}_L \qquad (1.30b)$$

Example 1.8

Determine the scalar and vector Cartesian components of the 300 lb force shown in Figure 1.17 which are parallel to the inclined plane AB.

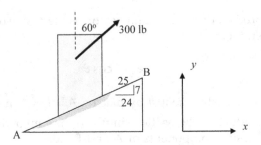

FIGURE 1.17
A force acting on a block.

The vector 300 lb force and a unit vector along AB are given by

$$F = 300 \sin 60°i + 300 \cos 60°j$$

$$e_{AB} = \frac{\overrightarrow{AB}}{|\overrightarrow{AB}|} = (24i + 7j)/25 \tag{1.31}$$

The scalar component along AB is

$$F_{AB} = F \cdot e_{AB} = 300 \sin 60°\left(\frac{24}{25}\right) + 300 \cos 60°\left(\frac{7}{25}\right) \tag{1.32}$$
$$\rightarrow \quad F_{AB} = 291.4 \text{ lb}$$

The vector component is

$$F_{AB} = 291.4 e_{AB} = 279.8i + 81.6j \text{ lb} \tag{1.33}$$

The solution in MATLAB, writing all the vectors in two dimensions, is:

```
F = [300*sind(60) 300*cosd(60)];  % the force F
AB = [24 7];
eAB = AB/norm(AB);               % the unit vector along AB
FAB = dot(F, eAB)                % the scalar component of F along AB
FAB = 291.4153
FV = FAB*eAB                     % the vector component of F along AB
FV = 279.7587  81.5963
```

Here, we used the built-in MATLAB function dot to obtain the dot product. There are two other ways we could do the dot product in

MATLAB. The first way multiplies the row vector F by a *column* vector, e', where " ' " —here indicates the transpose of a row vector (for a real vector). The matrix multiplication rules for this product (see Appendix A) give the dot product.

FAB = F*eAB'
FAB = 291.4153

A third way is to use the following approach:

FAB = sum(F.*eAB)
FAB = 291.4153

This method uses element-by-element multiplication of the row vectors F and eAB, denoted by the special multiplication operator .*. For vectors F = [Fx Fy] and [ex ey], this element-by-element multiplication gives a vector [Fx*ex Fy*ey] containing the products of the elements and applying the sum function adds up these components to give Fx*ex + Fy*ey, which is the dot product in Cartesian component form for this problem. Note that we must use element-wise multiplication here. If we try to use the matrix multiplication of two row vectors, we get an error, as shown below. MATLAB tells us we cannot multiply the two matrices (vectors in this case) with the multiplication operator " * " because our matrices do not satisfy the requirements for matrix multiplication (see Appendix A):

F*eAB
Error using *

Incorrect dimensions for matrix multiplication. Check that the number of columns in the first matrix matches the number of rows in the second matrix. To perform element-wise multiplication, use " .* ".

The choice of which of the three MATLAB methods you use to perform the dot product is up to you. The use of the dot function is nice in that it doesn't require the vectors in its argument to be of the same type (row or column vectors). It does require, however, that they have the same number of elements. Of course, you can always do the dot product by hand using the sum of the products of the Cartesian components.

1.8.2 Direction Cosines

Let's return to the Cartesian component representation of a vector (Figure 1.18(a)) and the unit vector that defines its direction (Figure 1.18(b)). The direction of a vector, **F**, and its associated unit vector, $\mathbf{e}_F$, can be defined by the angles $(\theta_x, \theta_y, \theta_z)$ with respect to the Cartesian axes. We have

$$\mathbf{F} = F_x\mathbf{i} + F_y\mathbf{j} + F_z\mathbf{k} = F\mathbf{e}_F$$

$$F = \sqrt{F_x^2 + F_y^2 + F_z^2} \qquad (1.34)$$

and

$$\mathbf{F} \cdot \mathbf{i} = F_x = F\mathbf{e}_F \cdot \mathbf{i} = F \cos \theta_x$$
$$\mathbf{F} \cdot \mathbf{j} = F_y = F\mathbf{e}_F \cdot \mathbf{j} = F \cos \theta_y \qquad (1.35)$$
$$\mathbf{F} \cdot \mathbf{k} = F_z = F\mathbf{e}_F \cdot \mathbf{k} = F \cos \theta_z$$

so that

$$\mathbf{e}_F = \frac{\mathbf{F}}{F} = \cos \theta_x\mathbf{i} + \cos \theta_y\mathbf{j} + \cos \theta_z\mathbf{k} \qquad (1.36)$$

where the direction cosines are

$$\cos \theta_x = \frac{F_x}{F}, \quad \cos \theta_y = \frac{F_y}{F}, \quad \cos \theta_z = \frac{F_z}{F} \qquad (1.37)$$

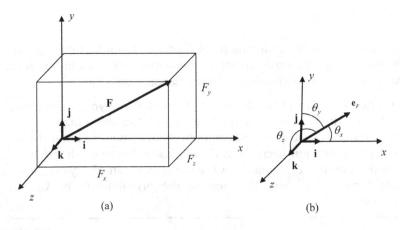

(a)   (b)

FIGURE 1.18
(a) The rectangular components of a vector, **F**. (b) The unit vector in the direction of **F** and the angles that the vector makes with the x-, y-, and z-axes.

and we can obtain the angles with the inverse cosine function

$$\theta_x = \cos^{-1}\left(\frac{F_x}{F}\right), \quad \theta_y = \cos^{-1}\left(\frac{F_y}{F}\right), \quad \theta_z = \cos^{-1}\left(\frac{F_z}{F}\right) \qquad (1.38)$$

Note: unlike computing the inverse sine function, there is never any problem in using the inverse cosine since the angles with respect to the axes always are given as "principal" values between 0° and 180° which is precisely the range needed to describe these angles.

Example 1.9

For the force shown in Figure 1.19(a),

    a. determine the (x, y, z) scalar components of the force,

    b. express the force in Cartesian vector form,

    c. determine the direction cosine angles of the force.

    a. We get the Cartesian components of the force by first breaking it into Cartesian components along the u- and z-directions and then breaking the u-component (in the x-y plane) into its x- and y-components. Here are the u- and z-components:

$$\begin{aligned} F_u &= 1000 \cos 25° = 906.3 \text{ lb} \\ F_z &= 1000 \sin 25° = 422.6 \text{ lb} \end{aligned} \qquad (1.39)$$

and here the x-, y-, and z-components:

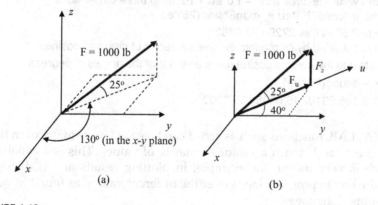

(a)

(b)

FIGURE 1.19

A force and the angles it makes in a Cartesian coordinate system. (b) Breaking the force into u- and z-components.

$$F_x = - F_u \sin 40° = -582.6 \text{ lb}$$
$$F_y = F_u \cos 40° = 694.3 \text{ lb} \qquad (1.40)$$
$$F_z = 1000 \sin 25° = 422.6 \text{ lb}$$

b. The vector Cartesian expression for **F** is thus:

$$\mathbf{F} = -582.6\mathbf{i} + 694.3\mathbf{j} + 422.6\mathbf{k} \text{ lb} \qquad (1.41)$$

c. The angles are obtained from the direction cosines:

$$\cos \theta_x = \frac{-582.6}{1000} = -0.5826, \quad \cos \theta_y = \frac{694.2}{1000} = 0.6942, \quad \cos \theta_z = \frac{422.6}{1000}$$
$$= 0.4226 \qquad (1.42)$$

as

$$\theta_x = \cos^{-1}\left(\frac{F_x}{F}\right) = 125.6°, \quad \theta_y = \cos^{-1}\left(\frac{F_y}{F}\right) = 46.0°, \quad \theta_z = \cos^{-1}\left(\frac{F_z}{F}\right) = 65.0°$$
$$(1.43)$$

The MATLAB solution is:

```
magF = 1000;                    %  magnitude of the force in lb
eu = [-sind(40) cosd(40) 0];  %  unit vector in the u - direction
ez = [0 0 1];                   %  unit vector in the z - direction
%  Now write the force as F = Fu*eu + Fz*ez in three-dimensions
F = magF*cosd(25)*eu +  magF*sind(25)*ez
F = -582.5634 694.2720 422.6183
%  Force divided by its magnitude gives a vector of direction cosines
%  so taking the inverse cosine gives a vector of the angles in degrees
angs = acosd(F/magF)
angs = 125.6310  46.0308  65.0000
```

In MATLAB, functions such as acosd can take vectors or matrices in their arguments and return a vector or matrix of values. This is a capability which is very useful, for example, in plotting results and it does not require one to program loops over the different values, as found in other computer languages.

1.9 Position Vectors

We may know the magnitude of a force and the fact that its direction is along a line between two points, as measured in a Cartesian coordinate system (see Figure 1.20(a)). To obtain the Cartesian components of the force, it is useful to use the concept of a *position vector*. By definition, a position vector is a vector that locates one point relative to another point. For example, the position vector of point A relative to the origin, O, of the coordinate system in Figure 1.20(b) is $\mathbf{r}_{OA} = x_A\mathbf{i} + y_A\mathbf{j} + z_A\mathbf{k}$, where (x_A, y_A, z_A) are the coordinates of point A. This position vector is a vector going from O to A; hence, we label its subscript as OA. Similarly, the position vector of point B relative to the origin O is $\mathbf{r}_{OB} = x_B\mathbf{i} + y_B\mathbf{j} + z_B\mathbf{k}$. The force of magnitude F in Figure 1.20(a) acts along the line from A to B. The position vector going from A to B is $\mathbf{r}_{AB}$ and from Figure 1.20(b):

$$\mathbf{r}_{OA} + \mathbf{r}_{AB} = \mathbf{r}_{OB} \tag{1.44}$$

or, solving for $\mathbf{r}_{AB}$ we have

$$\begin{aligned}\mathbf{r}_{AB} &= \mathbf{r}_{OB} - \mathbf{r}_{OA}\\ &= (x_B - x_A)\mathbf{i} + (y_B - y_A)\mathbf{j} + (z_B - z_A)\mathbf{k}\end{aligned} \tag{1.45}$$

We could also write this position vector as

$$\mathbf{r}_{AB} = x_{AB}\mathbf{i} + y_{AB}\mathbf{j} + z_{AB}\mathbf{k} \tag{1.46}$$

where (x_{AB}, y_{AB}, z_{AB}) are the coordinates of point B relative to point A, as shown in Figure 1.21. We can also use a notation where the position vectors are written directly as vector line segments such as $\mathbf{r}_{OA} = \overrightarrow{OA}$ and

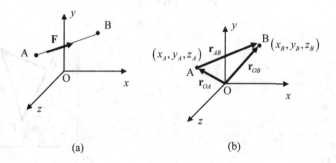

(a)   (b)

FIGURE 1.20
(a) A force acting along a line AB. (b) Geometry for defining the position vector from A to B.

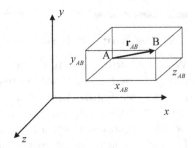

FIGURE 1.21
The components of a position vector.

$r_{OB} = \overrightarrow{OB}$. We have occasionally used this notation previously when expressing vectors (see Eq. (1.14)). Then we have in place of Eq (1.44) and Eq. (1.45):

$$\overrightarrow{OA} + \overrightarrow{AB} = \overrightarrow{OB} \tag{1.47a}$$

and, therefore,

$$\overrightarrow{AB} = \overrightarrow{OB} - \overrightarrow{OA} \tag{1.47b}$$

This alternate notation makes it clear that there is nothing special about using the origin O to calculate the position vectors of A and B. We could also locate points A and B relative to any point P (see Figure 1.22) we choose and write:

$$\overrightarrow{PA} + \overrightarrow{AB} = \overrightarrow{PB} \tag{1.48a}$$

and

$$\overrightarrow{AB} = \overrightarrow{PB} - \overrightarrow{PA} \tag{1.48b}$$

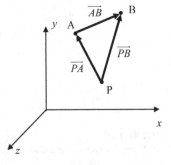

FIGURE 1.22
Using position vectors from a general point P to define the position vector from A to B.

Since the force in Figure 1.20 acts along the line from A to B, its direction is the same as the direction of the position vector $\mathbf{r}_{AB}$ and the direction of both of these vectors is defined by a unit vector, $\mathbf{e}_{AB}$, along the line from A to B, so we have

$$\mathbf{e}_{AB} = \mathbf{r}_{AB}/|\mathbf{r}_{AB}| = \overrightarrow{AB}/|\overrightarrow{AB}| \tag{1.49}$$

and

$$\mathbf{F} = F\mathbf{e}_{AB} = F\mathbf{r}_{AB}/|\mathbf{r}_{AB}| = F\,\overrightarrow{AB}/|\overrightarrow{AB}| \tag{1.50}$$

The components of the unit vector $\mathbf{e}_{AB}$ are the direction cosines that define the direction of both $\mathbf{r}_{AB}$ and the force $\mathbf{F}$.

In the next chapter, we will see that a position vector also plays an important role in defining the concept of the moment of a force about a point so it is essential that you can compute position vectors in various setups. Consider, for example, the following problem.

Example 1.10

A platform is held in the horizontal *x-z* plane by a cable that is attached to a vertical wall at point P with coordinates (2, 3, 0) ft. and to the platform at point R whose coordinates are (3.5, 0, 2.6) ft., as shown in Figure 1.23. The tension in the cable is 55 lb. Determine the vector force in the cable in terms of its Cartesian components. The coordinates are all measured in feet.

To get the direction of the force, we want a unit vector acting along RP. We have

$$\overrightarrow{OR} = 3.5\mathbf{i} + 0\mathbf{j} + 2.6\mathbf{k} \ \ \text{ft}$$

$$\overrightarrow{OP} = 2\mathbf{i} + 3\mathbf{j} + 0\mathbf{k} \ \ \text{ft}$$

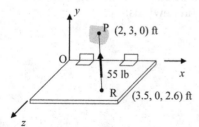

FIGURE 1.23
A platform supported by a cable.

so that

$$\overrightarrow{OR} + \overrightarrow{RP} = \overrightarrow{OP}$$
$$\rightarrow \overrightarrow{RP} = \overrightarrow{OP} - \overrightarrow{OR}$$

Placing the coordinates into the position vectors we find:

$$\overrightarrow{RP} = (2 - 3.5)\mathbf{i} + (3 - 0)\mathbf{j} + (0 - 2.6)\mathbf{k}$$
$$= -1.5\mathbf{i} + 3\mathbf{j} - 2.6\mathbf{k} \text{ ft}$$

and it follows that

$$|\overrightarrow{RP}| = \sqrt{(-1.5)^2 + (3)^2 + (-2.6)^2} = 4.2438 \text{ ft}$$
$$\mathbf{e}_{RP} = \overrightarrow{RP} / |\overrightarrow{RP}| = -0.353\mathbf{i} + 0.707\mathbf{j} - 0.613\mathbf{k}$$
$$\mathbf{F} = 55\mathbf{e}_{RP}$$
$$\rightarrow \mathbf{F} = -19.4\mathbf{i} + 38.9\mathbf{j} - 33.7\mathbf{k} \text{ lb}$$

In MATLAB, we have the same steps:

```
OR = [3.5 0 2.6];              % position vectors from O to R and from O to P
OP = [2 3 0];
RP = OP - OR;                  % position vector from R to P
eRP = RP/norm(RP);            % unit vector along RP
F = 55*eRP                     % force = magnitude*direction (in lb)
F = -19.4400  38.8801  -33.6961
```

1.10 Problems

P1.0 The mass density of water is one gram per cubic centimeter. Determine the weight of a volume $V = 1 \text{ m}^3$ of water (in newtons).

Choices (in N):

1. 9.81

2. 98.1

3. 981

4. 9810

5. 98100

P.1.1 A force of magnitude $F = 150$ lb acting along a bar (see Fig. P1.1) and a tensile horizontal spring force S produce a resultant force R acting in the negative x- and y-directions, with an angle $\beta = 75°$ as measured from the positive x-axis. Determine the magnitude of the spring force.

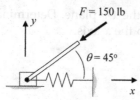

$F = 150$ lb

$\theta = 45°$

x

Fig. P1.1

Choices (in lb):

1. 40.1
2. 51.7
3. 63.4
4. 77.6
5. 81.5
6. 90.0

P1.2 Cables AB and AC help support a tower (Fig. P1.2). The tension in cable AC is $T_{AC} = 100$ lb. Determine the tension in cable AB if the resultant of the two cable forces is a force acting downwards (in the y-direction) at A.

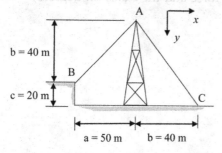

A

x

y

$b = 40$ m

B

$c = 20$ m

C

$a = 50$ m $b = 40$ m

Fig. P1.2

Choices (in lb):

1. 71.0
2. 58.0
3. 112.0

4. 64.0

5. 97.0

6. 32.0

P1.3 The 50 lb force shown in Fig. P1.3 acts in the *x-y* plane. Determine the direction cosine of this force with respect to the *x*-axis.

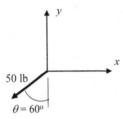

Fig. P1.3

Choices:

1. −0.5

2. 0.5

3. −0.866

4. 0.866

P1.4 A force has the *x*-, *y*-, and *z*-components shown in Fig. P1.4. Determine the angle between this force and the *z*-axis (in degrees).

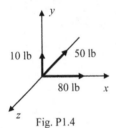

Fig. P1.4

Choices (in degrees):

1. −32.5

2. −58.2

3. 96.5

4. 121.8

5. 205.2

P1.5 The 60 lb force shown in Fig. P1.5 acts in the x-y plane. Determine the x- and y-components, (e_x, e_y), of a unit vector in the direction of this force.

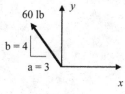

Fig. P1.5

Choices:

1. $(-0.8, -0.6)$
2. $(-0.6, -0.8)$
3. $(-0.6, 0.8)$
4. $(-0.8, -0.6)$

P1.6 The unit vector, e, shown in Fig. P1.6 acts in the x-y plane along a line between points A and B whose coordinates are given. Determine the components of e,(e_x, e_y), in the x- and y-directions, respectively.

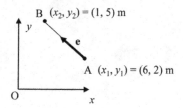

Fig. P1.6

Choices:

1. $(-0.781, 0.635)$
2. $(-0.858, 0.515)$
3. $(-0.825, 0.565)$
4. $(-0.785, 0.619)$

P1.7 Two forces (F_1, F_2) act in the x-y plane on the block shown in Fig. P1.7. Determine the angle θ that makes the resultant of these two forces act vertically down (in the negative y-direction) and determine the magnitude, R, of this resultant. Is there more than one answer?

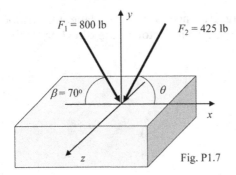

Fig. P1.7

P1.8 A 500 lb force **F** holds a triangular block against two walls (Fig. P1.8). Determine: (a) the scalar rectangular component of the force along OA, (b) the scalar rectangular component of the force along OB. (c) OA and OB are perpendicular to each other. Using your results from parts (a) and (b), show that we can recover the force through **F** = (**F** · **e**$_{OA}$)**e**$_{OA}$ + (**F** · **e**$_{OB}$)**e**$_{OB}$, where (**e**$_{OA}$, **e**$_{OB}$) are unit vectors along OA and OB, respectively.

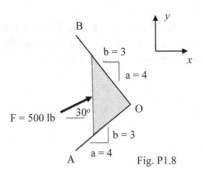

Fig. P1.8

P1.9 Write the force shown in Fig. P1.9 in Cartesian coordinates in terms of the unit vectors **i**, **j**, and **k**. The distances are a = 40 mm, b = 50 mm, and c = 110 mm.

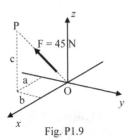

Fig. P1.9

Choices (in N):

1. 17.7i − 38.9j + 14.1k
2. 38.9i − 17.7j + 14.1k
3. 14.1i − 17.7j + 38.9k
4. 14.1i + 17.7j − 38.9k
5. −17.7i + 38.9j + 14.1k
6. 38.9i + 17.7j + 14.1k

P1.10 A 150 lb horizontal force acts on the frame, as shown in Fig. P1.10. Determine the magnitudes of the oblique components of the force acting along members AB and AC.

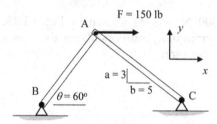

Fig. P1.10

P1.11 The three forces in Fig. P1.11 produce a resultant force of magnitude $R = 100$ N acting in the positive z-direction. Determine the magnitudes of these three forces which are the oblique components of the resultant force along OA, OB, and OC, respectively.

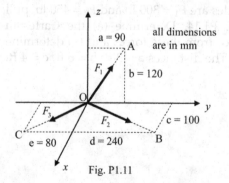

Fig. P1.11

P1.12 The force $F_1 = 45$ lb acts in the y-z plane at an angle of $\theta = 60°$ with respect to the y-axis, as shown in Fig. P1.12. Forces $F_2 = 60$ lb and $F_3 = 75$ lb act in the x-y plane at directions defined by the sides of triangles where a = 1 and b = 2. Determine the angles the resultant of these three forces make with respect to the (*x*, *y*, *z*) axes.

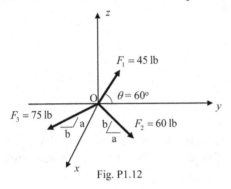

Fig. P1.12

P1.13 A force of magnitude F = 500 N acts as shown in Fig. P1.13. Determine the scalar and vector rectangular components of this force along the line OA, which lies in the *x-y* plane.

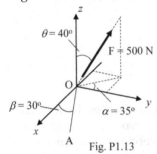

Fig. P1.13

P1.14 Two forces, whose magnitudes are F_1 = 300 lb and F_2 = 450 lb, pull on an eyebolt as shown in Fig. P1.14. Determine (a) the Cartesian representation of the resultant force from these forces and (b) determine the angle between the two forces. The distances a = 3 ft, b = 6 ft, c = 4 ft.

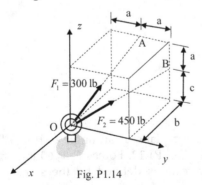

Fig. P1.14

P1.15 A door is held open at a $\theta = 30°$ angle by a cable that is attached to door at A and a vertical wall at B. If the tension in the cable is $T = 100$ N, determine the rectangular scalar component of the tension that acts along the diagonal CD of the door. The dimensions of the door are b = 1200 mm and c = 900 mm and the vertical distance a = 900 mm.

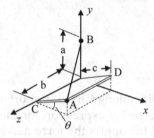

Fig. P1.15

1.10.1 Review Problems

These problems typically have the level of difficulty found on exams. They should be done by hand (i.e., with a calculator).

R1.1 A force of magnitude F = 300 lb acts along the diagonal BC of the block shown in Fig. R1.1. (a) Determine the vector expression for this force in terms of its Cartesian components. (b) Determine the angle (in degrees) that this force makes with respect to the positive y-axis. (c) Determine the rectangular (Cartesian) scalar component of this force along AB.

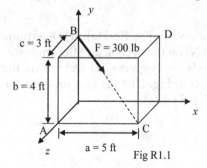

Fig R1.1

R1.2 A force of magnitude $F_1 = 100$ lb and a force of magnitude $F_2 = 200$ lb act at the origin O as shown in Fig. R1.2. The force $\mathbf{F}_1$ lies in the x-z plane. (a) Determine the vector resultant of these two forces in terms of its Cartesian components. (b) Determine the angles that the resultant force makes with respect to the positive x-, y-, and z-axes.

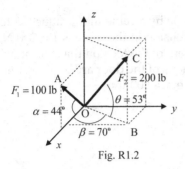

Fig. R1.2

R1.3 A force of magnitude T = 55 lb acts in a wire (along CE) to support a plate as shown in Fig. R1.3. Determine (a) the vector expression for the force in terms of its Cartesian components, (b) the angles the force make with respect to the x-, y-, and z-axes (in degrees), and (c) the scalar rectangular component of the force along the plate diagonal AD.

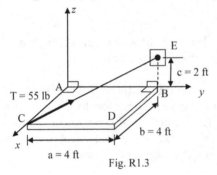

Fig. R1.3

R1.4 Two forces pull on an eyebolt as shown in Fig. R1.4 (a). (1) Determine the magnitude of the resultant force (in N) on the eyebolt and the angle this resultant makes with the positive x-axis (in degrees). (2) Determine the oblique force components (F_u, F_v) shown in Fig. R1.4 (b) that produce the same resultant force as the forces in Fig. R1.4 (a).

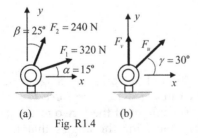

(a)   (b)
Fig. R1.4

R1.5 A force of magnitude F = 500 lb acts along the diagonal AD of the block as shown in Fig. R1.5. Determine the vector expression for this force.

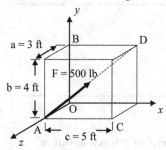

Figs. R1.5, R1.6, R1.7

Choices (in lb):

1. 283i + 212j −354k
2. 354i −212j −283k
3. 354i + 283j −212k
4. 212i + 283j −354k
5. −212i + 354j −283k

R1.6 Determine the angle that the force in Fig. R1.6 makes with the positive z-axis, approximately.

Choices (in degrees):

1. 55.6
2. 145.6
3. 45.0
4. 115.1
5. 64.9

R1.7 Determine the rectangular scalar component of the force in Fig. R1.7 along the line AB, approximately.

Choices (in lb):

1. 99.2
2. 339.4
3. −99.2
4. 353.6
5. −339.4

2

Moments and Couples

OBJECTIVES

- To introduce the concept of the moment of a force about a point.
- To define the vector cross product.
- To demonstrate calculations of the vector cross product with MATLAB®.
- To introduce the concept of a couple.
- To define the moment of a force about a line.

A force is a measure of a push or pull exerted on a body, but forces can also cause a body to rotate so we need to describe the strength of that rotational tendency. In this chapter, we will define the *moment of a force about a point* as the measure of the turning tendency the force has about that point. As we will see, the moment of a force is a vector which can be obtained through a *vector cross product*. Distributions of forces can also produce pure moments, without any net resultant force present. Such pure moments are called *couples*. Since a body may be designed to rotate about a particular axis that may not align with the direction of the moment, the concept of the *moment of a force about a line* (or axis), as opposed to about a point, will also be discussed, so that we can define the tendency that a force has to rotate a body about that axis.

2.1 Moment of a Force about a Point

The moment of a force about a point, as just indicated, is a measure of the turning effect that the force has about a particular point. For example, consider the force **F** which acts through a wrench on a bolt nut, as shown in Figure 2.1. From experience, we know that the strength of the turning

DOI: 10.1201/9781003372592-2

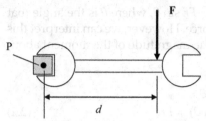

FIGURE 2.1
A force, **F**, applied to a wrench.

effect of the force to tighten (or loosen) the nut depends both on the magnitude of the force, F, and the distance, d, from the nut center, P, in Figure 2.1. A clue to how F and d enter into the magnitude of the moment can be found by considering the balance shown in Figure 2.2, where a pair of forces (F_1, F_2) are applied to the balance at different distances (d_1, d_2) from the pivot point, P. It can easily be demonstrated that the balance will not rotate if we have $F_1 d_1 = F_2 d_2$. Thus, we will take the product of the magnitude of a force, F, and the distance d as the definition of the magnitude of the moment of the force about point P, M_P, i.e.

$$M_P = Fd \tag{2.1}$$

This definition even works when the force is inclined to the wrench center line, as shown in Figure 2.3, if we choose d appropriately. In trying to apply Eq. (2.1) to this case (and to the balance of Figure 2.2 when the forces are acting at angles), we will find that we can still use Eq. (2.1) if we take the distance, d, in the product Fd to be the distance from point P along a line that is *perpendicular to the force*. We will call this perpendicular distance

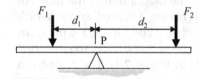

FIGURE 2.2
Two forces that have equal turning effects about the pivot point, P, of a balance.

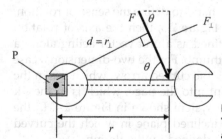

FIGURE 2.3
The more general case where a force is applied at an angle to the axis of the wrench.

$d = r_\perp$. Thus, as seen in Figure 2.3, $Fr_\perp = Fr \sin \theta$, where θ is the angle that the axis of the wrench makes with the force. However, we can interpret this result in two ways since we can write the magnitude of the moment about point P in the forms:

$$
\begin{aligned}
M_P &= Fr \sin \theta \\
    &= F(r \sin \theta) = Fr_\perp \\
    &= (F \sin \theta)r = F_\perp r
\end{aligned}
\tag{2.2}
$$

where the first form is the $Fr_\perp$ form just described, but the second form is $F_\perp r$, where r is the total distance to the force along a horizontal line and $F_\perp$ the component of the force that is perpendicular to the line along which r is measured. This second form also makes sense since any force component parallel to the position vector from point P to the force will have no tendency to turn a body about that point. In summary, we interpret the magnitude of the moment about a point as

---

magnitude of a moment = perpendicular distance × force

= perpendicular component of force × distance

---

since in either case we are computing the same product, $Fr \sin \theta$. A moment of a force about a point P will tend to turn a body about that point along a particular axis, so we will use that axis to define a direction for the moment. Thus, the moment will be a vector and have both a magnitude and a direction. For a two-dimensional problem like those we have been considering in Figures 2.1–2.3, we can give the magnitude of the moment as $Fd = Fr_\perp$ and define its direction either (1) using a curved arrow, which gives the sense in which the force tends to rotate a body in the x-y plane, or (2) using the *right-hand rule* to obtain a unit vector that defines the direction of the axis of that rotation (see an example in Figure 2.4). The right-hand rule is shown in Figure 2.5 for a general three-dimensional case. Suppose we have a force that tends to rotate a body about a point along a particular axis. If we curl the fingers of our right hand to define the sense of rotation, as indicated by the curved arrow in Figure 2.5, then the axis of rotation (direction of the moment) will be defined as a unit vector acting along a line that points in the direction of the thumb. For the two-dimensional case of Figure 2.4, the fingers would follow the curved arrow, which lies in the x-y plane, and the thumb would point into the page, which is in the $-\mathbf{k}$ direction for the Cartesian coordinate system shown in Figure 2.4. In the more general case of Figure 2.5, the inclined plane in which the curved arrow lies is perpendicular to the axis defined by the thumb.

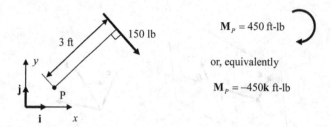

$M_p = 450$ ft-lb

or, equivalently

$M_p = -450\mathbf{k}$ ft-lb

FIGURE 2.4
A force acting in two dimensions, which generates a moment about point P. We can either define the direction of the moment by giving the sense of the rotational tendency it has about point P or by using the right-hand rule, defined in Figure 2.5, to obtain a unit vector that gives the direction. In this case, the unit vector is $-\mathbf{k}$.

FIGURE 2.5
The right-hand rule, where the turning tendency of a force is represented by the curved arrow (acting in the same sense as the curved fingers of the right hand), and the direction of the moment vector is given by a unit vector in the direction of the thumb. The plane in which the curved arrow lies is perpendicular to the axis along the thumb.

All the examples we have discussed so far have been for two-dimensional problems where the force and the line from a point P to the force lie in the x-y plane, but the same ideas also work in the 3-D case (Figure 2.6(a)) where the force $\mathbf{F}$ and a position vector, $\mathbf{r}$, from a point P to the force define a plane that contains both vectors, and we take the direction of the moment to be a unit vector, $\mathbf{e}$, perpendicular to that plane, as determined by the right-hand rule (Figure 2.6(b)).

2.1.1 The Vector Cross Product

If we are given any two vectors, $\mathbf{A}$ and $\mathbf{B}$, we can define the cross product of those vectors, $\mathbf{A} \times \mathbf{B}$ as the vector:

$$\mathbf{A} \times \mathbf{B} = AB \sin \theta \; \mathbf{e} \qquad (2.3)$$

where A and B are the magnitudes of those vectors, θ is the acute angle between the vectors, as measured in their common plane, and $\mathbf{e}$ is a unit vector perpendicular to both A and B and is in a direction given by the right-hand rule when $\mathbf{A}$ is rotated into $\mathbf{B}$ through the acute angle θ

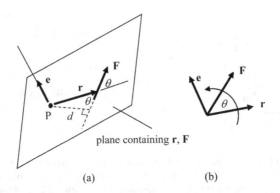

FIGURE 2.6

(a) The geometry for determining a moment in three dimensions. (b) Obtaining the direction of the moment by rotating **r** into **F** in their common plane, producing the direction along which the unit vector **e** acts by the right-hand rule.

FIGURE 2.7

Two vectors **A** and **B** and a unit vector, **e**, defined from them through the right-hand rule.

(Figure 2.7). It follows if we let **A** = **r**, a position vector from point P to a force **B** = **F**, that the cross product of **r** and **F** gives the moment of the force about point P:

$$\mathbf{M}_P = \mathbf{r} \times \mathbf{F} = rF \sin\theta \ \mathbf{e} \qquad (2.4)$$

and, again, we can interpret the magnitude of the moment in two ways as $rF \sin\theta = Fr_\perp = F_\perp r$. We can let the position vector in the calculation of the moment about point P to be the vector from P to any point along the line of action of **F** without changing the value of the moment of the force about P since the product $Fr_\perp$ is unchanged (Figure 2.8).

There are some important properties of vector cross products that we wish to highlight:

1. The order of the vector cross product does matter since **A** × **B** = −**B** × **A**.

2. The cross product of a vector with itself is zero, **A** × **A** = 0.

3. **A** × **B** = 0 if **A** = 0, or **B** = 0 (or both), or if **A** || **B** (**A** is parallel to **B**).

4. The unit vectors along the axes of a *right-handed Cartesian coordinate system* satisfy

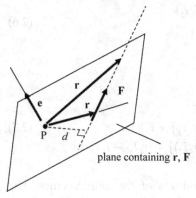

FIGURE 2.8

plane containing **r**, **F**

The vector **r** can be taken from P to any point along the line of action **F**, as seen by the two choices shown, without changing the moment about P since the perpendicular distance, *d*, is unchanged.

$$\mathbf{i} \times \mathbf{i} = \mathbf{j} \times \mathbf{j} = \mathbf{k} \times \mathbf{k} = 0$$
$$\mathbf{i} \times \mathbf{j} = \mathbf{k}, \quad \mathbf{j} \times \mathbf{k} = \mathbf{i}, \quad \mathbf{k} \times \mathbf{i} = \mathbf{j} \qquad (2.5)$$
$$\mathbf{j} \times \mathbf{i} = -\mathbf{k}, \quad \mathbf{k} \times \mathbf{j} = -\mathbf{i}, \quad \mathbf{i} \times \mathbf{k} = -\mathbf{j}$$

where a right-handed coordinate system can be defined as a system where, if we rotate the *x*-axis into the *y*-axis with the right hand, the thumb points in the direction of the *z*-axis.

It is important to always use a right-handed coordinate system since otherwise the cross products of the coordinate unit vectors will be inconsistent with the use of the right-hand rule in defining the direction of the cross product of any two vectors.

An easy way to remember all these cross products of different unit vectors is to place the **i**, **j**, **k** vectors around a clock dial in that order going clockwise (Figure 2.9). Then if we consider any two unit vectors in succession going clockwise, their cross product will be the third remaining unit vector. But if we consider any two unit vectors in succession going counterclockwise, their cross product will be the negative of the third remaining unit vector. For example, $\mathbf{i} \times \mathbf{j} = \mathbf{k}$ (clockwise) but $\mathbf{j} \times \mathbf{i} = -\mathbf{k}$, (counterclockwise), $\mathbf{k} \times \mathbf{j} = -\mathbf{i}$, etc. We can use property 4 to calculate the moment of a force in terms of the Cartesian components of **r** and **F**. If

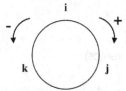

FIGURE 2.9

A "clock dial" construct for obtaining the cross products of the unit vectors **i**, **j**, **k**.

$$\mathbf{r} = r_x\mathbf{i} + r_y\mathbf{j} + r_z\mathbf{k}$$
$$\mathbf{F} = F_x\mathbf{i} + F_y\mathbf{j} + F_z\mathbf{k} \tag{2.6}$$

then

$$\begin{aligned}
\mathbf{M}_P &= M_{Px}\mathbf{i} + M_{Py}\mathbf{j} + M_{Pz}\mathbf{k} \\
&= (r_x\mathbf{i} + r_y\mathbf{j} + r_z\mathbf{k}) \times (F_x\mathbf{i} + F_y\mathbf{j} + F_z\mathbf{k}) \\
&= (r_yF_z - r_zF_y)\mathbf{i} + (r_zF_x - r_xF_z)\mathbf{j} + (r_xF_y - r_yF_x)\mathbf{k}
\end{aligned} \tag{2.7}$$

where we carried out all the cross products of the unit vectors. The Cartesian components of the moment are, therefore,

$$\begin{aligned}
M_{Px} &= (r_yF_z - r_zF_y) \\
M_{Py} &= (r_zF_x - r_xF_z) \\
M_{Pz} &= (r_xF_y - r_yF_x)
\end{aligned} \tag{2.8}$$

Another way to write the cross product is in the form of a 3 × 3 matrix determinant which we can expand in terms of smaller 2 × 2 matrix determinants, obtaining the same components seen in Eq. (2.8):

$$\mathbf{M}_P = \mathbf{r} \times \mathbf{F} = \begin{vmatrix} \mathbf{i} & \mathbf{j} & \mathbf{k} \\ r_x & r_y & r_z \\ F_x & F_y & F_z \end{vmatrix} = \mathbf{i}\begin{vmatrix} r_y & r_z \\ F_y & F_z \end{vmatrix} - \mathbf{j}\begin{vmatrix} r_x & r_z \\ F_x & F_z \end{vmatrix} + \mathbf{k}\begin{vmatrix} r_x & r_y \\ F_x & F_y \end{vmatrix} \tag{2.9}$$

$$= (r_yF_z - r_zF_y)\mathbf{i} + (r_zF_x - r_xF_z)\mathbf{j} + (r_xF_y - r_yF_x)\mathbf{k}$$

We place $\mathbf{i}, \mathbf{j}, \mathbf{k}$ in the first row of the determinant and the components of the position vector and force vector in the second and third rows, respectively. Thus, we are writing $\mathbf{r}$ and $\mathbf{F}$ here as row vectors within the 3 × 3 matrix.

In MATLAB, there are several ways we can compute cross products. The first is with the built-in MATLAB function cross. Suppose, for example, we have $\mathbf{r} = 4\mathbf{i} + 3\mathbf{j} - 2\mathbf{k}$ ft, $\mathbf{F} = 60\mathbf{i} + 30\mathbf{j} - 50\mathbf{k}$ lb. In MATLAB, the cross product, $\mathbf{r} \times \mathbf{F}$, then is obtained via:

```
r = [4 3 -2];        % the position vector
F = [60 30 -50];     % the force vector
M = cross(r, F)      % the cross product (moment vector)
M = -90  80  -60
```

There is nothing special about using row vectors. If we use column vectors instead:

```
rc = [4; 3; -2];          % the position vector
Fc = [60; 30; -50];       % the force vector
M = cross(rc, Fc)         % the cross product (moment vector)
M = -90
      80
     -60
```

To use the MATLAB function cross, we must have three-dimensional vectors. Thus, in two-dimensional problems where position vectors and force vectors only have two components, we must include a zero value in a third-dimensional component when we write them. This restriction makes sense since the cross product of two two-dimensional vectors always lies in a third dimension, so that third dimension must be included from the start.

We can also do the cross product symbolically, using the determinant expression and the built-in MATLAB function det for evaluating the determinant:

```
syms i j k                % define unit vectors symbolically
cross_mat = [i j k; r; F] % set up 3 × 3 matrix for the cross product
                          % note the use of row vectors r and F
cross_mat = i, j, k       % defined previously
             [4, 3, -2]
             [60, 30, -50]
M = det(cross_mat)        % evaluate the determinant
M = 80*j - 90*i - 60*k    % the moment of the force (symbolic result)
```

MATLAB has in the past reserved i and j to represent the imaginary value $\sqrt{-1}$. In statics, we do not use complex numbers so there will be no problem in defining (i, j, k) as symbolic Cartesian unit vectors. MATLAB more recently reserves 1i and 1j to represent the imaginary value $\sqrt{-1}$ precisely to avoid such conflicts. One can instead define these Cartesian unit vectors symbolically as (ex, ey, ez), but we will stay with the use of (i, j, k), which is commonly used in statics.

There is a third way to compute the cross product by writing it as a matrix-vector product where we place the position vector in a 3 × 3 antisymmetric matrix and multiply it by a *column* force vector in the form:

$$
\left\{ \begin{array}{c} M_x \\ M_y \\ M_z \end{array} \right\} = \left[\begin{array}{ccc} 0 & -r_z & r_y \\ r_z & 0 & -r_x \\ -r_y & r_x & 0 \end{array} \right] \left\{ \begin{array}{c} F_x \\ F_y \\ F_z \end{array} \right\} \tag{2.10}
$$

In our previous example, we would have

M = [0 2 3; -2 0 - 4; -3 4 0]*F.'   % matrix - vector form of cross product
M = -90
    80
    -60

where we see we have changed the row vector F into a column vector by use of the transpose. Using this form requires that we place the elements of the position vector in the appropriate places with the right signs to generate the 3 × 3 matrix, which can lead to potential errors, so it is usually better to either use the cross function or the determinant expression. We saw earlier that we could take the position vector in the cross product expression for the moment to be a position vector from point P to any point on the line of action of the force. Similarly, we can slide the force itself anywhere along its own line of action (Figure 2.10) and not change the cross product, since again the perpendicular distance and the magnitude of the force are unaffected. Thus, a force is sometimes called a *sliding vector* to indicate that the moment is unchanged by its placement along its line of action. It is important to realize, however, that although sliding a force along its line of action does not change the moment that the force produces, this does not mean that physical effects from sliding the force are not present. Figure 2.11(a), for example, shows a pair of equal and opposite forces acting on a block. For a rigid block, the conditions for it being in equilibrium depend on the resultant force and moment being zero, so

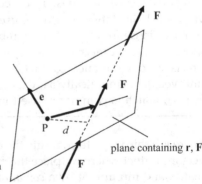

plane containing **r**, **F**

FIGURE 2.10
Sliding a force anywhere along the line of action
does not change the moment it produces.

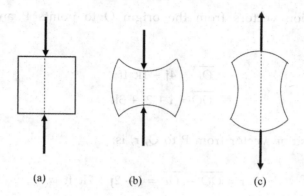

(a)   (b)   (c)

FIGURE 2.11
(a) A rigid block in equilibrium with a pair of equal and opposite forces, and (b), (c) the distortions of a highly deformable block when the forces are at different locations along their lines of action.

we are free to slide these forces along their common line of action and not violate that equilibrium. Similarly, when a rigid body moves, that motion only depends on the resultant force and moment. Thus, for rigid bodies, a force can indeed be treated as a sliding vector. However, if the block is deformable, sliding the forces along their lines of action can produce dramatic changes in the block, as seen in Figure 2.11(b) and (c). Statics typically deals with the equilibrium of rigid bodies where the force can be treated as a sliding vector, but in later courses, such as strength of materials, which consider deformable bodies, sliding a force vector may induce significant changes. In Chapter 10, we will also consider deformable bodies.

Example 2.1

The force $\mathbf{F} = \mathbf{i} + \mathbf{j} + \mathbf{k}$ lb acts at point Q whose coordinates are shown in Figure 2.12. Determine the moment of this about point P, whose coordinates are also given. All distances are measured in ft.

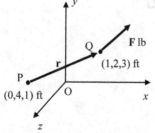

FIGURE 2.12
Geometry for determining the moment of the force **F** about point P.

The position vectors from the origin O to points P and Q are given by

$$\overrightarrow{OP} = 4\mathbf{j} + 1\mathbf{k} \ \ \text{ft}$$

$$\overrightarrow{OQ} = \mathbf{i} + 2\mathbf{j} + 3\mathbf{k} \ \ \text{ft}$$

(2.11)

so the position vector from P to Q, **r**, is:

$$\mathbf{r} = \overrightarrow{OQ} - \overrightarrow{OP} = \mathbf{i} - 2\mathbf{j} + 2\mathbf{k} \ \ \text{ft} \tag{2.12}$$

Let us start by evaluating the determinant expression for the moment by hand:

$$\mathbf{M}_P = \mathbf{r} \times \mathbf{F} = \begin{vmatrix} \mathbf{i} & \mathbf{j} & \mathbf{k} \\ 1 & -2 & 2 \\ 1 & 1 & 1 \end{vmatrix} = \mathbf{i}\begin{vmatrix} -2 & 2 \\ 1 & 1 \end{vmatrix} - \mathbf{j}\begin{vmatrix} 1 & 2 \\ 1 & 1 \end{vmatrix} + \mathbf{k}\begin{vmatrix} 1 & -2 \\ 1 & 1 \end{vmatrix}$$

$$= \mathbf{i}[-2 - 2] - \mathbf{j}[1 - 2] + \mathbf{k}[1 + 2]$$

$$= -4\mathbf{i} + \mathbf{j} + 3\mathbf{k} \ \ \text{ft-lb}$$

(2.13)

where the 2 × 2 sub-determinants are obtained by crossing out all the rows and columns in the original 3 × 3 determinant containing the unit vector under consideration. Note the sign change, however, on the **j** term. Evaluating 2 × 2 determinants is easy since we merely need to take the product of terms along the diagonal going down and to the right and subtract the product of the terms along a diagonal going down and to the left. If we want to do this cross product in MATLAB, we can use the built-in cross function:

```
OP = [0 4 1];
OQ = [1 2 3];
r = OQ–OP          % position vector from P to Q
r = 1 –2 2
F = [1 1 1];       % the force vector
M = cross(r, F)    % the cross product
M = –4  1  3
```

or we can do the problem symbolically with the determinant:

```
syms i j k                    % define unit vectors symbolically
cross_mat = [i j k;  r;  F]   % form matrix with r and F in 2nd and 3rd rows
   cross_mat = [i, j, k]
               [1, -2, 2]
               [1, 1, 1]
M = det(cross_mat)            % calculate determinant, which gives cross product
M = j - 4*i + 3*k
```

Example 2.2

Determine the moment of the 500 lb force shown in Figure 2.13 about point P in four different ways.

1. Use Figure 2.14(a) and $M_P = Fr_\perp$:

The angles α and β are given by

$$\alpha = \tan^{-1}(4/3) = 53.13° \quad \beta = \alpha - 45° = 8.13°$$

and we have

$$M_P = Fr_\perp = 500(2\sqrt{2} \sin 8.13°) = 200 \text{ ft-lb}$$

so $\mathbf{M}_P = -200\mathbf{k}$ ft-lb or $M_P = 200$ ft-lb ↻

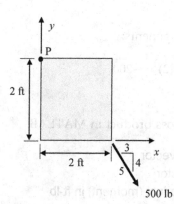

FIGURE 2.13
Geometry for determining the moment of a force about a point.

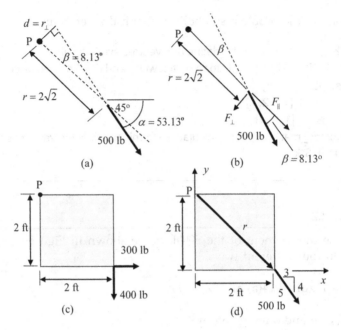

FIGURE 2.14
(a)–(d) Geometries for determining the moment of the 500 lb force about point P.

2. Use Figure 2.14(b) and $M_P = rF_\perp$:

$$F_\perp = 500 \sin 8.13° = 70.709$$
$$M_P = rF_\perp = (2\sqrt{2})(70.709) = 200 \text{ ft-lb}$$

so $\mathbf{M}_P = -200\mathbf{k}$ ft-lb or $M_P = 200$ ft-lb ↻

3. Use Figure 2.14(c) and rectangular components:

$$\sum M_{Pz} = (300)(2) - (400)(2) = -200$$

so $\mathbf{M}_P = -200\mathbf{k}$ ft-lb or $M_P = 200$ ft-lb ↻

4. Use of Figure 2.14(d) and the vector cross product in MATLAB:

```
r = [2 -2 0];        % the position vector
F = [300 -400 0];    % the force vector
M = cross(r, F)      % the cross product (moment) in ft-lb
M = 0 0 -200
```

2.2 Couples

A couple, **C**, is a vector quantity that represents a pure turning or twisting action on a body. It has the dimensions of a moment of a force (force × distance) and like a force is represented as an arrow but in three dimensions we normally also show the direction of the turning action of the couple, as determined with the right-hand rule, so we can distinguish couples from forces (Figure 2.15(a)). In two-dimensional problems where all the moments and couples are in the plus or minus z-direction, we either show the unit vector in the plus or minus z-direction or a curved arrow in the x-y plane to indicate the sense of turning.

A couple can be generated by a pair of equal and opposite forces separated by a perpendicular distance, D, as shown in Figure 2.16. Unlike

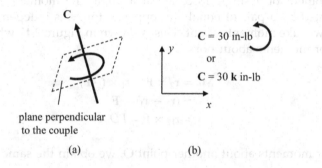

FIGURE 2.15

(a) A couple as represented in three-dimensional problems. (b) A couple as represented in two-dimensional problems.

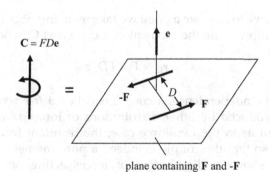

FIGURE 2.16

A couple produced by a pair of equal and opposite forces.

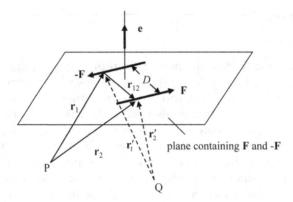

FIGURE 2.17
Geometry for showing that the moment of a couple is independent of the point we take moments about.

the moment of a single force about a point, the moment of a couple generated by a pair of equal and opposite forces is independent of the point we take moments about. This is shown in Figure 2.17 where we see that for moments about point P:

$$
\begin{aligned}
\mathbf{M}_P &= \mathbf{r}_2 \times \mathbf{F} + \mathbf{r}_1 \times (-\mathbf{F}) \\
&= (\mathbf{r}_2 - \mathbf{r}_1) \times \mathbf{F} \\
&= \mathbf{r}_{12} \times \mathbf{F} = FD\ \mathbf{e}
\end{aligned}
\tag{2.14}
$$

But for moments about another point Q, we obtain the same moment:

$$
\begin{aligned}
\mathbf{M}_Q &= (\mathbf{r}_2' - \mathbf{r}_1') \times \mathbf{F} \\
&= \mathbf{r}_{12} \times \mathbf{F} = \mathbf{M}_P
\end{aligned}
\tag{2.15}
$$

so we do not have to indicate a point we take moments about for a couple and we can simply write the moment of a couple as $\mathbf{C}$, where

$$
\mathbf{C} = \mathbf{r}_{12} \times \mathbf{F} = FD\ \mathbf{e}
\tag{2.16}
$$

Because of this independence, a couple is called a *free vector*. Couples can also be produced by other distributions of forces rather than just two forces, but as in the two-force case, the resultant force generated must be zero so that the couple is indeed a pure moment. Figure 2.18, for example, shows a distribution of forces acting on a pipe that generates a couple.

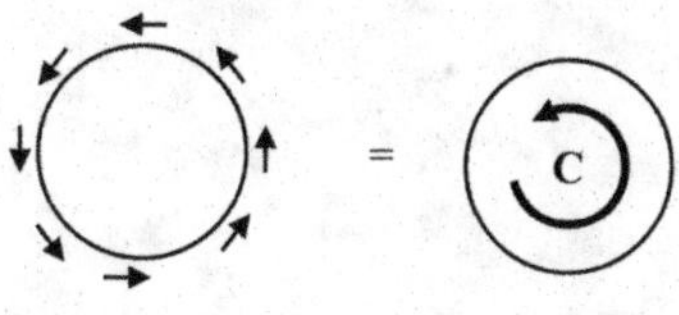

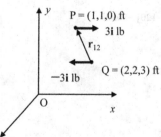

FIGURE 2.18
A distribution of forces that generates a couple.

Example 2.3

Determine the moment of the pair of forces shown in Figure 2.19, where the forces act at points P and Q. All distances and coordinates are measured in feet.

If we do this problem by hand, here are the steps:
First we get a position vector from one force to its opposite

$$\overrightarrow{OP} = i + j \ \text{ft}$$

$$\overrightarrow{OQ} = 2i + 2j + 3k \ \text{ft}$$

$$r_{12} = \overrightarrow{OP} - \overrightarrow{OQ} = -i - j - 3k \ \text{ft}$$

Then we can calculate the moment, which we do through a determinant

$$C = r_{12} \times F = \begin{vmatrix} i & j & k \\ -1 & -1 & -3 \\ 3 & 0 & 0 \end{vmatrix} = i \begin{vmatrix} -1 & -3 \\ 0 & 0 \end{vmatrix} - j \begin{vmatrix} -1 & -3 \\ 3 & 0 \end{vmatrix} + k \begin{vmatrix} -1 & -1 \\ 3 & 0 \end{vmatrix}$$

$$= 0i - 9j + 3k \ \text{ft-lb}$$

The same steps in MATLAB are

FIGURE 2.19
A pair of forces forming a couple.

```
syms i j k
OP = [1 1 0];
OQ = [2 2 3];
r12 = OP - OQ
r12 = -1 -1 -3
Cmat = [i j k; -1 -1 -3; 3 0 0]        % the matrix whose determinant we want

Cmat = [i, j, k]
       [-1, -1, -3]
       [3, 0, 0]
C = det(Cmat)                          % calculate determinant
C = 3*k -9*j                           % the couple in ft-lb
```

Here is another example where we can see the efficiency of the MATLAB solution.

Example 2.4

For the couple shown in Figure 2.20, determine the perpendicular distance between the two forces.

If we do the problem by hand, we must calculate the vector couple. Here are the steps:

$$r_1 = -4i + 4j + 4k \text{ ft}$$
$$r_2 = 4i + 2j + 0k \text{ ft}$$
$$C = (r_1 - r_2) \times F_1$$
$$= (-8i + 2j + 4k) \times F_1$$
$$= \begin{vmatrix} i & j & k \\ -8 & 2 & 4 \\ -70 & -120 & -80 \end{vmatrix} = i\begin{vmatrix} 2 & 4 \\ -120 & -80 \end{vmatrix} - j\begin{vmatrix} -8 & 4 \\ -70 & -80 \end{vmatrix} + k\begin{vmatrix} -8 & 2 \\ -70 & -120 \end{vmatrix}$$
$$= i[-160 + 480] - j[640 + 280] + k[960 + 140]$$
$$= 320i - 920j + 1100k \text{ ft-lb}$$

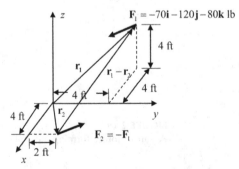

FIGURE 2.20
A couple in three dimensions.

Then, we must calculate the magnitude of the couple and the force, whose ratio gives us the perpendicular distance, D.

$$|\mathbf{C}| = \sqrt{(320)^2 + (-920)^2 + (1100)^2}$$
$$= 1469.29 \text{ ft-lb}$$
$$|\mathbf{F}_1| = \sqrt{(-70)^2 + (-120)^2 + (-80)^2}$$
$$= 160.31 \text{ lb}$$
$$D = \frac{|\mathbf{C}|}{|\mathbf{F}_1|} = \frac{1469.29}{160.31} = 9.17 \text{ ft}$$

The same steps in MATLAB using the cross function are:

```
r1 = [-4 4 4];          % position vectors
r2 = [4 2 0]
F1 = [-70 -120 -80];    % force in the couple
C = cross(r1 - r2, F1)  % moment of the couple
C = 320 -920 1100
D = norm(C)/norm(F1)    % distance in the couple (in ft)
D = 9.1652
```

where we see that all the algebra has been done automatically.

2.3 Moment of a Force about a Line

Figure 2.21(a) shows a force and the moment, $\mathbf{M}_P$, it produces about a point P. The direction of this moment is given by the unit vector $\mathbf{e}$. However, imagine if a body is only free to rotate about an axis (line) that is not along $\mathbf{e}$, such as the line AB seen in Figure 2.21(b). Then, we may need to find that part of the moment of the force that acts along AB to determine the turning action the force has about that line. The moment acting along AB, $\mathbf{M}_{AB}$, is called the *moment of the force about line AB*. In more precise terms, the moment of a force about a line AB is the Cartesian component along AB of the moment about a point P *that is located on that line*, so its scalar value, M_{AB}, is just the dot product of the moment about the point P with a unit vector, $\mathbf{e}_{AB}$, along AB:

$$M_{AB} = \mathbf{M}_P \cdot \mathbf{e}_{AB} \tag{2.17}$$

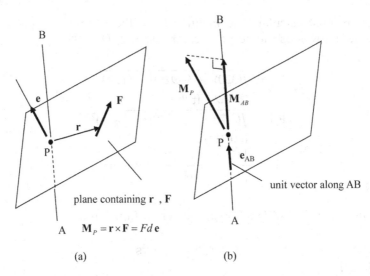

FIGURE 2.21
(a) The calculation of the moment of the force **F** about point P, and (b) the rectangular component of that moment along AB is the moment of the force about that line.

and the vector moment along AB is:

$$\mathbf{M}_{AB} = M_{AB}\mathbf{e}_{AB} \qquad (2.18)$$

To calculate the scalar value of the moment about a line AB in Cartesian coordinates, we have

$$\begin{aligned}
\mathbf{e}_{AB} &= e_x\mathbf{i} + e_y\mathbf{j} + e_z\mathbf{k} \\
\mathbf{r} &= r_x\mathbf{i} + r_y\mathbf{j} + r_z\mathbf{k} \\
\mathbf{F} &= F_x\mathbf{i} + F_y\mathbf{j} + F_z\mathbf{k}
\end{aligned} \qquad (2.19)$$

where $\mathbf{e}_{AB}$ is a unit vector along AB, and $\mathbf{r}$ is a position vector from a point, P, located on line AB, to the force **F**. Then, the scalar moment, M_{AB}, along AB in terms of those components is given either by a cross product followed by a dot product or, equivalently, by a determinant expression:

$$M_{AB} = (\mathbf{r} \times \mathbf{F}) \cdot \mathbf{e}_{AB} = \begin{vmatrix} e_x & e_y & e_z \\ r_x & r_y & r_z \\ F_x & F_y & F_z \end{vmatrix} = e_x\begin{vmatrix} r_y & r_z \\ F_y & F_z \end{vmatrix} - e_y\begin{vmatrix} r_x & r_z \\ F_x & F_z \end{vmatrix} + e_z\begin{vmatrix} r_x & r_y \\ F_x & F_y \end{vmatrix}$$

$$= (r_yF_z - r_zF_y)e_x + (r_zF_x - r_xF_z)e_y + (r_xF_y - r_yF_x)e_z$$

$$(2.20)$$

Example 2.5

A force $\mathbf{F} = 4\mathbf{k}$ lb acts at a point Q, as shown in Figure 2.22. Determine the moment of this force about line OP. All distances are in feet.

Doing this problem by hand, we first need to take moments about some point on line OP so we can choose O. The position vector to the force is then:

$$\mathbf{r} = 2\mathbf{i} + 3\mathbf{j} + 0\mathbf{k} \text{ ft}$$

We also need a unit vector along OP:

$$\mathbf{OP} = 3\mathbf{i} + 3\mathbf{j} + 3\mathbf{k} \text{ ft}$$

$$|\mathbf{OP}| = \sqrt{3^2 + 3^2 + 3^2} = 3\sqrt{3} \text{ ft}$$

$$\mathbf{e}_{OP} = \frac{\mathbf{OP}}{|\mathbf{OP}|} = \frac{1}{\sqrt{3}}(\mathbf{i} + \mathbf{j} + \mathbf{k})$$

Then, the scalar moment along OP is

$$M_{OP} = \frac{1}{\sqrt{3}}\begin{vmatrix} 1 & 1 & 1 \\ 2 & 3 & 0 \\ 0 & 0 & 4 \end{vmatrix} = \frac{1}{\sqrt{3}}\left[1\begin{vmatrix} 3 & 0 \\ 0 & 4 \end{vmatrix} - 1\begin{vmatrix} 2 & 0 \\ 0 & 4 \end{vmatrix} + 1\begin{vmatrix} 2 & 3 \\ 0 & 0 \end{vmatrix}\right]$$

$$= \frac{1}{\sqrt{3}}[12 - 8 + 0] = \frac{4}{\sqrt{3}} = 2.31 \text{ ft-lb}$$

and the vector moment is

$$\mathbf{M}_{OP} = M_{OP}\mathbf{e}_{OP} = \frac{4}{3}(\mathbf{i} + \mathbf{j} + \mathbf{k}) \text{ ft-lb}$$

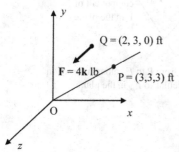

FIGURE 2.22
Geometry for determining the moment of the force F about the line OP.

Doing the problem in MATLAB we find

```
r = [2 3 0];              % position vector from O to the force
F = [0 0 4];              % the force
OP = [3 3 3];             % the vector OP
eOP = OP/norm(OP);        % unit vector along OP
MOP = cross(r, F)*eOP'    % using MOP = (r × F)·eOP
MOP = 2.3094              % scalar moment along OP
MV = MOP*eOP              % vector moment along OP
MV = 1.3333  1.3333  1.3333
```

We can interpret the moment of a force about a line in another way, as shown in Figure 2.23. If we let P be a point on a line AB, we can consider a plane that is perpendicular to the line at P. If we slide the force **F** along its line of action so that the point at its tail also lies in this plane, then we can take the position vector $\mathbf{r}_{pl}$ as the vector in the plane from P to **F**, as shown. Now consider breaking the force into two components – one that is parallel to AB, $\mathbf{F}_{AB}$, and one that lies in the plane, $\mathbf{F}_{pl}$. For the force component in the plane by the definition of the cross product, we have

$$\mathbf{r}_{pl} \times \mathbf{F}_{pl} = F_{pl}d_{pl}\mathbf{e} \qquad (2.21)$$

where F_{pl} is the magnitude of the component of the force that lies in the plane perpendicular to line AB, d_{pl} is the perpendicular distance in that

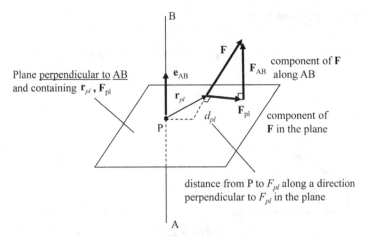

FIGURE 2.23
Geometry for examining the moment about a line.

plane from P to the force, and $\mathbf{e}$ is a unit vector along AB as determined by the right-hand rule from $\mathbf{r}_{pl}$ and $\mathbf{F}_{pl}$, so that it is either in the plus or minus $\mathbf{e}_{AB}$ direction. Now, from the expression for the moment of the force about line AB, we have

$$
\begin{aligned}
\mathbf{M}_{AB} &= (\mathbf{M}_P \cdot \mathbf{e}_{AB})\mathbf{e}_{AB} \\
&= [(\mathbf{r} \times \mathbf{F}) \cdot \mathbf{e}_{AB}]\mathbf{e}_{AB} \\
&= [(\mathbf{r} \times \mathbf{F}_{AB}) \cdot \mathbf{e}_{AB} + (\mathbf{r} \times \mathbf{F}_{pl}) \cdot \mathbf{e}_{AB}]\mathbf{e}_{AB} \qquad (2.22) \\
&= [(\mathbf{r}_{pl} \times \mathbf{F}_{pl}) \cdot \mathbf{e}_{AB}]\mathbf{e}_{AB} \\
&= [F_{pl}d_{pl}(\mathbf{e} \cdot \mathbf{e}_{AB})]\mathbf{e}_{AB}
\end{aligned}
$$

where we have used the fact that $(\mathbf{r} \times \mathbf{F}_{AB}) \cdot \mathbf{e}_{AB} = 0$ since the cross product is perpendicular to $\mathbf{F}_{AB}$, which is parallel to $\mathbf{e}_{AB}$, so the dot product must be zero. Similarly, $(\mathbf{r} \times \mathbf{F}_{pl}) \cdot \mathbf{e}_{AB} = (\mathbf{r}_{pl} \times \mathbf{F}_{pl}) \cdot \mathbf{e}_{AB}$ since any part of $\mathbf{r}$ which is not in the plane produces a zero dot product. Even though $\mathbf{e}$ and $\mathbf{e}_{AB}$ must lie along the same line, they may be in different directions along that line, i.e. $\mathbf{e} = \pm\mathbf{e}_{AB}$, but in either case, we have $(\mathbf{e} \cdot \mathbf{e}_{AB})\mathbf{e}_{AB} = \mathbf{e}$, so from Eq. (2.22) we find

$$
\mathbf{M}_{AB} = \mathbf{r}_{pl} \times \mathbf{F}_{pl} = F_{pl}d_{pl}\mathbf{e} \qquad (2.23)
$$

which shows that the moment of a force about a line can be expressed in the same form as the moment of a force about a point, with indicated changes to the distance and force involved.

There is a close relationship between moments about points and moments about lines. If we write the moment of a force about a point P in terms of its Cartesian components, then we have

$$
\mathbf{M}_P = M_{Px}\mathbf{i} + M_{Py}\mathbf{j} + M_{Pz}\mathbf{k} \qquad (2.24)
$$

where

$$
\begin{aligned}
M_{Px} &= (\mathbf{M}_P \cdot \mathbf{i}) \\
M_{Py} &= (\mathbf{M}_P \cdot \mathbf{j}) \qquad (2.25) \\
M_{Pz} &= (\mathbf{M}_P \cdot \mathbf{k})
\end{aligned}
$$

so that we have

$$
\mathbf{M}_P = (\mathbf{M}_P \cdot \mathbf{i})\mathbf{i} + (\mathbf{M}_P \cdot \mathbf{j})\mathbf{j} + (\mathbf{M}_P \cdot \mathbf{k})\mathbf{k} \qquad (2.26)
$$

But if we recall that the vector moment about a line AB is (see Eq. (2.17) and Eq. (2.18)):

$$\mathbf{M}_{AB} = (\mathbf{M}_P \cdot \mathbf{e}_{AB})\mathbf{e}_{AB} \qquad (2.27)$$

Letting $\mathbf{e}_{AB} = (\mathbf{i}, \mathbf{j}, \mathbf{k})$ successively we see that Eq. (2.26) says:

The moment of a force about a point P is just the sum of the moments of the force about three orthogonal lines (i.e., the x, y, and z axes) taken through P.

We can use this relationship and the expression of the moment of a force about a line in the form given by Eq. (2.23) to easily calculate the moment of a force about a point in terms of its rectangular components without having to memorize those rather complex-looking expressions (recall Eq. (2.9)). For example, consider Figure 2.24. Our relationship, Eq. (2.25), gives the same result we had before, namely,

$$M_{Px} = r_y F_z - r_z F_y$$
$$M_{Py} = r_z F_x - r_x F_z \qquad (2.28)$$
$$M_{Pz} = r_x F_y - r_y F_x$$

but which we now can interpret as moments about lines. For example, consider M_{Px}. This is the moment about an x-axis (line) through point P so that only force components (F_y, F_z) lie in a plane that is perpendicular to the x-line (the y-z plane), and the perpendicular distances from P to those forces in that plane are just (r_z, r_y). Using Eq. (2.23) and the right-hand rule tells us if the moment for a particular force and distance pair is in the plus or minus x-direction (see Figure 2.24 to apply the right-hand rule), we can obtain the signs seen in Eq. (2.28) for M_{Px}. Similar interpretations give us the other two components in Eq. (2.28).

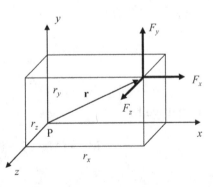

FIGURE 2.24
The Cartesian coordinates of the position vector and Cartesian components of a force vector appearing in a cross product, which can be used to calculate the moment of a force about a point in terms of moments about lines.

Example 2.6

Determine the moment of the 510 lb force in Figure 2.25 (the direction of the force is along line AD) with respect to (a) line BC and (b) with respect to point B. All dimensions are in inches.

First we need to describe the force vectorially:

$$\overrightarrow{AD} = -12i + 8j - 9k$$
$$|\overrightarrow{AD}| = \sqrt{12^2 + 8^2 + 9^2} = 17$$
$$e_{AD} = (-12i + 8j - 9k)/17$$
$$F = 510e_{AD} = -360i + 240j - 270k \ lb$$

Thus, the Cartesian components of the force are as shown in Figure 2.26.

a. Line BC is parallel to the y-axis so the only forces in the x-z plane are the 270 and 360 lb components. But the 270 lb component has no perpendicular distance from B, so it does not give a moment

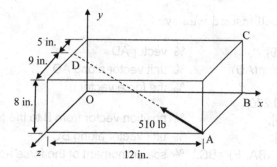

FIGURE 2.25
Geometry for determining the moment of the force about a line and point.

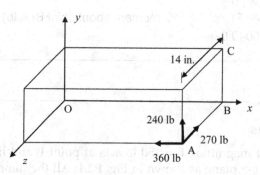

FIGURE 2.26
The Cartesian components of the force.

about BC and we have to only consider the 360 lb force, which gives, using $F_{pl} = 360$ lb, $d_{pl} = 14$in. and the right-hand rule:

$$\mathbf{M}_{BC} = (360)(14)(-\mathbf{j})$$
$$= -5040\mathbf{j} \text{ in} - \text{lb}$$

b. For the moment about point B, consider moments about x-, y-, and z-axes whose origin is at B. We have already calculated the moment about BC, which is this y-axis, so $M_{By} = -5040$ in-lb. For the other two components considering moments about the x- and z-axes (lines) at B, we have

$$M_{Bx} = -(240)(14) = -3360 \text{ in-lb}$$
$$M_{Bz} = 0$$

where the moment in the z-direction is zero since all components act through this z-axis at B. Thus, the moment about point B is

$$\mathbf{M}_B = -3360\mathbf{i} - 5040\mathbf{j} + 0\mathbf{k} \text{ in-lb}$$

Using MATLAB instead we have:

```
AD = [-12 8 9];          % vector AD
eAD = AD/norm(AD);       % unit vector along AD
F = 510*eAD              % the force vector
F = -360 240 270
rBA = [0 0 14];          % position vector from B to the force at A
eBC = [0 1 0];           % unit vector along BC
MBC = cross(rBA, F)*eBC.' % scalar moment of the force about line BC
MBC = -5040
MVBC = MBC*eBC           % vector moment of the force about line BC
MVBC = 0 -5040 0
MB = cross(rBA, F)       % moment about point B (in-lb)
MB = -3360 -5040 0
```

2.4 Problems

P2.1 A force of magnitude F = 150 lb acts at point B and lies in a plane parallel to the y-z plane as shown in Fig. P2.1. All the dimensions are in inches. Determine the moment of the force with respect to point A.

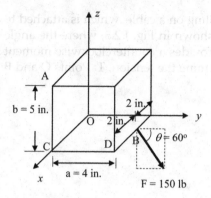

Fig. P2.1

Choices (inch-lb):

1. $-145i -260j -150k$
2. $-260i -145j -150k$
3. $-150i -260j -145k$
4. $145i +260j -150k$
5. $-145i +260j -150k$

P2.2 A force with a magnitude of F = 500 lb acts at point B as shown in Fig. P2.2. Determine the moment of the force about point A.

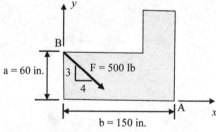

Fig. P2.2

Choices (measured in in-lb, where cw = clockwise, ccw = counterclockwise):

1. 45,000 cw
2. 45,000 ccw
3. 24,000 cw
4. 24,000 ccw
5. 21,000 cw
6. 21,000 ccw

P2.3 A flagpole is being raised by pulling on a cable, which is attached to the pole at A. When in the position shown in Fig. P2.3, where the angle $\alpha = 45°$, the tension, T, in the cable provides a counterclockwise moment about point O of M = 72 kN-m. Determine the tension, T. Points O and B are at the same vertical location.

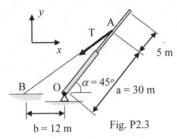

Fig. P2.3

Choices for T (in kN):

1. 13.67
2. 12.47
3. 11.15
4. 9.65
5. 6.75

P2.4 A force given by **F** = 200i+400j-300k lb is located at a point P whose coordinates are $(x, y, z) = (2, 6, 4)$ ft. The moment of this force about point A whose coordinates are $(x, y, z) = (3, 5, 2)$ ft is

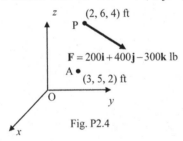

Fig. P2.4

Choices (in ft-lb):

1. 500i +100j + 600k
2. –600i –100j –500k
3. –100i –1100j + 200k
4. 1100i + 100j –600k
5. –1100i + 100j –600k
6. 100i –1100j – 600k

P2.5 A handle extension is placed over a wrench as shown in Fig. P2.5. Determine the moment exerted about O by a force of magnitude F = 7 lb. All dimensions are measured in inches.

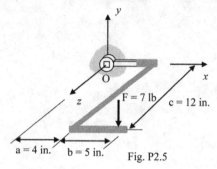

Fig. P2.5

Choices (measured in in-lb):

1. 84i + 0j –35k
2. –84i +35j +0k
3. 84i +0j –63k
4. 63i + 0j –84k
5. 0i +84j –63k
6. –35i +0j –84k

P2.6 A weight is being lifted by a hydraulic cylinder AB. When in the position shown in Fig. P2.6 the angle θ = 40° and force in the cylinder has a magnitude of 10 kN in compression, so the force in the cylinder pushes on the pin at A, in a direction acting from B to A. All dimensions are in mm. Note that AC is perpendicular to the arm OCD. Determine the moment of the force that the hydraulic cylinder exerts on the supporting arm at A with respect to point O (in kN-m).

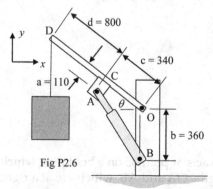

Fig P2.6

P2.7 Determine the magnitude of the scalar component of the moment of the force F = 200 lb about line OB. The force acts at A in a plane parallel to the y-z plane at an angle $\theta = 70°$, as shown in Fig. P2.7. The distances a = 4 in., b = 3 in., c = 2 in.

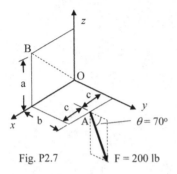

Fig. P2.7

Choices (approximately, in inch-lb):

1. 104
2. 201
3. 302
4. 414
5. 523

P2.8 A special lug wrench is used to tighten a lug nut on a tire (Fig. P2.8). The arm OA of the wrench is along the axis of the bolt holding the nut (parallel to the y-axis) and the wrench OAB lies in a plane parallel to the x-y plane. If a vertical force F = 50 lb (acting in the negative z-direction) is applied to the wrench, determine (a) the total moment exerted on the nut at O, (b) the magnitude of the component of the moment acting on the nut that is effective in tightening the nut. Let a = 7 in., b = 15 in.

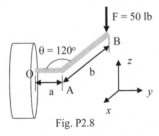

Fig. P2.8

P2.9 A force **F** = 10**i** + 50**j** − 100**k** lb acts at point C on a bent pipe which is screwed into a wall at O. The lengths OA and AB, which are at a right angle to each other, lie in the *x-y* plane, while the length BC is parallel to the *z*-axis. The length of OA is a = 10 in., while the length of AB is b = 5 in., and the length of BC is c = 3 in. Determine the moment that this force produces about point O and the scalar component of the moment of the force along OA which tends to unscrew the connection at the wall.

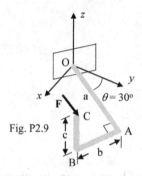

Fig. P2.9

P2.10 A pipe of radius *r* carries a uniform distributed force $f = w_0$ lb/in along its outer surface. Show that this distributed force produces a couple (i.e., a pure moment and no force) and determine the magnitude of that couple if $r = 5$ in., $w_0 = 2$ lb/in. [Hint: integrate the components of the distributed force and the moment it produces about O].

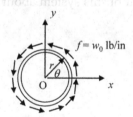

Fig. P2.10

2.4.1 Review Problems

These problems typically have the level of difficulty found on exams. They should be done by hand (i.e., with a calculator).

R2.1 Two disks are rigidly attached to a shaft at B and D as shown in Fig. R2.1. The disk at B lies in a plane parallel to the *y-z* plane, while the disk at D lies in a plane parallel to the *x-z* plane. Couples having force pairs $F_1 = 125$ lb and $F_2 = 250$ lb are applied to those disks. The radii of the disks are $r_1 = 6$ in. and $r_2 = 4$ in. and the distances a = 10 in., b = 18 in., c = 20 in., and d = 5 in. (a) Determine the total moment of this system of couples and (b) determine the total moment about the line AE.

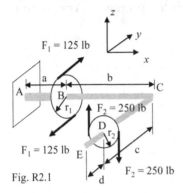

Fig. R2.1

R2.2 Two disks are rigidly attached to a shaft at B and D as shown in Fig. R2.2. The disk at B lies in a plane parallel to the y-z plane, while the disk at D lies in a plane parallel to the x-z plane. A force $F_1 = 450$ N acting in the y-direction is applied to the edge of the disk located at B and a force $F_2 = 200$ N acting in the negative z-direction is applied to the edge of the disk located at D. The radii of the disks are $r_1 = 125$ mm and $r_2 = 50$ mm and the distances a = 600 mm, b = 1500 mm, c = 850 mm, and d = 30 mm. Determine the total moment of this system about the line AE.

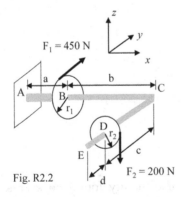

Fig. R2.2

R2.3 A bent pipe is attached to a wall at A and a force P = 350 lb and a force F = 130 lb are applied to the pipe at D as shown in Fig. R2.3. The distances a = 5 ft, b = 4 ft, and c = 2 ft. Determine the total moment of this system acting at A.

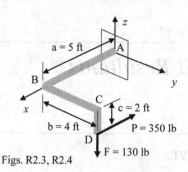

Figs. R2.3, R2.4

R2.4 A bent pipe is attached to a wall at A and a force P = 350 lb and a force F = 130 lb are applied to the pipe at D as shown in Fig. R2.4. The distances a = 5 ft, b = 4 ft, and c = 2 ft. Determine the magnitude of the moment of this system acting about line BC.

Choices (in ft-lb):

1. 260
2. 520
3. 650
4. 700
5. 1750

R2.5 A force couple acting at A and B on a triangular block shown in Fig. R2.5 has force pairs P = 75 lb acting parallel to the z-axis, while another force pair having F = 230 lb acts on the inclined surface of the block parallel to the x-axis. (a) Determine the total moment of this system and (b) determine the moment of this system acting about line AD. The distances a = 40 in, b = 30 in., and c = 60 in.

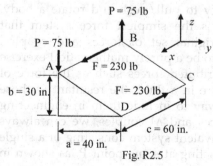

Fig. R2.5

3

Equivalent Systems and Resultants

OBJECTIVES

- To define statically equivalent force systems and their resultants.
- To determine force system resultants in two and three dimensions.
- To reduce a force system to a wrench.
- To demonstrate the use of MATLAB® in determining resultants and in performing symbolic integrations.
- To show the connection between resultants and the center of gravity.
- To obtain resultants for distributed forces.

If we have several forces and couples acting on a body, those forces and couples form a *force/couple system* (which we will simply call *a force system*). Two different force systems are said to be *statically equivalent* if they produce the same total force and moment with respect to any point. Some authors also refer to statically equivalent systems as *equipollent* systems. Thus, two different force systems that are statically equivalent have the same tendency to pull/push and rotate a body. The *resultant* of a force system is the simplest force system that is statically equivalent to the original set of forces and couples. Determining resultants may seem to be a purely mathematical exercise but later when we deal with distributed forces such as the force of gravity or the force due to pressure in a fluid, the resultant of those forces will give us a convenient way to include them in engineering analyses. Given any set of N forces and M couples, we can always reduce the force system to an equivalent system consisting of a single force, $\mathbf{R}$, and a single couple, $\mathbf{M}$, acting at some point P, as shown in Figure 3.1, if we have:

DOI: 10.1201/9781003372592-3

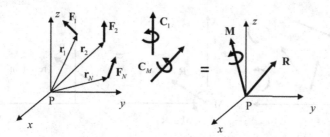

FIGURE 3.1
A force/couple system and an equivalent system of a resultant force and couple.

$$R = \sum_{n=1}^{N} F_n$$

$$M = \sum_{n=1}^{N} r_n \times F_n + \sum_{m=1}^{M} C_m \qquad (3.1)$$

(Note that the couple **M** will change if we change the point P about which we calculate the moments of the forces). For general systems, a single force and couple is the simplest equivalent system so it is the resultant (we will show later that we can reduce a general system to a resultant called a *wrench*, but it also contains a force and a couple).

In some special cases, we can go further and reduce the force system to only a single resultant force. We can do this if we can place the resultant force **R** at some location Q, where:

$$r \times R = M \qquad (3.2)$$

and where **r** is the position vector from P to Q (Figure 3.2). Equation (3.2) has a solution only if **R** and **M** are perpendicular to each other (which we can write as **M⊥R**) since the cross product in Eq. (3.2) must, by its definition, generate a vector that is perpendicular to **R**. One important force system where this reduction to a single resultant force is always possible is for coplanar (two-dimensional) force systems (see Figure 3.3)

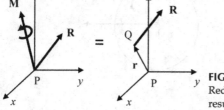

FIGURE 3.2
Reduction of a resultant force and couple to a resultant force.

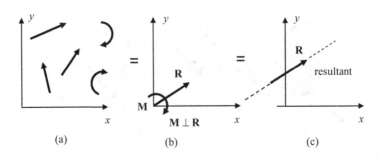

FIGURE 3.3
(a) A coplanar force/couple system and its reduction to (b) a force and couple, and (c) its further reduction to a single resultant force.

because the forces all lie in, say, the x-y plane and all moments and couples are in the plus or minus z-direction. Let's begin by examining some examples of such planar systems.

3.1 Resultants of Two-Dimensional Force Systems

Example 3.1

Determine the resultant of the coplanar force system shown in Figure 3.4 and locate it with respect to point O. The 400 lb force is tangent to the circle.

We will do this problem in three ways, which we will call the scalar method, the vector method, and the MATLAB method.

Scalar Method
In this method, we determine the scalar components of the forces, moments, and the resultant force, and locate the resultant force by its perpendicular distance from point O.

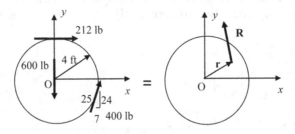

FIGURE 3.4
A force system and its equivalent resultant.

The scalar components of the resultant and its magnitude, R, are

$$R_x = \sum F_x = 212 + \left(\frac{7}{25}\right)(400) = 324$$

$$R_y = \sum F_y = -600 + \left(\frac{24}{25}\right)(400) = -216$$

$$R = \sqrt{R_x^2 + R_y^2} = 389 \text{ lb}$$

The net moment about O we can calculate with $M = Fd$, using perpendicular distances to the forces and taking positive moments acting in the z-direction:

$$M = (4)(400) - (4)(212)$$
$$= 752 \text{ ft-lb}$$

We can locate the resultant force by giving its perpendicular distance, D, from O to the resultant force:

$$RD = M = 752 \text{ ft-lb}$$
$$\rightarrow \quad D = 1.93 \text{ ft}$$

The resultant is shown in Figure 3.5. We can slide the force anywhere along its line of action which has a slope as shown since $R_x/R_y = -1.5$.

Vector Method
In this case, the resultant force is, in terms of Cartesian components,

$$\mathbf{R} = 212\mathbf{i} - 600\mathbf{j} + 400\left(\tfrac{7}{25}\mathbf{i} + \tfrac{24}{25}\mathbf{j}\right)$$
$$= 212\mathbf{i} - 600\mathbf{j} + (112\mathbf{i} + 384\mathbf{j})$$
$$= 324\mathbf{i} - 216\mathbf{j} \text{ lb}$$

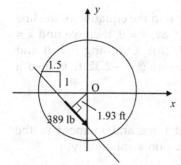

FIGURE 3.5
The resultant force for the force system given in Figure 3.4.

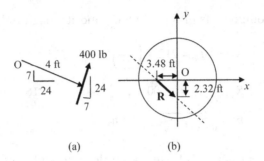

                                  (a)   (b)

FIGURE 3.6
(a) The geometry for defining the location of the 400 lb force of Figure 3.4. (b) The resultant force, **R**, for the system of Figure 3.4 and its location along the *x*- and *y*-axes.

and the resultant moment is (using the geometry of Figure 3.6(a)):

$$\mathbf{M} = \sum \mathbf{M}_O = 4\mathbf{j} \times (212\mathbf{i})$$
$$+4\left(\frac{24}{25}\mathbf{i} - \frac{7}{25}\mathbf{j}\right) \times (112\mathbf{i} + 384\mathbf{j})$$
$$= 752 \ \mathbf{k} \ \text{ft-lb}$$

(To get the position vector from O to the 400 lb force, we used the fact that the radial position vector is perpendicular to the force since the force is tangent to the circle. Thus, the angle of the force with respect to, say, the *x*-axis, is equal to the angle of the position vector with respect to the *y*-axis. This is why the sides of the two triangles seen in Figure 3.6(a) are exchanged.)

Letting the unknown position, **r**, of the resultant force have the coordinates (*x*, *y*) with respect to O, we need to solve the vector equation:

$$\mathbf{r} \times \mathbf{R} = \mathbf{M}$$
$$(x\mathbf{i} + y\mathbf{j}) \times (324\mathbf{i} - 216\mathbf{j}) = 752\mathbf{k}$$
$$\rightarrow -216x - 324y = 752$$

We do not get a definite location. Instead, we find the equation of the line of action of the resultant force. If we choose, say, *y* = 0, then we find *x* = −3.48 ft along the line of action (Figure 3.6(b)). Choosing *x* = 0 and locating the resultant force along the *y*-axis at *y* = −2.32 ft is also a legitimate solution.

MATLAB Method
MATLAB can help us set up the resultant force and moment in the equation **r** × **R** = **M** and then solve that equation symbolically:

```
F1 = [212 0 0];              % define vector forces and sum
F2 = [0 -600 0];             % to get the resultant force
F3 = [400*7/25 400*24/25 0]; % note that these are 3-D vectors
R = F1 + F2 + F3
R = 324 -216 0
r1 = [0 4 0];                % define the position vectors
                             % from O to the forces
r3 = [4*24/25 -4*7/25 0];    % note that r2 = 0.
M = cross(r1, F1) + cross(r3, F3)  % compute the moment
                             % (needs 3-D vectors)
M = 0 0 752
syms y                       % symbolic location along the y-axis
r = [0 y 0];                 % symbolic position vector
double(solve(cross(r, R) - M))  % solve r × R - M = 0
                             % and convert to numerical value
ans = -2.3210                % position along the y-axis
```

Here we chose to find the location along the *y*-axis, but we could have found the location along the *x*-axis in the same fashion. We used the MATLAB function solve to obtain the solution symbolically for the position *y* and then converted that symbolic result to a double precision numerical result with the MATLAB function double.

3.2 Resultants of Three-Dimensional Force Systems

In the last section, we obtained a single resultant force for two-dimensional force systems. This reduction was possible since the solution for a location of the resultant force, $\mathbf{r} \times \mathbf{R} = \mathbf{M}$, could be found (actually, only the line of action is obtained from this equation). This is only possible, as we indicated, if the net force, $\mathbf{R}$, is perpendicular to the net moment, $\mathbf{M}$. Concurrent force systems (Figure 3.7) always reduce to a single force resultant by simply sliding the forces along their lines of action to the common point of concurrency, P, and then summing the forces. Parallel force systems acting in three dimensions also can be reduced to single force resultant since the net moment always acts in a plane perpendicular to the direction of the resultant force (see Figure 3.8).

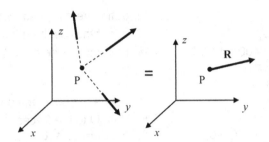

FIGURE 3.7
A concurrent force system reduces to a single equivalent resultant force.

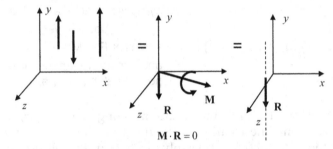

FIGURE 3.8
A parallel force system can be reduced to a single resultant force.

Example 3.2

Consider the parallel force system shown in Figure 3.9. Determine the resultant of this parallel force system and the intersection of the line of action of the resultant with the *x*-*z* plane.

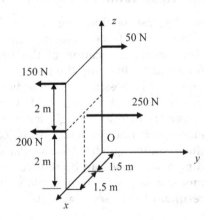

FIGURE 3.9
A parallel force system example.

Vector Method
First, we calculate the total force and the components of the moment
about point O:

$$R_y = \Sigma F_y = 50 + 250 - 150 - 200$$
$$= -50 \text{ N}$$

$$\Sigma M_{Ox} = (150)(4) + (200)(2) - (250)(2) - (50)(4)$$
$$= 300 \text{ N-m}$$
$$\Sigma M_{Oz} = (250)(1.5) - (150)(3) - (200)(3)$$
$$= -675 \text{ N-m}$$

This result is shown in Figure 3.10(a). To locate the force, we solve $\mathbf{r} \times$
$\mathbf{R} = \mathbf{M}$:

$$\mathbf{r} \times \mathbf{R} = 300\mathbf{i} - 675\mathbf{k}$$
$$(x\mathbf{i} + y\mathbf{j} + z\mathbf{k}) \times (-50\mathbf{j}) = 300\mathbf{i} - 675\mathbf{k}$$
$$-50x\mathbf{k} + 50z\mathbf{i} = 300\mathbf{i} - 675\mathbf{k}$$
$$x = \frac{-675}{-50} = 13.5 \text{ m}$$
$$z = \frac{300}{50} = 6 \text{ m}$$

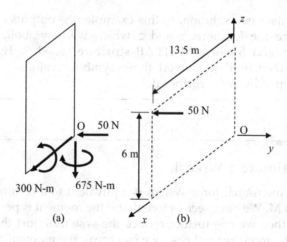

FIGURE 3.10
(a) The total force and moment of the parallel force system in Figure 3.9 at point O. (b) The
location of the force resultant.

We obtain a point (x, z) but no y-value since the term with y is absent in the cross product and we could place the force anywhere along the y-axis, including $y = 0$. The result with $y = 0$ is shown in Figure 3.10(b).

MATLAB Method
Here we use MATLAB to determine the resultant force and moment and then solve the equation $\mathbf{r} \times \mathbf{R} = \mathbf{M}$ symbolically:

```
ey = [0 1 0];                          % unit vector in y-direction
F1 = 250*ey;                           % the forces
F2 = 50*ey;
F3 = -150*ey;
F4 = -200*ey;
R = F1 + F2 + F3 + F4;                 % resultant force
r1 = [1.5 0 2];                        % the position vectors
r2 = [0 0 4];
r3 = [3 0 4];
r4 = [3 0 2];
M = cross(r1, F1) + cross(r2, F2)  + cross(r3, F3) + cross(r4, F4);
M = 300   0  -675                      % the total moment
syms x z                               % symbolic locations
r = [x 0 z];                           % symbolic position vector
S = solve(cross(r, R) - M);            % solve rxR - M = 0 and place in structure S
double([S.x S.z])                      % obtain numerical values for x and z(in m)
ans = 13.5000   6.0000
```

which gives the same solution. In this example, the outputs of the solve function were the locations x and z which were symbolic values in the fields S.x and S.z of a MATLAB structure called S. The function double was then used to convert those symbolic values to numerical ones. See Appendix A for further details.

3.3 Reduction to a Wrench

Any three-dimensional force system is equivalent to a resultant force $\mathbf{R}$ and moment $\mathbf{M}$. We have seen before that if the moment is perpendicular to the force then we can further reduce the system to just the resultant force $\mathbf{R}$. In the more general case, we can break the moment $\mathbf{M}$ into two components – one that is parallel to the resultant force, $\mathbf{R}$, which we write as $\mathbf{M}_{\parallel}$, and one that is perpendicular to the force, $\mathbf{M}_{\perp}$ (see Figure 3.11). We can solve the equation

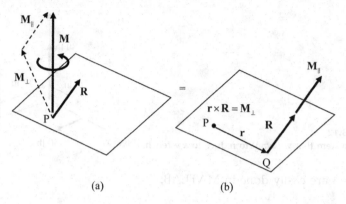

(a)   (b)

FIGURE 3.11
(a) A force-couple system where we decompose the couple into components along and perpendicular to the force **R**, and (b) the reduction of the system to a wrench.

$$\mathbf{r} \times \mathbf{R} = \mathbf{M_\perp} \tag{3.3}$$

which will move the force **R** so as to eliminate $\mathbf{M_\perp}$, leaving the resultant force **R** and the moment $\mathbf{M_{\|}}$, as shown in Figure 3.11. This combination of force and a moment that is parallel to it is called a *wrench* since when we try to loosen or tighten a nut on a shaft with a wrench, we are usually applying a moment along the shaft as well as pushing on the shaft. The steps in the reduction to a wrench are:

1. Calculate the total force **R**, in the force system and a couple **M** equal to the total moment about a point P, i.e., $\mathbf{M} = \mathbf{M}_P$.
2. Determine a unit vector along **R**, by dividing the force by its magnitude, $|\mathbf{R}|$, $\mathbf{e}_R = \mathbf{R}/|\mathbf{R}|$.
3. Determine the moment along **R**, $\mathbf{M_{\|}} = (\mathbf{M \cdot e}_R)\mathbf{e}_R$.
4. The remainder of **M** is the part that is perpendicular to **R**, $\mathbf{M_\perp} = \mathbf{M} - (\mathbf{M \cdot e}_R)\mathbf{e}_R$.
5. Solve $\mathbf{r} \times \mathbf{R} = \mathbf{M_\perp}$ to eliminate $\mathbf{M_\perp}$, leaving the wrench **R** and $\mathbf{M_{\|}}$.

The next example illustrates these steps.

Example 3.3

Reduce the force system of Figure 3.12 to a wrench and locate the wrench in (1) the *x-z* plane, and (2) the *x-y* plane.

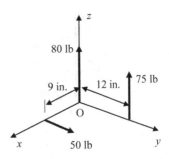

FIGURE 3.12
A force system that we want to reduce to a wrench.

Steps 1–4 are easily done in MATLAB:

```
R = [0 50 (75 + 80)];                 % total force
M = [75*12 0 50*9];                   % total moment about O
eR = R/norm(R);                       % unit vector along R
Mpar = dot(M, eR)*eR                  % vector component of M along R
Mpar = 0 131.4797 407.5872
Mperp = M - Mpar                      % vector component perpendicular to R
Mperp = 900.0000 -131.4797 42.4128
```

Now we want to solve $\mathbf{r} \times \mathbf{R} = \mathbf{M}_\perp$. Let's do this by hand:
Let $\mathbf{r} = x\mathbf{i} + y\mathbf{j} + z\mathbf{k}$, giving

$$(x\mathbf{i} + y\mathbf{j} + z\mathbf{k}) \times (50\mathbf{j} + 155\mathbf{k}) = 900\mathbf{i} - 131.4797\mathbf{j} + 42.4128\mathbf{k}$$
$$= 50x\mathbf{k} - 155x\mathbf{j} + 155y\mathbf{i} - 50z\mathbf{i}$$

This yields the three equations

$$-50z + 155y = 900$$
$$-155x = -131.4797$$
$$50x = 42.4128$$

The second and third equations both yield $x = 0.8483$ in. If we set $y = 0$ in the first equation to locate the wrench in the x-z plane, we find $z = -18$ in. This solution is shown in Figure 3.13. If instead, we set $z = 0$ in the first equation to locate the wrench in the x-y plane, we find $y = 5.807$, as shown in Figure 3.14. Note that the wrench is given explicitly by $\mathbf{R} = 50\mathbf{j} + 155\mathbf{k}$ lb, and $\mathbf{M}_\| = 131.48\mathbf{j} + 407.59\mathbf{k}$ in-lb.

If we try to find the location with a MATLAB symbolic solution procedure, as done in previous examples, we will be unsuccessful since we see that two of the equations in $\mathbf{r} \times \mathbf{R} = \mathbf{M}_\perp$ are not independent. In Appendix C of the addendum to this text (see www.eng-statics.org), we define a MATLAB function R_loc that can be used to overcome this

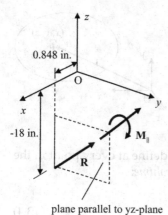

FIGURE 3.13
Reduction of the force system in Figure 3.12 to a wrench located in the *x-z* plane.

plane parallel to yz-plane

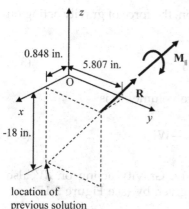

location of
previous solution

FIGURE 3.14
Reduction of the force system of Figure 3.12 to a wrench located in the *x-y* plane.

problem and solve for a particular location of the resultant. However, it is not necessary to use that function since we can always easily solve $\mathbf{r} \times \mathbf{R} = \mathbf{M}_\perp$ by hand, as just shown.

3.4 Resultants and the Center of Gravity

Parallel force systems can always be reduced to a single resultant force. This fact has significant practical implications for distributed parallel forces like the force of gravity. Gravity acts on every part of a body (Figure 3.15) so imagine if we take a small volume element, dV, of a body and divide its weight, dW, by that volume. If we take the limit of this

Engineering Statics with MATLAB®

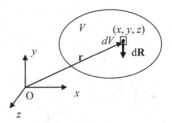

FIGURE 3.15
Gravity acting on a small volume element of a body.

ratio as the volume dV shrinks to zero, we can define at every point in the body a quantity, γ, called the *weight per unit volume*:

$$\gamma = \lim_{dV \to 0} \frac{dW}{dV} \tag{3.4}$$

Assuming it acts in the negative y-direction, the force of gravity acting on a small volume dV is then

$$d\mathbf{R} = -\gamma dV \mathbf{j} \tag{3.5}$$

and the total force, $\mathbf{R}$, acting on the entire volume, V, is

$$\mathbf{R} = -\mathbf{j} \int_V \gamma dV = -W\mathbf{j} \tag{3.6}$$

where W is the total weight of the body. Gravity acting on dV also produces a moment about the origin O given by (see Figure 3.15)

$$\begin{aligned} d\mathbf{M}_0 &= \mathbf{r} \times d\mathbf{R} \\ &= (x\mathbf{i} + y\mathbf{j} + z\mathbf{k}) \times (-\gamma dV\mathbf{j}) \end{aligned} \tag{3.7}$$

so the total moment of the force of gravity about O is

$$\begin{aligned} \mathbf{M}_0 &= \int_V (x\mathbf{i} + y\mathbf{j} + z\mathbf{k}) \times (-\gamma dV\mathbf{j}) \\ &= -\mathbf{k} \int_V x\gamma dV + \mathbf{i} \int_V z\gamma dV \end{aligned} \tag{3.8}$$

Since the total force, Eq. (3.6), is perpendicular to the total moment, Eq. (3.8), we can replace this distributed gravity force by a single force acting at a point G called the *center of gravity* (Figure 3.16). To find the center of gravity, we must satisfy

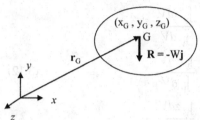

FIGURE 3.16
The center of gravity.

$$\mathbf{r}_G \times \mathbf{R} = \mathbf{M}_0$$

$$\begin{vmatrix} \mathbf{i} & \mathbf{j} & \mathbf{k} \\ x_G & y_G & z_G \\ 0 & -W & 0 \end{vmatrix} = -\mathbf{k}\int_V xydV + \mathbf{i}\int_V zydV \tag{3.9}$$

$$z_G W\mathbf{i} - x_G W\mathbf{k} = -\mathbf{k}\int_V xydV + \mathbf{i}\int_V zydV$$

where we have written the cross product as determinant which then was expanded. It follows that:

$$x_G = \frac{\int_V xydV}{W} = \frac{\int_V xydV}{\int_V ydV}$$

$$z_G = \frac{\int_V zydV}{W} = \frac{\int_V zydV}{\int_V ydV} \tag{3.10}$$

Because we assumed gravity acted along the y-axis, we do not get a location along that axis, but if we assumed that gravity acted in a direction parallel to one of the other axes, we would obtain a y-location also given by

$$y_G = \frac{\int_V yydV}{W} = \frac{\int_V yydV}{\int_V ydV} \tag{3.11}$$

If the body is homogeneous so that y is a constant, then the weight/unit volume cancels out and we find $(x_G, y_G, z_G) = (x_C, y_C, z_C)$, which is a point called the *centroid* of the volume V, where

$$x_C = \frac{\int_V x\, dV}{V} = \frac{\int_V x\, dV}{\int_V dV}$$

$$y_C = \frac{\int_V y\, dV}{V} = \frac{\int_V y\, dV}{\int_V dV} \qquad (3.12)$$

$$z_C = \frac{\int_V z\, dV}{V} = \frac{\int_V z\, dV}{\int_V dV}$$

In Chapter 7, centroids will be discussed in more detail. From Eq. (3.12), it is obvious that the centroid depends only on the geometry of the body.

Note that if we use the *mass/unit volume*, ρ, in these calculations instead of the weight per unit volume, then we can define the *center of mass* location as

$$x_m = \frac{\int_V x\rho\, dV}{M} = \frac{\int_V x\rho\, dV}{\int_V \rho\, dV}$$

$$y_m = \frac{\int_V y\rho\, dV}{M} = \frac{\int_V y\rho\, dV}{\int_V \rho\, dV} \qquad (3.13)$$

$$z_m = \frac{\int_V z\rho\, dV}{M} = \frac{\int_V z\rho\, dV}{\int_V \rho\, dV}$$

where M is the total mass of the body. When the body is homogeneous, this again reduces to the location of the centroid.

When the geometry of a homogeneous body has a plane of symmetry, then the center of gravity, center of mass, and the centroid will all lie in that plane. Bodies such as spheres, cylinders, and rectangular cuboids have three intersecting orthogonal planes of symmetry so the centroid will lie at the point of intersection of all three planes, which will be at the geometric center (Figure 3.17). The centroid of a body of revolution such as a cone or hemisphere will lie on the axis of revolution (see also Figure 3.17).

Example 3.4

Prove that the volume and the vertical location of the centroid are the values given in Figure 3.17 for the cone.

Take a set of (x, y, z) axes as shown in Figure 3.18(a), whose origin is at the tip of the cone rather than at the base as shown in Figure 3.17. Then the radius, ρ, of the cone is given by $\rho = rz/h$. The volume element dV can be taken as that of a thin cylinder of radius ρ and height dz, where $dV = \pi\rho^2 dz$ (Figure 3.18(b)). Then we have the volume given by

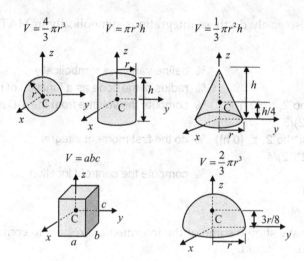

FIGURE 3.17
Centroid locations and volumes for a sphere, cylinder, and rectangular cuboid where the centroids are at the geometric centers. Centroid locations and volumes for a cone and a hemisphere, which are bodies of revolution where the centroid lies on the axis of revolution.

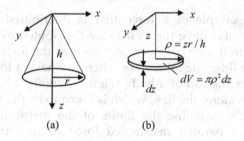

FIGURE 3.18
(a) The coordinate system. (b) A volume element of the cone.

$$V = \int_{z=0}^{z=h} \pi \rho^2 dz = \frac{\pi r^2}{h^2} \int_{z=0}^{z=h} z^2 dz = \frac{\pi r^2 h}{3} \tag{3.14}$$

And the so-called first moment integral is

$$z_c V = \int_{z=0}^{z=h} z dV = \frac{\pi r^2}{h^2} \int_0^h z^3 dz = \frac{\pi r^2 h^2}{4}$$

$$\rightarrow \quad z_c = \frac{3}{4}h \tag{3.15}$$

which agrees with Figure 3.17 where the distance given is measured from the base of the cone instead.

We can also easily do these integrations symbolically in MATLAB. We have:

```
syms r z h                 % define variables symbolically
rho = r*z/h;               % radius of the cone as a function of r, z, h
V = int(pi*rho^2, z, [0 h])   % compute the volume from the 1-D integration
V = (pi*h*r^2)/3
zV = int(z*pi*rho^2, z, [0 h])  % do the first moment integral
zV = (pi*h^2*r^2)/4
zc = zV/V                  % compute the centroid location
zc = (3*h)/4
```

where we have shown some of the intermediate results to compare with our previous solution.

3.5 Resultants of Distributed Forces

Gravity is an example of a force that is distributed throughout a volume. We can also have forces that are distributed over areas (such as the pressure in a fluid) or over lines such as the forces acting along beams. Let's consider the latter case where we have a load distributed along a line over the top of a body (Figure 3.19(a)), where $w(x)$ is the load/unit length along the line, which is taken to be the x-axis. We will let $x = x_1$ and $x = x_2$ define the limits of the distributed load. Like gravity, this is a parallel distributed force so we can determine a resultant vertical force, R, and its location, x_R, that gives the same net

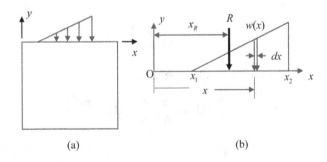

(a)   (b)

FIGURE 3.19
(a) A force distributed over a line on a body. (b) Geometry for determining the resultant force from the distributed line load.

force and moment as the load distribution (Figure 3.19(b)). The force due to a small length, dx, of the distributed load located at x is $w(x)dx$ so the total resultant force due to the sum of all such small segments is in the limit, as these small segments go to zero, an integral over the extent of the load, i.e.

$$R = \int_{x=x_1}^{x=x_2} w(x)dx \tag{3.16}$$

Similarly, the moment about point O of the force $w(x)dx$ acting at x is $xw(x)$ dx so the total moment about point O for the distributed load is the integral

$$M_O = \int_{x=x_1}^{x=x_2} xw(x)dx \tag{3.17}$$

This moment must be equal to the moment produced about O by the resultant force R so we can use that relationship to locate the distance, x_R, from O to the resultant:

$$x_R R = M_0 = \int_{x=x_1}^{x=x_2} xw(x)dx$$

$$\rightarrow x_R = \frac{\int_{x=x_1}^{x=x_2} xw(x)dx}{R} \tag{3.18}$$

If we view the intensity of the distributed load, $w(x)$, as the height of a curve above the x-axis that describes this distribution, then we can interpret Eq. (3.16) as saying that the resultant force is just the total area under that curve. Similarly, as we will see in Chapter 7, the location of the resultant can be interpreted as saying that the location of the resultant is at the *centroid* of the area under the loading curve. For simple shapes such as a uniform load or a linearly increasing load, as shown in Figure 3.20, the value and location of the resultant are easy to determine. Consider first the constant load case. The resultant force and its location are found as

$$R = \int_0^L wdx = w\int_0^L dx = wL$$

$$\int_0^L xwdx = w\int_0^L xdx = \frac{wL^2}{2} \tag{3.19}$$

$$\rightarrow x_R = \frac{1}{R}\int_0^L xwdx = \frac{L}{2}$$

while for the linearly increasing load where $w(x) = wx/L$, we have

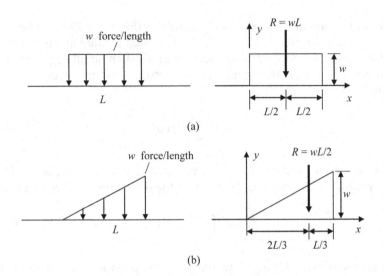

FIGURE 3.20
(a) A distributed force whose intensity varies quadratically. (b) The resultant force and its location.

$$R = \int_0^L w\,dx = \frac{w}{L}\int_0^L x\,dx = \frac{wL}{2}$$

$$\int_0^L xw\,dx = \frac{w}{L}\int_0^L x^2\,dx = \frac{wL^2}{3} \tag{3.20}$$

$$\rightarrow x_R = \frac{1}{R}\int_0^L xw\,dx = \frac{2L}{3}$$

and these locations are shown in Figure 3.20.

Example 3.5

Consider a distributed load whose intensity varies quadratically, as shown in Figure 3.21(a). Determine the resultant force and its location.
  For the resultant force, we have

$$R = \int_{x=0}^{x=L} w\,dx = \frac{w_0}{L^2}\int_0^L x^2\,dx = \frac{w_0 L}{3} \tag{3.21}$$

and its location is given as

$$x_R R = \int_{x=0}^{x=L} xw\,dx = \frac{w_0}{L^2}\int_0^L x^3\,dx = \frac{w_0 L^2}{4}$$

$$\rightarrow x_R = \frac{3L}{4} \tag{3.22}$$

These results are shown in Figure 3.21(b). If we do the problem in MATLAB, we have

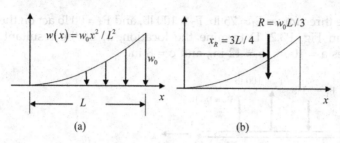

FIGURE 3.21
(a) A constant distributed load and its resultant. (b) A linearly increasing distributed load and its resultant.

```
syms w0 L x                          % define symbolic variables
R = int(w0*x^2/L^2, x, [0 L])        % resultant force
R = (L*w0)/3
xR = int(w0*x^3/L^2, x, [0 L])       % first moment integral
xR = (L^2*w0)/4
xc = xR/R                            % location of centroid of area
                                     % under the distributed load
xc = (3*L)/4
```

3.6 Problems

P3.1 A force P = 500 lb acting at an angle $\theta = 40°$ and a force F = 175 lb act on a bracket as shown in Fig. P3.1. Determine the magnitude of the resultant force and the angle it makes with respect to the positive x-axis. Locate the resultant by giving the perpendicular distance from A to the line of action of the resultant and indicating if the perpendicular distance is to the right or left of A. The distances a = 2 in., b = 7 in., and c = 12 in.

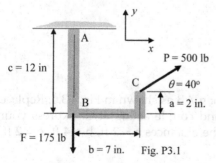

Fig. P3.1

P3.2 The three forces $F_1 = 75$ lb, $F_2 = 100$ lb, and $F_3 = 90$ lb act on the beam shown in Fig. P3.2. Determine the location, x, of the resultant if the distances a = 10 in., b = 12 in., and c = 8 in.

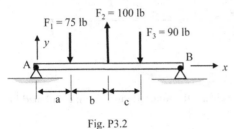

Fig. P3.2

Choices (in inches):

1. 16.5
2. 17.8
3. 18.9
4. 19.2
5. 20.1

P3.3 Forces P = 5 kN, F = 3 kN, and T = 10 kN act on the truss shown in Fig. P3.3. Determine the resultant of these forces and locate the resultant by giving the perpendicular distance from A to the line of action of the resultant and indicating if the perpendicular distance is to the right or left of A. The distances a = 2 m and b = 3 m.

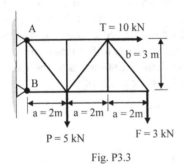

Fig. P3.3

P3.4 A force F = 300 lb acts at A along AB as shown in Fig. P3.4. Replace this force by an equivalent force and couple acting at C. Express your answer in Cartesian vector form. The distances a = 2 ft, b = 4 ft, c = 2 ft, and d = 3 ft.

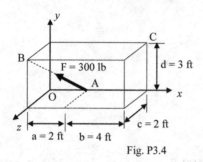

Fig. P3.4

P3.5 Three forces $P_1 = 450$ lb, $P_2 = 70$ lb, and $P_3 = 30$ lb act on a post as shown in Fig. P3.5. Determine the y-coordinate location of the resultant of these forces along the post. The distances a = 2 ft, b = 3 ft, and c = 4 ft and the angle $\theta = 30°$.

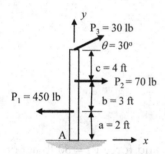

Fig P3.5

Choices (in ft):

1. 0.93
2. 2.85
3. 1.96
4. 0.78
5. 0.89

P3.6 The three forces $P_1 = 4$ kN, $P_2 = 6$ kN, and $P_3 = 3$ kN all act parallel to the z-axis on a U-shaped frame as shown in Fig. P3.6. Determine the resultant of this force system and its location in the x-y plane. The distances a = 6 m, b = 3 m, and c = 2 m.

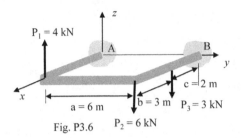

Fig. P3.6 $P_2 = 6$ kN

P3.7 The force $P_1 = 75$ lb, force $P_2 = 90$ lb, and force $P_3 = 150$ lb act on the T-shaped bar in Fig. P3.7. Replace this force system with a force and a couple acting at O. The distances a = 20 in., b = 9 in., and c = 8 in.

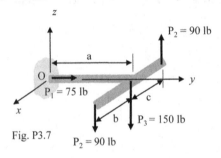

Fig. P3.7 $P_2 = 90$ lb

P3.8 The force $P_1 = 125$ lb, forces $P_2 = 60$ lb, and force $P_3 = 80$ lb act on the T-shaped bar in Fig. P3.8. Determine the resultant of this force system and locate it along the z-axis. The distances a = 20 in. and b = 18 in.

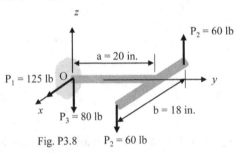

Fig. P3.8 $P_2 = 60$ lb

P3.9 Replace the system of forces $F_1 = 75$ lb, $F_2 = 90$ lb, and $F_3 = 60$ lb shown in Fig. P3.9 by an equivalent wrench and determine the location of this wrench in the x-z plane. The distances a = 6 in. and b = 12 in.

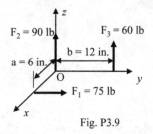

Fig. P3.9

P3.10 Replace the system of forces $P_1 = 30$ lb, $P_2 = 300$ lb, and $P_3 = 200$ lb shown in Fig. P3.10 by an equivalent wrench and determine the location of this wrench in the x-y plane. The distances a = 6 ft. and b = 4 ft., c = 3 ft, and d = 5 ft.

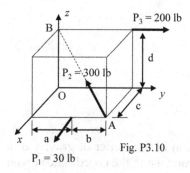

Fig. P3.10

P3.11 A distributed load $w = 100\sqrt{x}$ lb/ft acts on the beam in Fig. P3.11 which has a total length L = 12 ft. Determine the resultant force and its location along the x-axis.

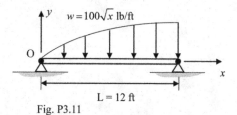

Fig. P3.11

P3.12 The beam of Fig. P3.12 carries a distributed load that varies linearly from zero to a maximum intensity of w = 25 lb/ft and then back to zero. Determine the resultant force and its location along the x-axis. The distances a = 3 ft and b = 7 ft.

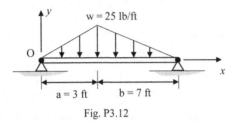

Fig. P3.12

P3.13. Determine by integration the location of the center of gravity of a homogeneous hemisphere of radius $r = 4$ in. using the coordinates given in Fig. P3.13. Show that your results agree with values obtained from Figure 3.17.

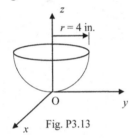

Fig. P3.13

P3.14 Determine by integration the location of the center of gravity of a quarter of a homogeneous right circular cone using the coordinates given in Fig. P3.14.

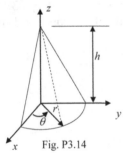

Fig. P3.14

3.6.1 Review Problems

These problems typically have the level of difficulty found on exams. They should be done by hand (i.e., with a calculator).

R3.1 The three forces $P_1 = 80$ N, $P_2 = 30$ N, and $P_3 = 100$ N together with the couple $C = 15$ N-m act on a plate as shown in Fig. R3.1. Determine the resultant of this coplanar force system and its location along the *x*-axis. The distances a = 2 m, b = 3 m, and the angle θ = 30°.

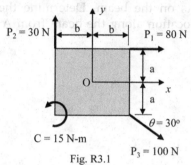

Fig. R3.1

R3.2 Replace the three parallel forces P_1 = 30 lb, P_2 = 50 lb, and P_3 = 100 lb in Fig. R3.2 by an equivalent single resultant force and determine its location in the *x-y* plane. The distance a = 2 ft and b = 3 ft.

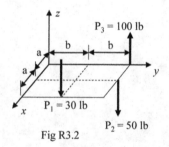

Fig R3.2

R3.3 A force system consists of a couple C_1 = 50 ft-lb acting in the negative *z*-direction, a couple C_2 = 30 ft-lb acting in the *y*-direction, and a force P_1 = 10 lb acting in the *y*-direction, as shown in Fig. R3.3. Replace this force system with a wrench and determine the location of this wrench along the *x*-axis. The distances a = 6 ft and b = 8 ft.

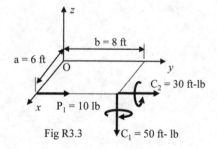

Fig R3.3

R3.4 A distributed load on a beam has an intensity that varies linearly from $w_1 = 10$ lb/ft to $w_2 = 15$ lb/ft, as shown in Fig. R3.4. A concentrated force P =150 lb also acts on the beam. Determine the resultant of this force system and its location along the beam from A. The distances a = 6 ft and b = 3 ft.

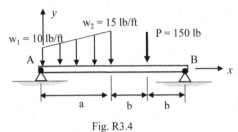

Fig. R3.4

4

Equilibrium

OBJECTIVES

- To define the equations of equilibrium of rigid bodies in 2-D and 3-D problems.
- To demonstrate the importance of free body diagrams.
- To define common types of reaction forces and moments.
- To illustrate the use of free body diagrams and the equations of equilibrium.
- To apply various sets of equilibrium equations.

The main objective of a statics course is to examine the condition under which a rigid body remains at rest, or in a state of *"equilibrium"*, where there is no tendency of the body to translate or rotate. Requiring that the body be rigid means that the shape of the body does not change when forces or moments are applied to it so that we can draw those forces and moments acting on the original unloaded shape. The condition for equilibrium is that the resultant of the force system must be zero. For a general system of forces and couples, equilibrium, therefore, requires that:

$$\mathbf{R} = \sum \mathbf{F}_i = 0$$
$$\mathbf{M} = \sum \mathbf{r}_i \times \mathbf{F}_i + \sum \mathbf{C}_j = 0 \tag{4.1}$$

where $\mathbf{R}$ and $\mathbf{M}$ are the resultant force and couple acting on the system, respectively. These vector equilibrium equations are equivalent to six scalar equations:

$$\sum F_{ix} = 0 \quad \sum (\mathbf{r}_i \times \mathbf{F}_i)_x + \sum C_{jx} = 0$$
$$\sum F_{iy} = 0 \quad \sum (\mathbf{r}_i \times \mathbf{F}_i)_y + \sum C_{jy} = 0 \tag{4.2}$$
$$\sum F_{iz} = 0 \quad \sum (\mathbf{r}_i \times \mathbf{F}_i)_z + \sum C_{jz} = 0$$

DOI: 10.1201/9781003372592-4

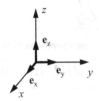

FIGURE 4.1
Unit vectors along a Cartesian coordinate system.

where the components of the moments of the forces about the (x, y, z) axes are:

$$(\mathbf{r}_i \times \mathbf{F}_i)_x = (\mathbf{r}_i \times \mathbf{F}_i) \cdot \mathbf{e}_x$$
$$(\mathbf{r}_i \times \mathbf{F}_i)_y = (\mathbf{r}_i \times \mathbf{F}_i) \cdot \mathbf{e}_y \qquad (4.3)$$
$$(\mathbf{r}_i \times \mathbf{F}_i)_z = (\mathbf{r}_i \times \mathbf{F}_i) \cdot \mathbf{e}_z$$

and $(\mathbf{e}_x, \mathbf{e}_y, \mathbf{e}_z)$ are unit vectors along the axes of a Cartesian coordinate system (Figure 4.1). For two-dimensional (coplanar) systems, where all the forces lie in an x-y plane and all the moments and couples are in the plus or minus z-direction, these six equations reduce to three scalar equations:

$$\sum F_{ix} = 0$$
$$\sum F_{iy} = 0 \qquad (4.4)$$
$$\sum (\mathbf{r}_i \times \mathbf{F}_i)_z + \sum C_{jz} = 0$$

The point that we use to compute the moments of the forces in either two-dimensional or three-dimensional problems is arbitrary and if the resultant force and couple in Eq. (4.1) is zero when we use a point P to calculate the moments of the forces, the resultant force and couple will also be zero if we use any other point Q.

4.1 Free Body Diagrams

In applying equilibrium equations, we must include all forces and couples acting on a body. These usually are separated into three types:

1. applied forces or couples – some, or all of which may be known.
2. reaction forces or couples which arise from the body interacting with other bodies. These are generally unknown.
3. internal forces/couples, which arise from interactions internal to a body. They also are typically unknown.

To ensure that we have included all the forces and couples in the equilibrium equations, we remove all the physical interactions of a body with other bodies and replace them with equivalent reaction forces and couples. If a cut is made through a body to expose internal forces, they must also be included. *A diagram of all the forces acting on the body is called a free body diagram.* Generally, we also place all pertinent dimensions on free body diagrams so that we can apply the equilibrium equations directly from the diagrams.

Consider, for example, the beam of Figure 4.2(a), which has a 300 lb applied force acting on it and is supported by two rollers. Such a beam is called a simply supported beam. If we want to find all the forces acting on the beam, we must imagine removing the supports and replacing them with reactions coming from those supports. In this case of a simply supported beam, those reactions are just vertical forces A and B acting at the supports, as shown in Figure 4.2(b). If there are no other forces (such as the weight of the beam, for example, which is neglected here), then the diagram of Figure 4.2(b) is a free body diagram. Once we have a free body diagram, we can use the equations of equilibrium. Here, summing forces and taking moments about point A, we have

$$
\begin{aligned}
\sum F_x &= 0 & 0 &= 0 \\
\sum F_y &= 0 & A + B - 300 &= 0 \\
\sum M_A &= 0 & 10B - (5)(300) &= 0 \\
&\rightarrow A = 150 \text{ lb}, & B &= 150 \text{ lb}
\end{aligned}
\tag{4.5}
$$

and we can solve for the reaction forces at the support points A and B. They are both equal to half the applied force, which we could also have deduced from the symmetry of the problem. Reaction forces and couples

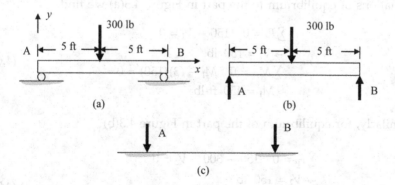

FIGURE 4.2
(a) A simply supported beam. (b) A free body diagram of the beam showing the reaction forces at A and B. (c) The equal and opposite reaction forces acting on the ground.

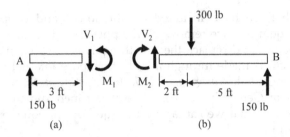

FIGURE 4.3
(a), (b) Free body diagrams of parts of the beam of Figure 4.2(a).

always occur in equal and opposite pairs. Here, A and B are the forces
that the ground exerts on the beam through the roller supports and there
are equal and opposite forces that are exerted on the ground, as shown in
Figure 4.2(c). Unlike Figure 4.2(b), however, Figure 4.2(c) is not a free
body diagram since there are other forces present in the ground (that are
not shown) needed to keep it in equilibrium.

Now consider internal forces for the problem of Figure 4.2. Just as we
removed the beam from its supports to reveal the reactions at the rollers,
if we imagine cutting through the beam at some location along its length
and examine the cut parts, we see the two free body diagrams in
Figure 4.3(a) and (b), where we have given the reaction forces at A and B
their previously determined values. In this case, we took the cut at 3 ft
from the left end of the beam so we will see the internal forces/couples at
that cut but note that these internal forces/couples will depend on where
we make the cut so they will vary throughout the beam. We must have
both forces and couples acting at the cut to keep those parts in
equilibrium. We showed the internal forces and couples as having the
different values (V_1, M_1) and (V_2, M_2) in Figure 4.3 but if we apply the
equations of equilibrium to the part in Figure 4.3(a) we find

$$\sum F_y = 0 \quad 150 - V_1 = 0$$
$$\rightarrow V_1 = 150 \ \text{lb}$$
$$\sum M_A = 0 \quad M_1 - (3)(150) = 0 \tag{4.6}$$
$$\rightarrow M_1 = 450 \ \text{ft-lb}$$

Similarly, for equilibrium of the part in Figure 4.3(b)

$$\sum F_y = 0 \quad 150 - 300 + V_2 = 0$$
$$\rightarrow V_2 = 150 \ \text{lb}$$
$$\sum M_B = 0 \quad (300)(5) - (150)(7) - M_2 = 0 \tag{4.7}$$
$$\rightarrow M_2 = 450 \ \text{ft-lb}$$

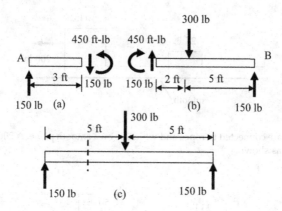

FIGURE 4.4

(a), (b) All the forces and couples acting on the parts of the beam shown in Figure 4.3(a), and (c) the forces seen when we place the beam back together at the cut.

so that we have shown that $V_1 = V_2$ and $M_1 = M_2$, and from Figure 4.4(a) and (b) we see that these internal forces and couples occur in equal and opposite pairs. For beam problems, the internal forces are called *shear forces* and the internal couples are called *bending moments*. Having internal forces/couples always occurring in equal and opposite pairs makes sense, since if we imagine putting the parts of the beam back together, they *must* cancel because there is no net force or moment acting on the beam at the location of the imaginary cut (see Figure 4.4(c)). By the same argument, external reaction forces such as those at A and B also always occur in equal and opposite pairs. This leads to the following general rule:

When drawing multiple free body diagrams, where reaction forces and/or internal forces appear in more than one free body diagram, because each part must be in equilibrium those reaction and/or internal forces always occur in equal and opposite pairs and must be drawn consistently as such on the free body diagrams.

As an illustration of this rule, consider the following example.

Consider the problem where a 200 lb horizontal force is applied to one of two blocks that are connected to each other and to a wall by horizontal ropes and are sitting on a smooth (frictionless) surface (Figure 4.5). Draw free body diagrams of (1) the system of both blocks and (2) the individual blocks and the connecting rope.

A free body diagram of the system of both blocks is shown in Figure 4.6. Since the surface is frictionless it can only exert a vertical reaction force (normal to the surface) on either block. The free body diagram of the

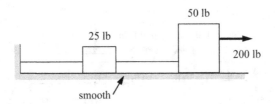

FIGURE 4.5
Two blocks that are connected to ropes and which lie on a smooth plane. A 200 lb horizontal force is applied as shown.

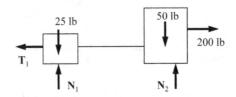

FIGURE 4.6
A free body diagram of the two blocks together.

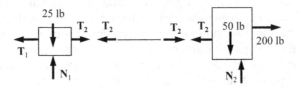

FIGURE 4.7
Free body diagrams of the two blocks separately and a free body diagram of the rope connecting them.

system of both blocks shows these normal forces, the reaction force T_1, which the rope exerts on the 25 lb block, and the two weights. Note that the rope is in tension so it must always pull to the left on the 25 block. It cannot push on the block (unless it was frozen rigid like a brick, of course).

When we look at the individual blocks and the connecting rope (Figure 4.7), we also see the tensions that the middle string exerts on the blocks. These again occur in equal and opposite pairs which we can see clearly by also showing the free body diagram of the rope itself, where the forces acting at the rope ends by equilibrium must also be equal and opposite.

The next example shows a more complex case where there are internal forces and couples.

Here is a slightly more complex example of free body diagrams for an L-shaped beam where the beam is "built-in", i.e., fixed, to a wall (Figure 4.8). In this case, the wall must exert both forces on the beam (to prevent its motion in the horizontal and vertical directions) as well as a couple to prevent the beam from rotating. The free body diagram for the L-beam removed from the wall is shown in Figure 4.9. We could solve for those wall reactions by equilibrium. If we imagine taking a cut through the beam at B, as seen in Figure 4.8, we will also expose internal reaction forces and a couple at B, which are shown for both parts of the cut beam in Figure 4.10. Again, these occur in equal and opposite pairs, and we can also find those internal reactions from equilibrium of the two parts.

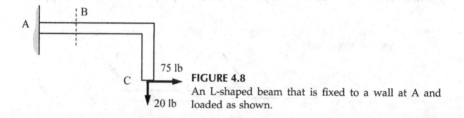

FIGURE 4.8
An L-shaped beam that is fixed to a wall at A and loaded as shown.

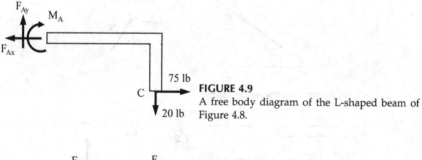

FIGURE 4.9
A free body diagram of the L-shaped beam of Figure 4.8.

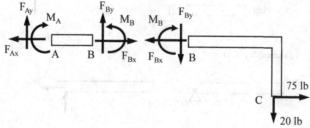

FIGURE 4.10
Free body diagrams of the two parts of the L-shaped beam that is cut at B as shown in Figure 4.8.

4.2 Reaction Forces at Supports

In the previous examples, we have seen some of the reaction forces and internal forces that can exist for structures. In this section, we want to summarize the nature of the reaction forces and couples that appear in commonly occurring support conditions. Figure 4.11 shows some of those cases. A wire or rope must be in tension so it will always pull (not push) on a body as shown in Figure 4.11(a). A smooth surface must always push on a body in a direction normal to the surface, as shown in Figure 4.11(b). If friction was present at the surface, we would also have a tangential frictional force along the surface. The smooth supports shown in Figure 4.11(c) will likewise only support a normal (vertical) force that pushes on the body at the support. A smooth pin, however, can support

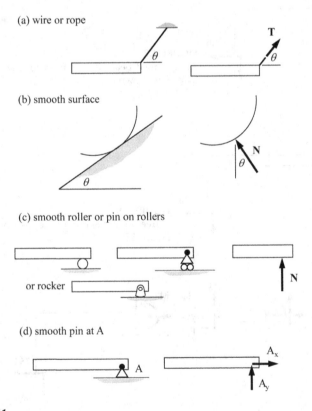

FIGURE 4.11
Reaction forces for (a) a wire or rope support, (b) contact with a smooth surface, (c) contact with a smooth roller or a pin on rollers or a rocker, and (d) a smooth pin support.

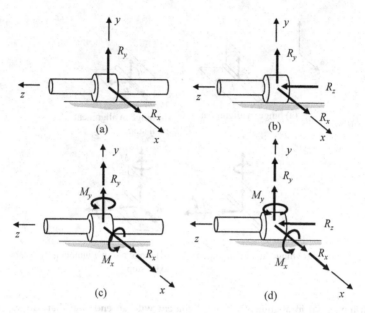

FIGURE 4.12

Smooth bearing (a) in alignment, (b) in alignment with an end stop that can support a force along the bearing axis, (c) not in alignment, where couples can exist along the x- and y-directions that will tend to "bind" the bearing (no couple can exist along the z-axis since the bearing is smooth), and (d) not in alignment with an end stop force present.

both horizontal and vertical force components. If there was friction in the pin, a couple could be generated by the frictional forces distributed around the surface of the pin. All the examples in Figure 4.11 were shown for two-dimensional problems. Figure 4.12 shows a smooth bearing, and the forces/couples present on the shaft in three dimensions for various conditions. Figure 4.12(a) shows a bearing that is used in conjunction with other supports so that it only needs to apply forces in the x- and y-directions. In this condition, the bearing is said to be in alignment. No force can exist in the z-direction since a smooth bearing cannot hold onto the shaft in that direction. However, if the bearing has an end stop, as shown in Figure 4.12(b), then a force along the shaft can exist. When a bearing is also required to support couples in addition to the forces, it is said to be not in alignment, as shown in Figure 4.12(c) and (d). In this case, the shaft tends to "bind" in the bearing and to make contact with the bearing at different locations in the bearing, creating couples along the x- and y-axes, so it is good to avoid this condition by providing other supports that can supply these moments. No couple can exist along the z-axis, which is along the shaft, since the bearing is assumed to be smooth. With an end stop, however, a force can exist in

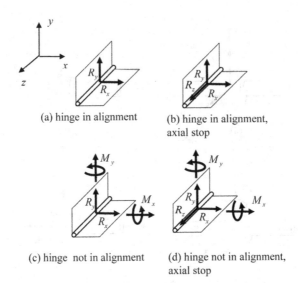

(a) hinge in alignment   (b) hinge in alignment,
                                  axial stop

(c) hinge not in alignment   (d) hinge not in alignment,
                                  axial stop

FIGURE 4.13
Smooth hinges (a) in alignment, (b) in alignment with an end stop that can support a
force along the hinge axis, (c) not in alignment, where couples can exist along the *x*- and
y-directions that will tend to "bind" the hinge (no couple can exist along the *z*-axis since the
hinge is smooth), and (d) not in alignment but with an end stop force present.

the *z*-direction. These various bearing cases are also present for a smooth
hinge, as shown in Figure 4.13.

In a problem, how do we know if we should draw the bearings or hinges as
if they were in alignment or not? There is not a comprehensive answer to
this question since the answer depends on how the bearing/hinge is used
in conjunction with other supports. Bearings and hinges are often used in
pairs along a common axis so that they can support external moments
without having couples develop at the hinges themselves. One way to
proceed is to assume that the hinges are in alignment and see if it is
possible to satisfy all the force and moment equilibrium equations under
that assumption. If not, then we will need to include couples.

Figure 4.14 shows several more supports in three dimensions. A smooth
ball support, for example, can only support a force normal to the body, as
shown in Figure 4.14(a). The smooth ball and socket of Figure 4.14(b) can
support forces in all three coordinate directions but no couples since the
ball and socket are assumed to be smooth.

If we have a beam that is fixed (built into a wall) and is loaded in three
dimensions, then since the support does not allow any displacement

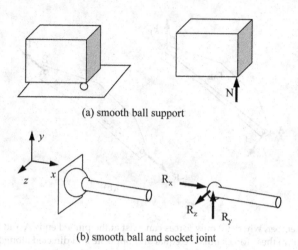

(a) smooth ball support

(b) smooth ball and socket joint

FIGURE 4.14
(a) The normal force at a smooth ball support, and (b) the forces that a smooth ball and socket joint can support.

in the x-, y-, and z-directions and does not allow any rotation of the beam in those directions, there are three force components and three couples that are developed at the support, as shown in Figure 4.15(a). If the loading on the beam, however, is only in two dimensions, then there are

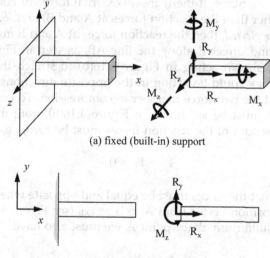

(a) fixed (built-in) support

(b) fixed (built-in) support
for coplanar force system

FIGURE 4.15
(a) The forces and couples at a fixed support for a beam loaded in three dimensions and (b) a fixed support when the loading is two-dimensional.

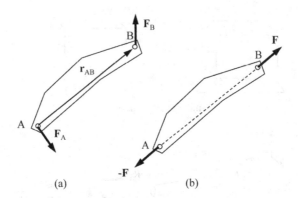

FIGURE 4.16
A two-force member, where (a) only forces can exist at the pinned ends A and B, but where (b) by equilibrium these forces must be equal and opposite and directed along the line from A to B.

only two force components and one couple component present, as we have discussed previously and which is shown in Figure 4.15(b).

Another case that we will see often in later examples is a *two-force member* (also called a *weightless link*), as seen in Figure 4.16(a). Let there be smooth pins at A and B which connect the link to other bodies. Then, in two-dimensional problems, there can be only forces at A and B acting on the link in the x-y plane. If there are no external forces or couples acting on the link (other than the reaction forces at A and B) and *if the weight of the link can be neglected*, then the reaction forces at A and B must be equal and opposite and directed along the line AB, as seen in Figure 4.16(b). The forces are shown acting in Figure 4.16(b) to stretch the two-force member, but they could be acting in the opposite directions from those shown and put the two-force member in compression. To prove that the reaction forces must be as shown in Figure 4.16(b), note that by force equilibrium the sum of the reaction forces must be zero, i.e.

$$\mathbf{F}_A + \mathbf{F}_B = 0 \tag{4.8}$$

which shows that the forces must be equal and opposite where $\mathbf{F}_A = -\mathbf{F}_B$. If we let the position vector from A to B be $\mathbf{r}_{AB}$ (see Figure 4.16(a)), then by moment equilibrium about point A we must also have

$$\mathbf{r}_{AB} \times \mathbf{F}_B = 0 \tag{4.9}$$

which can only be satisfied for a non-zero position vector and force, if $\mathbf{r}_{AB}$ is parallel to $\mathbf{F}_B$. Thus, if we let $\mathbf{F}_B = \mathbf{F}$, we will have the reaction forces for the two-force member, as shown in Figure 4.16(b).

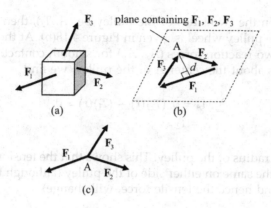

FIGURE 4.17
(a) Three forces acting on a body. (b) The common plane of the three forces when they are in equilibrium. (c) The three forces forming a concurrent force system.

There is also the case of a so-called *three-force member* where there are three non-parallel forces acting on a structure, as shown in Figure 4.17(a). If the three forces are in equilibrium, then they must lie in a common plane defined by the sum of those forces from the parallelogram law of addition, as shown in Figure 4.17(b). If the net moment about point A is zero, then from Figure 4.17(b), it is clear that $F_1d = 0$, where d is the perpendicular distance to the force $\mathbf{F}_1$ from point A. Thus, we must have $d = 0$ and so the force system must be a concurrent force system at A, as shown in Figure 4.17(c).

The last example we will consider in this section is the case where a wire or rope runs over a pulley, where the wheel of the pulley is in smooth contact with the pulley shaft, as shown in Figure 4.18(a). If we let

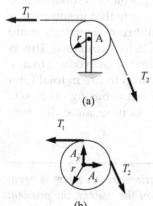

FIGURE 4.18
(a) A rope running over a smooth pulley, and (b) a free body diagram of the pulley.

the tensions on the two sides of the pulley be (T_1, T_2), then a free body diagram of the pulley wheel is given in Figure 4.18(b). At the shaft, there can only be two reaction forces (A_x, A_y) for smooth contact. Thus, if we sum moments about the center A of the pulley, we find

$$\sum M_A = 0 \ (T_2)(r) - (T_1)(r) = 0$$
$$\rightarrow T_1 = T_2$$
(4.10)

where r is the radius of the pulley. This shows that the tension in the rope must remain the same on either side of the pulley (although the direction of the rope, and hence the tensile force, will change).

4.3 Solving Equilibrium Problems – I

Let us begin by solving some two-dimensional (coplanar) problems where the equations of equilibrium are:

$$\sum F_{ix} = 0$$
$$\sum F_{iy} = 0$$
$$\sum M_{Pz} = 0$$
(4.11)

It may appear that we can only solve for three unknown forces or couples because we have three equations. However, we can break a system into parts and apply these three equations to the free body diagrams of each part, so in fact, we can solve for many more than three unknowns. If the total number of unknowns is equal to the total number of independent equilibrium equations, then the problem is solvable and we say the problem is *statically determinate*. Also, note that in some cases it may be desirable to use the moment equilibrium equation more than once, taking moments about different points in a structure. This is acceptable if the equations we use are independent of each other. In any case, we still have only three independent equations to use in total for a given free body diagram. We will say more about choosing other sets of equilibrium equations later. For now, let's examine some different problems with these equations.

When the number of independent equilibrium equations for a system is equal to the number of unknown forces/moments acting on the system, the problem is statically determinate. In statically indeterminate problems we cannot solve for all the unknowns, although we may be able to solve for some of them. Statics courses traditionally have only dealt with statically determinate problems but in chapter 10 we will show you how to also solve statically indeterminate problems.

Example 4.1

Consider the two-member structure shown in Figure 4.19(a) where there are two bars, each weighing 100 lb, and where a 200 lb force is applied to a smooth pin at B. The bars are attached to each other and to the floor through smooth pins. Determine all the forces acting between the bars and the external reaction forces on the entire structure.

Free Body Diagrams

Examine first a free body diagram of the entire structure (Figure 4.19(b)). We will assume that the bars are uniform and place their weights at their geometrical centers, which are the centers of gravity for the bars. There are four unknown reaction forces at the smooth pins A and C but we only have three equilibrium equations so it would appear we cannot solve for all these unknowns. However, if we break the system into parts, as shown in Figure 4.20, we have three equations of equilibrium for each of the two bars shown in Figure 4.20(a) and (c) and two equations for the pin connecting those parts, as shown in Figure 4.20(b) since the force system for the pin is a concurrent force system where the moment equilibrium equation will automatically be satisfied if the forces are in equilibrium. Thus, there are a total of eight equations for these parts and eight unknowns, and the problem is statically determinate. Examine Figure 4.20 carefully to see

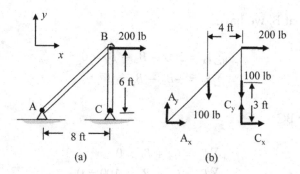

FIGURE 4.19
(a) A frame structure and (b) its free body diagram.

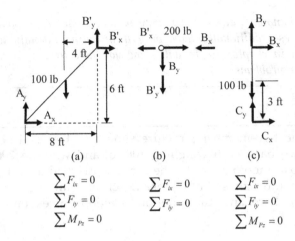

$$\sum F_{ix} = 0$$
$$\sum F_{iy} = 0$$
$$\sum M_{Pz} = 0$$

$$\sum F_{ix} = 0$$
$$\sum F_{iy} = 0$$

$$\sum F_{ix} = 0$$
$$\sum F_{iy} = 0$$
$$\sum M_{Pz} = 0$$

FIGURE 4.20
(a), (b), (c) Free body diagrams of the two bars and the pin and their equilibrium equations.

how we have drawn the forces at B for the two bars and the pin connecting them in a consistent manner, which shows these forces in equal and opposite pairs. Drawing free body diagrams in such a consistent manner is essential.

Equations of Equilibrium
The equations of equilibrium for bar AB are

$$_{AB}\sum F_x = A_x + B'_x = 0$$
$$_{AB}\sum F_y = A_y + B'_y - 100 = 0 \tag{4.12}$$
$$_{AB}\sum M_{Az} = 8B'_y - 6B'_x - 4(100) = 0$$

For the pin at B, we have

$$_{B}\sum F_x = 200 - B_x - B'_x = 0$$
$$_{B}\sum F_y = -B_y - B'_y = 0 \tag{4.13}$$

and for bar BC

$$_{BC}\sum F_x = C_x + B_x = 0$$
$$_{BC}\sum F_y = C_y + B_y - 100 = 0 \tag{4.14}$$
$$_{BC}\sum M_{Cz} = -6B_x = 0$$

We could solve these eight equations for the eight unknowns, which we will do shortly, but in many problems like this one, we can sequentially work through the system and solve for all the unknowns without considering simultaneous equations. For example, let's place the equations in the following order:

$$
\begin{aligned}
&\text{(1)} \quad \sum M_{Cz} = -6B_x = 0 \\
&\text{(2)} \quad \sum F_x = C_x + B_x = 0 \\
&\text{(3)} \quad \sum F_x = 200 - B_x - B_x' = 0 \\
&\text{(4)} \quad \sum F_x = A_x + B_x' = 0 \\
&\text{(5)} \quad \sum M_{Az} = 8B_y' - 6B_x' - 4(100) = 0 \qquad\qquad (4.15) \\
&\text{(6)} \quad \sum F_y = A_y + B_y' - 100 = 0 \\
&\text{(7)} \quad \sum F_y = -B_y' - B_y' = 0 \\
&\text{(8)} \quad \sum F_y = C_y + B_y - 100 = 0
\end{aligned}
$$

From the first equation, we find $B_x = 0$, the second equation gives $C_x = 0$, and the third equation then yields $B_x' = 200$. The fourth equation gives $A_x = -200$, while the fifth equation gives $B_y' = 200$. The sixth equation gives $A_y = -100$, while the seventh equation yields $B_y = -200$, and finally the eighth equation gives $C_y = 300$. All these forces are in pounds. We can now place all these forces on the parts as shown in Figure 4.21. You should be able to identify the equal and opposite pairs of forces. If we put these parts all back together and examine the entire structure, as shown in Figure 4.22, we see by inspection that the forces are balanced. If we take moments about point B, we find

$$
\sum M_{Bz} = (100)(4) + (100)(8) - (200)(6) = 0
$$

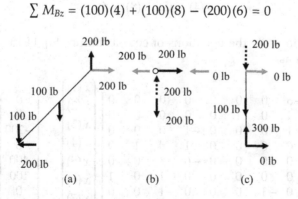

FIGURE 4.21
(a), (b), (c) All the forces acting on the parts of the structure.

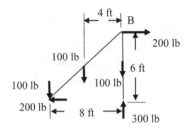

FIGURE 4.22
The forces acting on the entire structure.

so the moments are balanced as well. This shows:

If we guarantee that all parts of a structure are in equilibrium,

then the entire structure must also be in equilibrium.

This statement is true in general, not just for this problem.

Solving Equilibrium Equations Numerically in MATLAB
Although we can solve this problem in the sequential manner shown, in other problems this may not be possible, and we will have to deal with a solution of simultaneous equations. Using MATLAB, that is relatively easy. We can do the problem in a purely numerical manner, or we can use a symbolic algebra approach. First, consider a numerical solution where we define the unknowns as components of an x-vector as

$$
\begin{aligned}
x(1) &= B_x & x(5) &= B_x' \\
x(2) &= B_y & x(6) &= B_y' \\
x(3) &= C_x & x(7) &= A_x \\
x(4) &= C_y & x(8) &= A_y
\end{aligned}
\tag{4.16}
$$

Then, we can write the equations of equilibrium of Eq. (4.15) in matrix-vector form as

$$
\begin{bmatrix}
-6 & 0 & 0 & 0 & 0 & 0 & 0 & 0 \\
1 & 0 & 1 & 0 & 0 & 0 & 0 & 0 \\
-1 & 0 & 0 & 0 & -1 & 0 & 0 & 0 \\
0 & 0 & 0 & 0 & 1 & 0 & 1 & 0 \\
0 & 0 & 0 & 0 & -6 & 8 & 0 & 0 \\
0 & 0 & 0 & 0 & 0 & 1 & 0 & 1 \\
0 & -1 & 0 & 0 & 0 & -1 & 0 & 0 \\
0 & 1 & 0 & 1 & 0 & 0 & 0 & 0
\end{bmatrix}
\begin{Bmatrix}
x(1) \\
x(2) \\
x(3) \\
x(4) \\
x(5) \\
x(6) \\
x(7) \\
x(8)
\end{Bmatrix}
=
\begin{Bmatrix}
0 \\
0 \\
-200 \\
0 \\
400 \\
100 \\
0 \\
100
\end{Bmatrix}
\tag{4.17}
$$

and we can solve this system of equations in MATLAB. First, we write
the matrix E of coefficients of the equilibrium equations:

```
E = [-6    0    0    0    0    0    0    0;
      1    0    1    0    0    0    0    0;
     -1    0    0    0   -1    0    0    0;
      0    0    0    0    1    0    1    0;
      0    0    0    0   -6    8    0    0;
      0    0    0    0    0    1    0    1;
      0   -1    0    0    0   -1    0    0;
      0    1    0    1    0    0    0    0];
```

Next, we write the known force vector on the right-hand side of these
equations (which contains the negative values of the applied loads, as
mentioned earlier) as a column vector:

P = [0; 0; −200; 0; 400; 100; 0; 100];

Then, we can solve these equations in MATLAB as

```
    E\P
ans = 0          % Bx = 0
     -200        % By = −200 lb
      0          % Cx = 0
      300        % Cy = 300 lb
      200        % B'x = 200 lb
      200        % B'y = 200 lb
     -200        % Ax = −200 lb
     -100        % Ay = −100 lb
```

Solving Equilibrium Equations Symbolically in MATLAB
The previous MATLAB solution required that we turn the equilibrium
equations into an equivalent equilibrium matrix and a known force vector
so that we could solve a set of simultaneous equations numerically. But we
can also enter the equations symbolically in much the same form as they
are written in Eq. (4.15) and solve them in MATLAB. This is a very simple
way to solve equilibrium problems that can avoid the errors that may
occur when writing the equilibrium matrix from the equations. We have
seen some use of a similar symbolic approach in previous chapters. First,
define the eight unknowns as symbols and symbolically enter the eight
equations as the eight components of a symbolic vector Eq. Here Bpx, Bpy
are B'x and B'y.

```
syms Bx By Cx Cy Bpx Bpy Ax Ay      % define unknowns symbolically
Eq(1) = -6*Bx == 0;                 % form equations of equilibrium,
                                    % == is symbolic equals sign
Eq(2) = Cx + Bx == 0;
Eq(3) = 200 - Bx - Bpx == 0;
Eq(4) = Ax + Bpx == 0;
Eq(5) = 8*Bpy - 6*Bpx - 400 == 0;
Eq(6) = Ay + Bpy - 100 == 0;
Eq(7) = -By - Bpy == 0;
Eq(8) = Cy + By - 100 == 0;
```

Next, we solve these eight symbolic equations with the MATLAB function solve:

```
S = solve(Eq)            % solve the equilibrium equations
S =
struct with fields:      Ax: -200
                         Ay: -100
                         Bpx: 200
                         Bpy: 200
                         Bx: 0
                         By: -200
                         Cx: 0
                         Cy: 300
```

We see that output of the solve function is placed in a MATLAB structure, S, that contains eight fields with the values of the symbolic variables. (You can look up the MATLAB documentation if you want to understand more about MATLAB structures. For our purposes we only need to know how to access the elements which we will now address.) Each symbolic variable in the S structure can be accessed by writing S followed by a dot and the symbolic name of one of the unknown variables. For example, typing S.Ax will return a symbolic value of –200, which is the value of Ax. To change these symbolic values to the ordinary numerical values you are accustomed to working with, we can use the double function:

```
NS = double([ S.Ax S.Ay S.Bpx S.Bpy S.Bx S.By S.Cx S.Cy])
NS = -200 -100 200 200 0 -200 0 300
```

The double function converts all these symbolic values to double precision numerical values, which is why it is called double. The vector NS is now a vector of numerical values.

Note the directness of this approach in MATLAB. We can simply write the equilibrium equations in a symbolic form which is very close to how

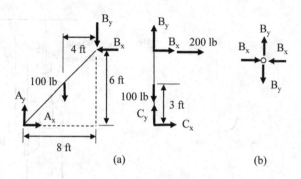

FIGURE 4.23
(a) The free body diagrams of the bars when the 200 lb force acts on BC. (b) The free body diagram of the pin.

we wrote the equations by hand, and then solve those symbolic equations with the solve function. We can effortlessly do many statics problems in this symbolic manner that would be more tedious if done by hand or even if done numerically in MATLAB.

Before we close this problem, we want to mention some details about drawing the free body diagrams. We showed the two bars and the pin in Figure 4.20 because we said the 200 lb was applied to the pin, which in practice is likely how such a pinned structure is loaded. The free body diagram of the pin shows this force as well as the forces coming from bars AB and BC. If instead the 200 lb force was applied to bar BC, then the free body diagrams would be as shown in Figure 4.23(a). In this case, the pin at B simply transfers the force on AB to BC (see Figure 4.23(b)) and we do not have to consider the pin by itself, reducing the number of unknown forces by two. The solution procedure, however, otherwise stays the same.

Many equilibrium problems will require us to draw multiple free body diagrams. The pulley system in the next example is one of them.

Example 4.2

Determine the tension, T, required to hold the weight W (whose mass is given) in equilibrium in Figure 4.24. The strings on pulleys 1 and 2 at C (which are joined) are attached to C at points 1 and 2 indicated by the dots. The radii of the two pulleys at C are shown in the figure.

Free Body Diagrams
Consider the free body diagrams for the three pulleys seen in Figure 4.25. In drawing those free body diagrams, we have used the fact that the

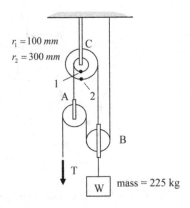

FIGURE 4.24
A pulley system.

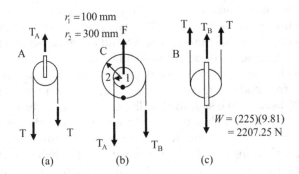

FIGURE 4.25
Free body diagrams for (a) pulley A, (b) pulley C, and (c) pulley B.

tension T is unchanged as it goes over pulleys A and B. There are four unknown forces and four equations of equilibrium involving the forces in all three free body diagrams, but we do not need to solve directly for the force F.

Equations of Equilibrium
First consider force equilibrium for pulley A:

$$\sum F_y = 0$$
$$T_A - 2T = 0 \quad \rightarrow T_A = 2T \tag{4.18}$$

For the fixed pulley C, we can take moments about the center O of that pulley to find the tension T_B in terms of T_A, and hence the tension T.

$$\sum M_O = 0$$

$$(100)(T_A) - (300)(T_B) = 0$$

$$\rightarrow T_B = \frac{T_A}{3} = \frac{2T}{3}$$

(4.19)

Finally, we can sum forces for pulley B and obtain the tension T from T_B and the known weight. Here, we use $g = 9.81$, so $W = (225)(9.81) = 2207.25$ N

$$\sum F_y = 0$$

$$2T + \frac{2T}{3} - 2207.25 = 0$$

(4.20)

$$\rightarrow T = 827.7 \ N$$

The remaining force equilibrium equation for pulley C can be used to find the force F, if wanted. Since we have enough equations to solve for all the unknowns, the problem is *statically determinate*. It took three equations to obtain the tension T. If we did the problem in MATLAB, we could solve for all the forces with four equations and very little additional effort:

```
syms T Ta Tb F              % the symbolic unknown forces
W = 225*9.81;               % the known weight
Eq(1) = 2*T - Ta == 0;      % the four equilibrium equations
Eq(2) = Ta + Tb  - F == 0;
Eq(3) = 100*Ta - 300*Tb == 0;
Eq(4) = 2*T + Tb  - W == 0;
S = solve(Eq);              % the solution
double([S.T S.Ta S.Tb S.F]) % numerical values for T, Ta, Tb, F(in N)
ans = 1.0e+03 *
0.8277   1.6554   0.5518   2.2073
```

The equilibrium examples considered so far have all written the equilibrium equations in terms of a set of horizontal and vertical Cartesian axes. The next example shows that we may want to use a set of inclined axes instead when most of the forces act along such axes.

Example 4.3

Determine the magnitude of the force in the cable in Figure 4.26(a) and the magnitudes of the forces exerted on the wheels at A and B by the inclined smooth surface. The mass of the cart is 3500 kg and its weight acts as the center of gravity, G.

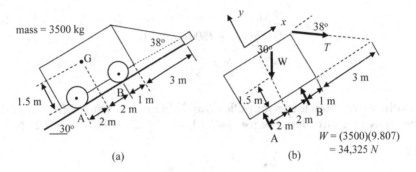

FIGURE 4.26
(a) A cart held on an inclined smooth surface. (b) The free body diagram.

Free Body Diagram
By using inclined x- and y-axes we see in the free body diagram of Figure 4.26(b) that we only must break up two of the four forces into components. We have three equations of equilibrium for the three unknown forces, so the problem is *statically determinate*.

Equations of Equilibrium
Here, we will use $g = 9.807$ so $W = (3500)(9.807) = 34,325$ N. Summing forces in the inclined x-direction, we find:

$$\sum F_x = 0$$
$$T \cos 38° - W \sin 30° = 0 \tag{4.21}$$
$$\rightarrow T = 21,780 \text{ N} = 21.8 \text{ kN}$$

If we sum moments about the point where the force A acts, then:

$$\sum M_A = 0$$
$$(B)(4) + (W \sin 30°)(1.5) - (W \cos 30°)(2)$$
$$- (T \cos 38°)(3 \tan 38°) - (T \sin 38°)(5) = 0 \tag{4.22}$$
$$\rightarrow B = 35,245 \text{ N} = 35.2 \text{ kN}$$

Summing forces in the inclined y-direction give, finally:

$$\sum F_y = 0$$
$$A + B - W \cos 30° - T \sin 38° = 0 \tag{4.23}$$
$$\rightarrow A = 7890 \text{ N} = 7.89 \text{ kN}$$

The equations in MATLAB are:

```
syms A B T                              % symbolic unknown forces
W = 3500*9.807;                         % the known weight(in N)
Eq(1) = T*cosd(38) − W*sind(30) == 0;   % the equations of equilibrium
Eq(2) = 4*B + W*sind(30)*1.5...
−W*cosd(30)*2 − T*cosd(38)*(3*tand(38))...
−T*sind(38)*5 == 0;
Eq(3) = A + B − W*cosd(30)...
−T*sind(38)== 0;
S = solve(Eq);                          % the solution
double([S.A S.B S.T])                   % numerical values for A, B, T (in N)
ans = 1.0e + 04 *
0.7890   3.5244   2.1779
```

which gives the same values.

The examples so far have been for two-dimensional problems. Now, consider some three-dimensional ones.

Example 4.4

Consider a weight W = 3000 lb that is supported by three inextensible cables as shown in Figure 4.27(a). Determine the tensions in the cables.

Free Body Diagram
The free body diagram of this concurrent force system is given in Figure 4.27(b). Since we have a concurrent force system, there are three

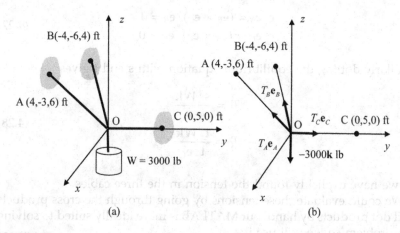

FIGURE 4.27
(a) A weight supported by three wires. (b) The free body diagram of the connection point O.

force equilibrium equations for the three unknown cable tensions and the problem is statically determinate.

Equations of Equilibrium

Since there are three equations of equilibrium, we can find the three tensions (T_A, T_B, T_C) by solving three simultaneous equations:

$$\sum \mathbf{F} = 0 \quad T_A \mathbf{e}_A + T_B \mathbf{e}_B + T_C \mathbf{e}_C - W\mathbf{k} = 0 \qquad (4.24)$$

where $(\mathbf{e}_A, \mathbf{e}_B, \mathbf{e}_C)$ are unit vectors along the cables. However, we can avoid those simultaneous equations and solve the problem directly. This is possible if we first define the vectors $(\mathbf{r}, \mathbf{s}, \mathbf{t})$ through the cross products:

$$
\begin{aligned}
\mathbf{r} &= \mathbf{e}_B \times \mathbf{e}_C \\
\mathbf{s} &= \mathbf{e}_A \times \mathbf{e}_C \\
\mathbf{t} &= \mathbf{e}_A \times \mathbf{e}_B
\end{aligned}
\qquad (4.25)
$$

By the definition of the cross product, these are three vectors (but not unit vectors) that are orthogonal to the unit vectors that define them. For example, $\mathbf{r}$ is orthogonal to $\mathbf{e}_B$ and $\mathbf{e}_C$. Thus, if we dot the equilibrium equation, Eq. (4.24), with $\mathbf{r}$, we find

$$
\begin{aligned}
T_A \mathbf{r} \cdot \mathbf{e}_A &= \mathbf{r} \cdot W\mathbf{k} \\
\rightarrow T_A &= \frac{\mathbf{r} \cdot W\mathbf{k}}{\mathbf{r} \cdot \mathbf{e}_A}
\end{aligned}
\qquad (4.26)
$$

since

$$
\begin{aligned}
\mathbf{r} \cdot \mathbf{e}_B &= (\mathbf{e}_B \times \mathbf{e}_C) \cdot \mathbf{e}_B = 0 \\
\mathbf{r} \cdot \mathbf{e}_C &= (\mathbf{e}_B \times \mathbf{e}_C) \cdot \mathbf{e}_C = 0
\end{aligned}
\qquad (4.27)
$$

Similarly dotting the equilibrium equation with $\mathbf{s}$ and $\mathbf{t}$ gives

$$
\begin{aligned}
T_B &= \frac{\mathbf{s} \cdot W\mathbf{k}}{\mathbf{s} \cdot \mathbf{e}_B} \\
T_C &= \frac{\mathbf{t} \cdot W\mathbf{k}}{\mathbf{t} \cdot \mathbf{e}_C}
\end{aligned}
\qquad (4.28)
$$

so we have explicitly found the tension in the three cables.

We could evaluate these tensions by going through the cross products and dot products by hand, but MATLAB is more ideally suited to solving the problem so we will use it:

```
OA = [4 -3 6];                  % define unit vectors along cables
ea = OA/norm(OA);
OB = [-4 -6 4];
eb = OB/norm(OB);
OC = [0 5 0];
ec = OC/norm(OC);
W = 3000*[0 0 1];               % define Wk (note: no minus sign)
r = cross(eb, ec);              % define r, s, t vectors
s = cross(ea, ec);
t = cross(ea, eb);
Ta = dot(W, r)/dot(r, ea)       % tensions(in lb)
Ta = 2.3431e+003
Tb = dot(W, s)/dot(s, eb)
Tb = 2.4739e+003
Tc = dot(W, t)/dot(t, ec)
Tc = 2700
```

This MATLAB solution was a purely numerical one. We could also solve this problem in MATLAB symbolically by solving the three equilibrium equations in Eq. (4.24) directly.

```
syms Ta Tb Tc                        % define tensions symbolically
R = Ta*ea + Tb*eb + Tc*ec + [0 0 -3000];  % define resultant force
                                     % (which is zero)
S = solve(R);                        % solve for the tensions and
                                     % obtain them numerically
double([S.Ta, S.Tb S.Tc])
ans = 1.0e+03 *
2.3431   2.4739   2.7000
```

This solution is highly efficient, showing the power of the use of MATLAB's symbolic capabilities. Note that in using the solve function, we did not place the symbolic == 0 in the equilibrium equations. The function solve will assume that we are trying to solve a set of equations that are equal to zero. Thus, we can omit "== 0" in the equations.

Here is another three-dimensional example.

Example 4.5

A three-foot radius table (Figure 4.28(a)) is supported symmetrically by three legs (each separated by 120° as shown in Figure 4.28(b))

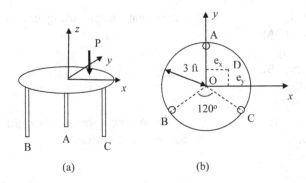

FIGURE 4.28
(a) A force P acting on a three-leg table. (b) The geometry looking down the z-axis. The force acts at point D.

1. If a vertical force P is applied at point D, determine the vertical forces in legs A, B, and C as a function of the distance e_x and e_y. Let the weight of the table be W (the table is assumed to be homogeneous so that its weight acts at its center).

2. Determine the negative value of e_y which will cause the table to tip about legs B and C if (a) the weight of the table is neglected, and (b) if the weight is included.

Free Body Diagram
This is a parallel force system acting on the table, as shown in the free body diagram of Figure 4.29(a), so there are only three non-trivial equations of equilibrium that can be used to solve for the three unknown forces in the table legs. The problem is statically determinate.

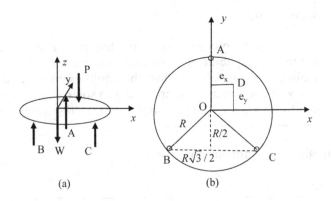

FIGURE 4.29
(a) The parallel force system acting on the table. (b) The distances involved in the x-y plane.

Equations of Equilibrium

Using the forces and distances shown in Figure 4.29(a) and (b), the equations of equilibrium are

$$\sum F_z = 0 \quad A + B + C - P - W = 0$$
$$\sum M_{x0} = 0 \quad AR - Pe_y - BR/2 - CR/2 = 0 \quad\quad (4.29)$$
$$\sum M_{z0} = 0 \quad Pe_x + CR\sqrt{3}/2 - BR\sqrt{3}/2 = 0$$

We will set this problem up and solve it symbolically in MATLAB:

```
syms A B C P R W ex ey          % define symbolic variables
Eq(1) = A + B + C - P - W;      % the three equations in a symbolic
                                % vector (omitting == 0)
Eq(2) = A*R - P*ey - B*R/2 - C*R/2;
Eq(3) = P*ex + C*R*sqrt(3)/2 - B*R*sqrt(3)/2;
S = solve(Eq, A, B, C)          % solve for the forces A, B, C and
                                % place in a structure
S = struct with fields:
```

A: (P*R + R*W + 2*P*ey)/(3*R)
B: (P*R + R*W − P*ey + 3^(1/2)*P*ex)/(3*R)
C: −(3^(1/2)*(3*P*ex − 3^(1/2)*P*R − 3^(1/2)*R*W + 3^(1/2)*P*ey))/(9*R)

In this case, we see that the structure S contains the symbolic solution for the three forces. We can rewrite them as

$$A = \frac{W}{3} + \frac{P}{3R}[R + 2e_y]$$
$$B = \frac{W}{3} + \frac{P}{3R}[R - e_y + e_x\sqrt{3}] \quad\quad (4.30)$$
$$C = \frac{W}{3} + \frac{P}{3R}[R - e_y - e_x\sqrt{3}]$$

which gives the solution for part (1). Setting $W = 0$ in these expressions, we find for that case:

$$A = \frac{P}{3R}[R + 2e_y]$$
$$B = \frac{P}{3R}[R - e_y + e_x\sqrt{3}] \qu\quad (4.31)$$
$$C = \frac{P}{3R}[R - e_y - e_x\sqrt{3}]$$

We can also do the substitution $W = 0$ in MATLAB with the subs function. The expressions for the three forces are in the structure fields S.A, S.B, and S.C:

subs(S.A, W, 0)
ans = (P*R + 2*P*ey)/(3*R)
subs(S.B, W, 0)
ans = (P*R − P*ey + 3^(1/2)*P*ex)/(3*R)
subs(S.C, W, 0)
ans = −(3^(1/2)*(3*P*ex − 3^(1/2)*P*R + 3^(1/2)*P*ey))/(9*R)

which give the same solution. Now, consider part (2) for tipping. Let's examine the $W = 0$ case first. From Eq. (4.31), we see that as the eccentricity e_y gets more and more negative, the force in leg A will eventually go to zero. But the force at A cannot be negative so when the force $A = 0$, the table must be on the verge of tipping about legs B and C. For $A = 0$, we find:

$$e_y = \frac{-R}{2} \qquad (4.32)$$

Since this condition for tipping is independent of e_x, a force P applied at any point in the shaded region $e_y < -R/2$, shown in Figure 4.30, will cause the table to tip. Because of the symmetry of this problem, we will have similar regions between A-B and A-C where tipping can occur at the lifting of legs C and B. When the weight W is not zero, we have

$$A = \frac{W}{3} + \frac{P}{3R}[R + 2e_y] \qquad (4.33)$$

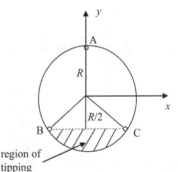

FIGURE 4.30
The region on the table where applying the force P
will cause the table to tilt when W = 0.

region of
tipping
between B-C

so that tipping occurs at larger magnitudes of e_y where

$$e_y = -\frac{R}{2}\left(1 + \frac{W}{P}\right) \qquad (4.34)$$

with similar results to make legs B or C to be involved with tipping.

Three-dimensional equilibrium problems can be challenging because of the 3-D geometry involved. MATLAB can help us deal with that complexity as the next example shows.

Example 4.6

A 50 kg block is held on a smooth inclined plane ABC by a force **P** that acts parallel to the plane (Figure 4.31(a)). The lines AB and AC in the plane make angles of 30° and 45° with the x- and y-axes, respectively, as shown in that figure. The vector **n** is a unit vector normal to the plane. Determine the force **P**.

Free Body Diagram
The free body diagram is shown in Figure 4.31(b). There is an unknown normal force $N\mathbf{n}$ acting on the block from the plane and an unknown tangential vector force **P**. If we can find the unit normal **n**, then only the magnitude of the normal force, N, is unknown and the vector **P** can be

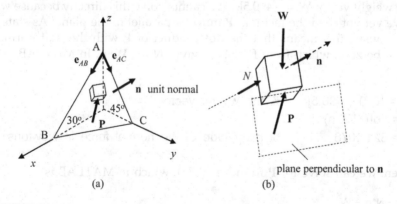

(a)   (b)

FIGURE 4.31
(a) A block held on a smooth plane by a force **P** that is tangential to the surface. (b) A free body diagram of the forces on the block, where the normal force acts along **n** and the force **P** lies in a plane perpendicular to **n**.

described in terms of only two unknowns since $\mathbf{P} \cdot \mathbf{n} = 0$ if that force is to lie in a plane parallel to the smooth plane. Thus, there are three force equations for three unknowns for this concurrent three-force-system and the problem is statically determinate.

Let us first find the unit normal vector $\mathbf{n}$. Since lines AB and AC lie in the plane, if we define unit vectors ($\mathbf{e}_{AB}$, $\mathbf{e}_{AC}$) along those lines (see Figure 4.31(a)) and take their cross product, we can determine a unit vector, $\mathbf{n}$, in a direction normal to the plane. First, let's find that unit normal vector. It is easy in MATLAB:

```
eab = [cosd(30) 0 - sind(30)];    % define the unit vectors
eac = [0 cosd(45) - sind(45)];
v = cross(eab, eac);              % take the cross product and
                                  % divide by magnitude to
                                  % get a unit vector
n = v/norm(v)
n = 0.3780   0.6547   0.6547
```

Equations of Equilibrium
We can write the equations of equilibrium vectorially for this concurrent force system as

$$\sum \mathbf{F} = 0 \quad N\mathbf{n} + \mathbf{P} + \mathbf{W} = 0 \qquad (4.35)$$

where the magnitude of the weight of the block is $(50)(9.81) = 490.5$ N, so the weight vector $\mathbf{W} = -490.5\mathbf{k}$. We cannot solve this directly because we have yet not used the fact that $\mathbf{P}$ must be parallel to the plane. As stated previously, this means that the dot product of $\mathbf{P}$ with the unit normal must be zero, which from Eq. (4.35) gives $N = -\mathbf{W} \cdot \mathbf{n}$. In MATLAB, we have

```
W = [0 0 - 490.5];    % the weight vector
N = dot(-W, n)
N = 321.1076          % magnitude of the normal force in Newtons
```

Then, we can solve for $\mathbf{P}$ from Eq. (4.35), which in MATLAB is

```
P = -N*n - W
P = -121.3673 - 210.2143 280.2857    % the force components of P in Newtons
```

4.4 Alternate Systems of Equilibrium Equations

Satisfying the force and moment equilibrium equations guarantees that there is no net force or moment acting on a structure. However, these are not the only sets of equations we can use. We can, for example, replace one or more of the force equilibrium equations with moment equations that might make the solution easier to obtain. However, such substitutions have some restrictions. It is best to examine this behavior with a simple example.

Example 4.7

Consider the two-dimensional example shown in Figure 4.32(a) where a homogeneous block of weight W is supported by a horizontal weightless link at D, a cable at C, and a smooth roller at A (a rather unusual support arrangement, indeed!). Neglect the small distance of the link to the top of the block at D.

Free Body Diagram
The free body diagram of the block is shown in Figure 4.32(b).

Equations of Equilibrium (1)
If we use the force and moment equations for this two-dimensional problem, we have

$$\sum F_x = 0 \quad D_x + 4F/5 = 0$$
$$\sum F_y = 0 \quad A_y + 3F/5 - W = 0 \tag{4.36}$$
$$\sum M_{zA} = 0 \quad -6D_x + (3F/5)(4) - (4F/5)(6) - 2W = 0$$

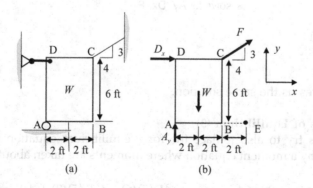

(a)   (b)

FIGURE 4.32
(a) A problem for examining the use of different sets of equilibrium equations. (b) The free body diagram.

We can write this as a set of linear equations in matrix-vector form as [E] {x} = {P}, where explicitly we have:

$$\begin{bmatrix} 0 & 1 & 0.8 \\ 1 & 0 & 0.6 \\ 0 & -6 & -2.4 \end{bmatrix} \begin{Bmatrix} A_y \\ D_x \\ F \end{Bmatrix} = \begin{Bmatrix} 0 \\ W \\ 2W \end{Bmatrix} \tag{4.37}$$

Let's solve this system symbolically in MATLAB:

```
syms Ay Dx F W
E = [0 1 0.8; 1 0 0.6; 0 -6 -2.4];   % using force equations and moment
                                     % equation

P = [0; W; 2*W];
E\P                                  % solve for Ay, Dx, F
ans = W/2
      -(2*W)/3
      (5*W)/6
```

Equations of Equilibrium (2)
Now, consider replacing the force equilibrium equation in the y-direction by a moment equation where the moments are taken about point B. This equation is

$$\sum M_{zB} = 0 \quad -4A_y -6D_x -(4/5F)(6) + 2W = 0 \tag{4.38}$$

Making this substitution in MATLAB and solving again:

```
E(2,:) = [-4 -6 -4.8]; % replace force equation in y-direction by moment about B
P(2) = -2*W;
E\P                    % solve for Ay, Dx, F
ans = W/2
      -(2*W)/3
      (5*W)/6
```

which gives us the same solution.

Equations of Equilibrium (3)
Now, let's try to also replace the force equilibrium equation in the x-direction by a moment equation where moments are taken about point E.

$$\sum M_{zE} = 0 \quad -6A_y -6D_x -(4F/5)(6) - (3F/5)(2) + 4W = 0 \tag{4.39}$$
$$\rightarrow \quad -6A_y - 6D_x - 6F + 4W = 0$$

Making this additional replacement in MATLAB:

```
E(1,:) = [-6 -6 -6];              % replace force equation in x-direction
                                  % by moment about E
P(1) = -4*W;
E\P;                              % solve
Warning: Solution is not unique because the system is rank - deficient.
In symengine
In sym/privBinaryOp (line 1136)
In \ (line 497)
```

The solution in this case fails. If we evaluate the determinant of the equilibrium matrix [E], we find it is nearly zero:

```
det(E)
ans = 4.2633e - 15
```

and when the determinant of the matrix coefficients is zero for a system of linear equations, that system either has no solution or infinitely many solutions.

Taking moments about three points (A, B, E) that lie along a line does not give us a set of three independent equilibrium equations, so we do have infinitely many solutions with that choice.

Equations of Equilibrium (4)
However, if we had replaced the force equation in the x-direction by a moment equation where moments are taken about point C, we have

$$\sum M_{zC} = 0 \quad -4A_y + 2W = 0 \tag{4.40}$$

and make that replacement instead in MATLAB we find

```
E(1,:) = [-4 0 0];    % replace force equation in x - direction by moment about C
P(1) = -2*W;
E\P                   % solve for Ay, Dx, and F
ans = W/2
      -(2*W)/3
       (5*W)/6
```

and we do obtain the solution again. This is an example where choosing the moment equation about point C to replace a force equation is actually a very good choice since it allows us to solve for the force A_y directly.

There are other cases in two- and three-dimensional problems where replacing force equations by moment equations can lead to not having a unique solution. If the determinant of the system of equilibrium equations is zero (or nearly zero), you know that the choice of replacement was not a good one. One can, of course, either make a different replacement (i.e., choose a different point to take moments about) or simply use the original set of force and moment equations.

In this chapter, we have solved a few equilibrium problems of different types. We will solve additional examples in the following chapters. Statics is primarily about solving equilibrium problems so you should attempt to solve as many kinds of problems as possible. The homework problems will give you a selection to choose from. Try solving problems by hand (i.e., with a pocket calculator) to prepare for tests and solve them using MATLAB, as that is a very valuable engineering tool. As we have seen in several problems, a MATLAB solution can often avoid intermediate calculations that can lead to errors, so also obtaining a MATLAB solution to a problem you have previously done by hand can help ensure your calculations are correct.

4.5 Problems

Free body diagrams typically show all the forces/moments and pertinent distances and angles present in a problem so that we can solve an equilibrium problem directly from the free body diagrams. Some distances in the free body diagram problems given below remain unspecified so that you can focus on the forces/moments. Thus, in all the free body diagram problems below assume that the necessary distances and angles in the problem are known.

P4.1 A beam is pinned at A and is supported by a wire that runs over a smooth pulley at D. There is a known applied force P as shown in Fig. P4.1. Neglecting the weight of the beam, draw a free body diagram for it. Is the problem statically determinate?

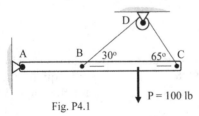

Fig. P4.1

P4.2 Two bars are connected to fixed pins at A and B and connected to each other by a pin at C (Fig. P4.2). Two known forces (P_1, P_2) are applied to bar BC. Draw consistent free body diagrams for the entire system and for the individual bars AC and AB. Neglect the weights of the bars. How do the free body diagrams needed change if the force P_2 is applied to the pin at C? Is the problem statically determinate in either case?

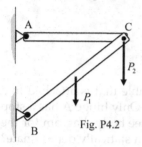

Fig. P4.2

P4.3 An angle bracket is connected at A to a fixed pin and is in contact with a roller at B. The bracket holds two smooth homogeneous cylinders, as shown in Fig. P4.3, whose weights are known. Draw a free body diagram of just the bracket. Neglect the weight of the bracket. Is this the "best" free body diagram one could draw? Explain. Is the problem statically determinate?

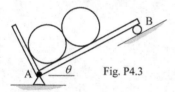

Fig. P4.3

P4.4 A bent bar is fixed to a wall at A and carries the three known forces (P_1, P_2, P_3) shown in Fig. P4.4. Neglect the weight of the bar. Draw a free body diagram for the bar. Is this problem statically determinate?

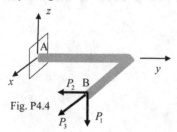

Fig. P4.4

P4.5 A door of known weight W is held by a cable that is attached to a vertical wall at D and a hinge at A. Draw a free body diagram for the door, assuming it is homogeneous. Is this problem statically determinate?

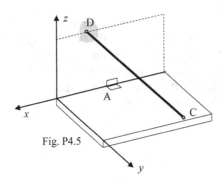

Fig. P4.5

P4.6 A door of known weight W is held by a cable that is attached to a vertical wall at D and by two hinges at A and B. Only hinge A has a stop that can support a force along its axis. Draw a free body diagram for the door, assuming it is homogeneous. Is the problem statically determinate? If not, can you make a different set of assumptions about the reactions needed at the hinges that will make it statically determinate?

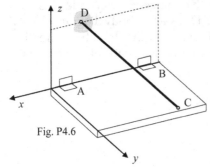

Fig. P4.6

P4.7 A pulley system consisting of three smooth pulleys, A, B, C, as shown in Fig. P4.7, supports a known weight, W, and an applied force P, which is unknown. The ropes are continuous over the pulleys. Draw consistent free body diagrams for the three pulleys. Is the problem statically determinate?

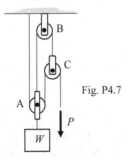

Fig. P4.7

P4.8 An X-frame consists of two bars pinned together at C and connected by a cable at A and B. The X-frame sits on a smooth floor and supports a homogeneous cylinder of known weight W. All contacts are smooth. Draw consistent free body diagrams of the entire system, bars ACE and BCD individually, and the cylinder. Neglect the weights of the bars. Is the problem statically determinate?

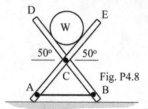

Fig. P4.8

P4.9 A rod assembly is supported by a ball and socket joint at A, a fixed bearing at C, and a smooth roller at D. A known force P is applied to an arm of the assembly in a plane parallel to the y-z plane. The weight of the assembly can be neglected. Draw a free body diagram of the rod assembly. What reactions are needed at the bearing to keep the assembly in equilibrium? Is this problem statically determinate?

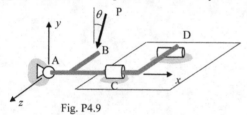

Fig. P4.9

P4.10 A beam sits on three rollers and is loaded by a known concentrated force and distributed load as shown in Fig. P4.10. Neglect the weight of the beam. Draw a free body diagram of the beam. Is the problem statically determinate?

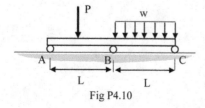

Fig P4.10

P4.11 A U-shaped frame is supported by two pins and carries a known applied force, P, as shown in Fig. P4.11. Neglect the weight of the frame. Draw a free body diagram of the frame. Is the problem statically determinate?

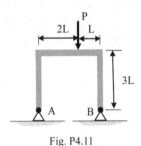

Fig. P4.11

P4.12 The three-bar truss shown in Fig. P4.12 is pinned together at D and is supported by pins at A, B, and C. The truss supports the known forces P and F, which are applied to pin D. Neglect the weights of the truss members. Draw a free body diagram for pin D. Is the problem statically determinate? Does your answer change if you consider pins A, B, and C?

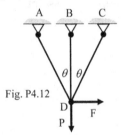

Fig. P4.12

P4.13 A uniform beam weighs $W = 250$ lb and is acted upon by a $P = 1500$ lb force, as shown in Fig. P4.13. The beam is pinned at A and in contact with an inclined smooth surface at B which makes an angle $\theta = 45°$ with respect to the x-axis. The distances $a = 4$ ft, $b = 3$ ft. Determine the magnitude of the force at B acting on the beam.

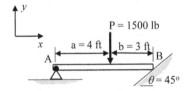

Fig. P4.13

Choices (in lb):

1. 874
2. 919
3. 1016
4. 1212
5. 1320

P4.14 A beam whose own weight can be neglected is pinned at A and supported by a cable at B and C that runs over a smooth pulley at D. An applied force P = 100 lb acts on the beam, as shown in Fig. P4.14. Cable portion BD makes an angle of θ = 30° with respect to the x-axis and cable portions BD and DC are at right angles to each other. Determine the tension in the cable (and the reactions at A). The distances a = 4 ft, b = 4 ft, and c = 2 ft.

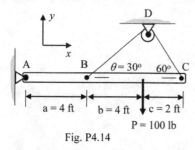

Fig. P4.14

Choices for the tension (in lb):

1. 37.5
2. 54.6
3. 25.7
4. 60.3
5. 75.0

P4.15 A force P = 450 N is applied to the L-shaped frame shown in Fig. P4.15 which is supported by a pin at A and a smooth roller at B. Determine the forces acting on the frame. The distances a = 2 m and b = 4 m. Neglect the weight of the frame.

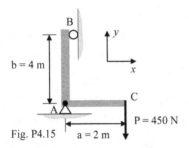

Fig. P4.15   a = 2 m

P4.16 Pulleys A and B of the chain hoist shown in Fig. P4.16 have radii given by a = 3.6 in. and b = 4 in., respectively. These pulleys are connected and rotate as a unit. A force P is applied to the chain at the right to hold a block whose weight is W = 500 lb. The chain is continuous and prevented from slipping by cogs at all the pulleys. Determine the force P.

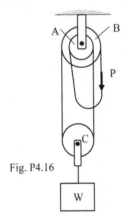

Fig. P4.16

Choices (in lb)

1. 125
2. 75
3. 50
4. 25
5. 15

P4.17 A uniform bar of length L = 4 ft that has a weight W = 25 lb and is pinned to rollers at A and B. The bar assembly rests on the inclined planes, as shown in Fig. P4.17. Determine the angle θ at equilibrium.

Neglect the radii of the rollers and their weights. The angles $\alpha = 30°$ and $\beta = 45°$.

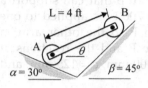

Fig. P4.17

P4.18 A uniform triangular plate has a weight W = 350 lb and is supported by two vertical cables at A and B and sits on a smooth ball support at C, as shown in Fig. P4.18. The distances a = 24 in., b = 30 in. Determine the tensions in the cables and the reaction at the ball support.

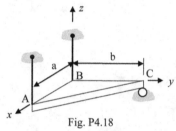

Fig. P4.18

P4.19 A uniform plate has a weight W = 150 lb and is supported by hinges at A and B and a cable at C which is attached to a vertical wall at D, as shown in Fig. P4.19. The hinges are in alignment and only hinge B can support a force along its axis. The distances a = 5 in., b = 26 in., c = 20 in., and the angle of the cable is $\theta = 30°$. Determine the reactions at A and B and the tension in the cable.

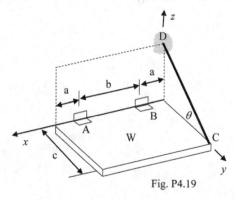

Fig. P4.19

P4.20 A shaft with two arms is supported by two smooth bearings at A and B. Assume the bearings are in alignment and only support forces and assume that only bearing B has a stop that can support a force along the axis of the shaft. A force F = 700 N is applied to one arm in the negative x-direction and a force P to the other arm in the negative z-direction. Determine the force P and the reactions at the bearings. The distances a = 150 mm, b = 100 mm, c = 125 mm, and d = 150 mm.

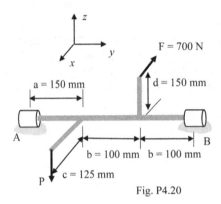

Fig. P4.20

4.5.1 Review Problems

These problems typically have the level of difficulty found on exams. They should be done by hand (i.e., with a calculator).

R4.1 A bent bar whose own weight can be neglected is supported by a pin at A and a cable at B, as shown in Fig. R4.1. A force P = 750 N acts at D. Determine the forces acting at A and the tension in the cable. The distances a = 325 mm, b = 250 mm, and the angles θ = 45° and β = 25°.

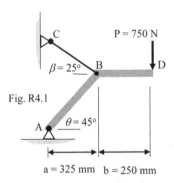

Fig. R4.1

R4.2 The uniform cantilever beam in Fig. R4.2 has a mass of m = 300 kg
and carries a linearly varying distributed load whose intensity varies
from zero to w = 3500 N/m. Determine the reactions acting at A. The
distances a = 3 m and b = 4 m.

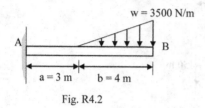

w = 3500 N/m

A

B

a = 3 m   b = 4 m

Fig. R4.2

R4.3 The three-pulley assembly shown in Fig. R4.3 supports a block
whose mass is m = 240 kg. The ropes are continuous over the pulleys.
Determine the force P required to hold the block in equilibrium.

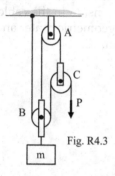

Fig. R4.3

R4.4 A belt runs over a drum which is rigidly attached to a shaft at D that
is supported by smooth bearings at A and B, as shown in Fig. R4.4. A
force P acting in the negative z-direction and a F = 150 lb force acting in
the y-direction are both applied to an arm of the shaft at C. The radius
of the drum is r = 6 in. and there is friction between the belt and the drum
so that the tensions in the belt on either side of the drum are not equal.
(a) If those tensions are T_1 = 300 lb and T_2 = 50 lb, both acting in the
negative x-direction, determine the force P needed to keep the assembly
in equilibrium. Assume the bearings are in alignment and only support
forces and assume that only bearing B has a stop that can support a force
along the axis of the shaft. (b) Without solving for the bearing reactions,
identify the additional equilibrium equations (after P is known) you
could use to determine those reactions that do not require the solution of
any simultaneous equations.

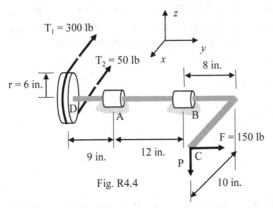

Fig. R4.4

R4.5 The homogeneous bar AB is supported by a pin at A and a cable at B. The bar weighs W_b = 25 lb and is in smooth contact at D with a cylindrical drum which weighs W_d = 125 lb. The drum is also in contact with a smooth vertical surface at C. Determine the tension in the cable. Assume the center of gravity G of the drum is at its geometric center and that the distances a = 32 in., b = 50 in., and c = 15 in.

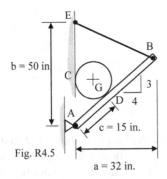

Fig. R4.5

5

Trusses

OBJECTIVES

- To learn how to solve truss problems with the method of pins (joints).
- To learn how to solve truss problems with the method of sections.
- To identify zero-force members.
- To analyze spatial (3-D) trusses.

One very old type of structure that can be analyzed with statics is a truss. A truss consists of an interconnected system of weightless members that are pinned together at their ends with smooth pins. Thus, all the members of the truss are two-force members. The applied loads and support forces are all assumed to be applied to the pins. Early trusses were built in a fashion that matched these connection assumptions, but most trusses are now constructed with bolted, riveted, or welded gusset plate connections and the weight of the truss members may be significant, in which case the members are not two-force members. We will consider only weightless, pinned trusses in this chapter. If all the members lie in a single plane, as is the case for the truss shown in Figure 5.1(a), the truss is called a *planar truss*. If a truss consists of a series of interconnected triangular sections (Figure 5.1(a)), then the truss is also called a *simple truss*.

Trusses can be statically determinate, statically indeterminate (statically underdetermined), or statically overdetermined. To determine which of these cases is appropriate, recall that in a planar truss, the forces at the pins of the truss form a two-dimensional concurrent force system, so there are two equilibrium equations for every pin. Since the internal members of a planar truss are two-force members that carry axial loads, there are as many unknown forces in the members as there are members. The only other unknown forces on the truss are the reaction forces at the supports. To determine the appropriate case (statically

DOI: 10.1201/9781003372592-5

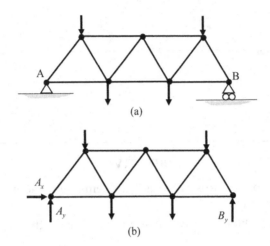

FIGURE 5.1
(a) A simple planar truss that is supported by a pin at A and a smooth roller at B, and (b) a free body diagram of the truss.

determinate, etc.), we need to compare the total number of unknown forces with the total number of equilibrium equations.

If there are *m* members in the planar truss, *n* reaction forces on the entire truss, and *j* pins, then a simple truss is *statically determinate* if

$$m + n = 2j \tag{5.1a}$$

number of unknown forces = number of equilibrium equations

The truss is *statically indeterminate* if

$$m + n > 2j \tag{5.1b}$$

number of unknown forces > the number of equilibrium equation

and the truss is a *statically overdetermined* truss if

$$m + n < 2j \tag{5.1c}$$

number of unknown forces < number of equilibrium equations

If we assume that the members of the truss are rigid, then we can only solve for all the forces if the truss is statically determinate or statically overdetermined. However, statically overdetermined trusses may be unstable or not sufficiently constrained to prevent motion, so they are usually not present in practice. Thus, our primary focus in this chapter will be on the solution of statically determinate trusses. In Chapter 10, however, we will allow the truss members to deform slightly and show how under those conditions we can solve for the forces in statically indeterminate trusses. The truss in Figure 5.1(a) is a statically determinate simple truss if all the applied forces shown in Figure 5.1(a) are known. There are 11 bars that contain 11 forces in the members and three reaction forces at A and B, as seen in the free body diagram of Figure 5.1(b), for a total of 14 unknown forces. There are seven pins and so there also are 14 equilibrium equations.

There are two methods for analyzing the equilibrium of trusses. The first, called the *method of pins*, examines, as its name implies, equilibrium at the pins. The method of pins is also called the *method of joints*. The second method, called the *method of sections*, imagines cutting through (sectioning) the truss into parts and drawing free body diagrams of the parts to examine the equilibrium of those parts and obtain one or more of the internal forces in the truss members. We will examine both methods. Let us begin with the method of pins.

5.1 The Method of Pins

Example 5.1

Consider the three-member truss shown in Figure 5.2. It is a statically determinate truss since there are three reaction forces and three forces in the members for a total of six unknown forces and there are three pins at which we have a total of six equilibrium equations. If we obtain this

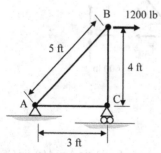

FIGURE 5.2
A simple truss.

solution via computer, we simply set up and solve all six equations at the pins simultaneously. If we obtain this solution by hand, it is easier to use the three equilibrium equations for the overall structure first and then we will have to solve only three equations from a subset of the six equilibrium equations at the pins. Let's consider first solving this problem by hand.

Free Body Diagrams
The free body diagrams of the entire structure and two of the pins we will use are shown in Figure 5.3(a)–(c). The forces at A and C will be found from equilibrium of the entire structure so those forces are shown as known on the two pins.

Equations of Equilibrium
For the entire truss ABC, we have three equilibrium equations for this coplanar force system, which are

$$_{ABC}\Sigma F_x = 0 \quad A_x + 1200 = 0$$
$$\rightarrow \quad A_x = -1200 \text{ lb}$$
$$_{ABC}\Sigma F_y = 0 \quad A_y + 1600 = 0$$
$$\rightarrow \quad A_y = -1600 \text{ lb}$$
$$_{ABC}\Sigma M_A = 0 \quad 3C_y - (1200)(4) = 0$$
$$\rightarrow \quad C_y = 1600 \text{ lb}$$

(5.2)

Now, consider pin A (Figure 5.3(b)). From the two equilibrium equations at this pin, we obtain

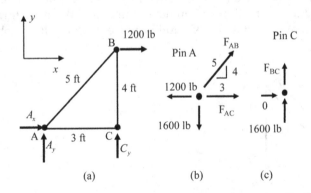

(a)   (b)   (c)

FIGURE 5.3
(a) A free body diagram of the entire truss, and free body diagrams for (b) pin A and (c) pin C.

$$_A\Sigma F_y = 0 \quad \frac{4}{5}F_{AB} - 1600 = 0$$
$$\rightarrow \quad F_{AB} = 2000 \ \text{lb}$$
$$_A\Sigma F_x = 0 \quad -1200 + \left(\frac{3}{5}\right)(2000) + F_{AC} = 0 \tag{5.3}$$
$$\rightarrow \quad F_{AC} = 0$$

Finally, from pin C (Figure 5.3(c)), we have

$$_C\Sigma F_y = 0 \quad F_{BC} + 1600 = 0$$
$$\rightarrow \quad F_{BC} = -1600 \ \text{lb} \tag{5.4}$$

Note that *if we always draw the unknown forces in the members as pulling on the pin, this assumes the members are in tension.* The member forces in Figure 5.3(b) and (c) are all drawn in this manner. If a member force, therefore, comes out to be negative, we know the member is in compression. When giving the solution for the forces in the members we usually just give the magnitude of the force together with a T (for tension) or C for compression. For the three members for the truss in Figure 5.3(a), therefore, we would write the total solution as

$$A_x = -1200 \ \text{lb}$$
$$A_y = -1600 \ \text{lb}$$
$$C_y = 1600 \ \text{lb}$$
$$F_{AB} = 2000 \ \text{lb} \ \text{T} \tag{5.5}$$
$$F_{AC} = 0$$
$$F_{BC} = 1600 \ \text{lb} \ \text{C}$$

The complete solution for this truss is given in Figure 5.4, where we show both the forces at all the pins and the forces in the members. In obtaining this solution, we did not have to consider the x-force equation for pin C or either the x- or y-force equations for pin B to solve for all the unknowns, since we used the force and moment equilibrium equations for the entire structure instead. One could examine the equilibrium equations not used as checks on the results obtained from the other pins and the entire structure. Thus, this solution was not purely a method of pins.

Now, let's reconsider this problem where we do use all the pins instead.

Free Body Diagrams
Free body diagrams for the three pins are shown in Figure 5.5.

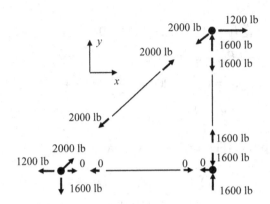

FIGURE 5.4
All the member forces and reaction forces for the three-member truss of Figure 5.2.

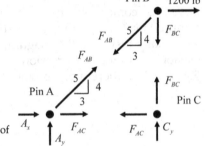

FIGURE 5.5
Free body diagrams of the pins for the truss of
Figure 5.2.

Equations of Equilibrium
For pin A

$$_A\sum F_x = 0 \quad A_x + \frac{3}{5}F_{AB} + F_{AC} = 0 \quad (1)$$

$$_A\sum F_y = 0 \quad \frac{4}{5}F_{AB} + A_y = 0 \quad\quad\quad (2)$$

(5.6)

where we have numbered these equations also. For pin C

$$_C\sum F_x = 0 \quad -F_{AC} = 0 \quad\quad (3)$$

$$_C\sum F_y = 0 \quad C_y + F_{BC} = 0 \quad (4)$$

(5.7)

and for pin B

$$_B\Sigma F_x = 0 \quad 1200 - \frac{3}{5}F_{AB} = 0 \quad (5)$$

$$_B\Sigma F_y = 0 \quad -\frac{4}{5}F_{AB} - F_{BC} = 0 \quad (6)$$

(5.8)

We want to solve simultaneously these six equations. This is not strictly necessary since some forces are obtained directly from individual equilibrium equations, but it is very easy to directly enter and solve these six equations symbolically in MATLAB, as we have demonstrated in other problems. First, we declare the unknowns symbolically and then enter the six equations as components of a symbolic vector Eq:

```
syms Ax Ay Cy Fab Fac Fbc
Eq(1) = Ax + 0.6*Fab + Fac == 0;
Eq(2) = 0.8*Fab + Ay == 0;
Eq(3) = -Fac == 0;
Eq(4) = Cy + Fbc == 0;
Eq(5) = -0.6*Fab + 1200. == 0;
Eq(6) = -0.8*Fab - Fbc == 0;
```

Here, we used the == 0 in the symbolic equations but, as discussed previously, that is not necessary. [In trusses where there are many pins to consider, by not including the == 0 we can save some effort in writing the equations.] We can solve this system of equations symbolically with the MATLAB function solve. The output of solve here is in a variable S which represents a MATLAB structure with six fields:

```
S = solve(Eq)
S = struct with fields:
     Ax: -1200
     Ay: -1600
     Cy: 1600
     Fab: 2000
     Fac: 0
     Fbc: -1600
```

These values are still in symbolic form. If we use the double function, we can convert them to numeric values:

```
Sn = double([S.Ax S.Ay S.Cy S.Fab S.Fac S.Fbc])
Sn =
[-1200, -1600, 1600, 2000, 0, -1600]
```

If you do not have the MATLAB symbolic toolkit, you can still do this problem with solely numerical calculations, as we have shown previously. We write these six equilibrium equations in matrix vector form as $[E]\{x\} = \{P\}$, where $[E]$ is a 6×6 matrix of coefficients of the unknowns, $\{P\}$ is a 6×1 column vector of the known applied forces in these equations (actually $\{P\}$ contains the negative of the applied loads since they are placed on the left side of the equilibrium equations), and $\{x\}$ is a 6×1 column vector of the unknown forces, whose components we have to identify with the various unknown forces. Here we take this x-vector to represent the following unknowns:

$$
\begin{matrix}
A_x \\
A_y \\
C_y \\
F_{AB} \\
F_{AC} \\
F_{BC}
\end{matrix}
\quad
\begin{Bmatrix}
x(1) \\
x(2) \\
x(3) \\
x(4) \\
x(5) \\
x(6)
\end{Bmatrix}
\tag{5.9}
$$

Examining our six equations, we have

$$
[E] = \begin{bmatrix}
1 & 0 & 0 & 0.6 & 1 & 0 \\
0 & 1 & 0 & 0.8 & 0 & 0 \\
0 & 0 & 0 & 0 & -1 & 0 \\
0 & 0 & 1 & 0 & 0 & 1 \\
0 & 0 & 0 & -0.6 & 0 & 0 \\
0 & 0 & 0 & -0.8 & 0 & -1
\end{bmatrix}, \quad
\{P\} = \begin{Bmatrix}
0 \\
0 \\
0 \\
0 \\
-1200 \\
0
\end{Bmatrix}
\tag{5.10}
$$

which we can solve in MATLAB. Here is the entry of the coefficient matrix and P vector in MATLAB and the solution:

```
E = [1  0  0   0.6   1   0;
     0  1  0   0.8   0   0;
     0  0  0    0   -1   0;
     0  0  1    0    0   1;
     0  0  0  -0.6   0   0;
     0  0  0  -0.8   0  -1];
```

P = [0; 0; 0; 0; −1200; 0];
x = E\P
x = −1200
    −1600
     1600
     2000
        0
    −1600

Unlike the solution by hand, we never considered equilibrium of the entire structure since, as we have shown before in Chapter 4, if each part of the structure (each pin) is in equilibrium, then the entire structure will be in equilibrium.

5.1.1 Summary

Solving truss problems with the method of pins in a symbolic manner is relatively easy. Most statics texts instead concentrate on the method of sections (which we will discuss shortly) where we do not have to solve as many equilibrium equations to get results for specific forces. However, with the use of software such as MATLAB and its symbolic capabilities, we see that a method of pins solution is very direct. Our numerical MATLAB solution to this problem was less efficient than the symbolic approach since we had to set up a rather large matrix for the equilibrium equations even though many elements of that matrix were zero.

5.2 Zero-Force Members

A truss may have members that do not carry any loads (zero-force members) but that may be needed for stability of the truss or for other reasons. For example, the force in member AC for the truss in Figure 5.2 was zero but if we remove that member from the truss, we end up with the configuration of Figure 5.6(a), which is clearly not stable. We proved that the force in AC was zero by summing forces in the *x*-direction at pin C (Figure 5.6(b)) but if we simply examine the original three-member truss, we see that since both the force in member BC and the force from the roller must be vertical, there cannot be any force in member AC. In the same fashion, we can identify zero-force members by inspection. For example, there is a zero-force member at pin A in Figure 5.7(a) since if we sum forces in the *n*-direction, there is no other force with a component in that direction. In Figure 5.7(b), we see that at pin A we can again identify a zero-force member along AB, but at pin B summing forces in the

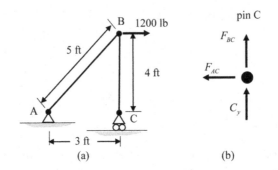

FIGURE 5.6
(a) The truss of Figure 5.2 with member AC removed. (b) The free body diagram of pin C for the truss of Figure 5.2.

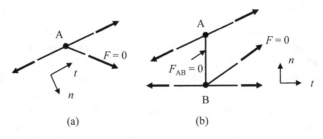

FIGURE 5.7
(a) A zero-force member acting at pin A. (b) A zero-force member acting at pin B in conjunction with a zero-force member in AB.

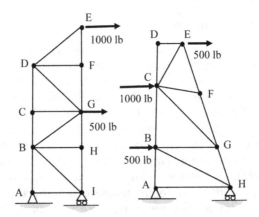

FIGURE 5.8
Two trusses where you should be able to identify a few zero-force members by inspection.

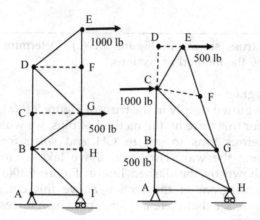

FIGURE 5.9
The zero-force members (dashed lines) for the trusses in Figure 5.7.

n-direction means that the inclined member is also a zero-force member. Similarly, we can identify zero-force members by inspection in other trusses. Consider, for example, the trusses shown in Figure 5.8. You should be able to identify the zero-force members directly through examination of that figure. The zero-force members are shown in Figure 5.9.

5.3 The Method of Sections

Unlike the method of pins, the method of sections is not very amenable to computer solution. However, it is a very efficient way of determining a few of all the truss member forces by hand. In the method of sections, we take an imaginary cut through the truss to expose the member whose force we want and then use the equations of equilibrium for the section of the truss we have cut off. In some cases, we must take more than one cut. In general, the forces for a cut section will represent a coplanar system of forces, so we will have three equations of equilibrium:

$$\sum F_x = 0$$
$$\sum F_y = 0 \qquad (5.11)$$
$$\sum M_{Pz} = 0$$

The moment equation is especially useful since we can choose P, the point which we take moments about, to eliminate as many unknowns as possible. In some cases, taking moments about two different points can also be helpful.

Example 5.2

Consider the truss shown in Figure 5.10(a). Determine the force in member GH by the method of sections.

Free Body Diagram

Member GH is buried deeply in the truss of Figure 5.10(a). If we tried to get the value for this force by the method of pins, we would have to go through numerous pins to get to GH and solve for many other unknowns along the way. However, if we take a cut through the structure as shown by the dashed line in Figure 5.10(a) and look at the free body diagram of the section above this cut, as shown in Figure 5.10(b), we see that there are three unknown forces acting in this free body diagram.

Equations of Equilibrium

We have three equilibrium equations for this section so we could solve those equations for all three forces, but if we simply take moments of the forces about point B, we could get the force in GH directly:

$$\sum M_{Bz} = 0 \quad (1000)(12) + (500)(4) + 6F_{GH} = 0$$
$$\rightarrow \quad F_{GH} = -2333.3 \text{ lb} \quad or \quad F_{GH} = 2333.3 \text{ lb C} \tag{5.12}$$

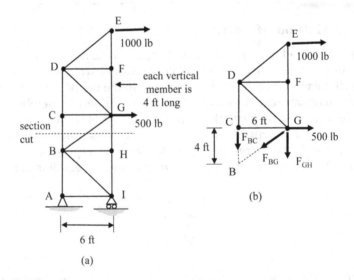

(a)

(b)

FIGURE 5.10
(a) A truss structure. (b) Free body diagram of a section above the cut.

Note that if we had looked at the section below the cut, then we would also have the reaction forces at A and I in that free body diagram and we could not find the force in GH until we first found the reaction forces (which we could obtain from equilibrium of the entire truss). Thus, using the section above the cut is the most efficient choice.

5.4 Space Trusses

Up to this point, all the trusses we have considered have been two-dimensional trusses. Trusses where the members do not all lie in a single plane are called *space trusses*. The two-force members in the space truss are again connected by smooth pins, so the member forces (and any applied forces) form a concurrent force system at each pin where there are three equations of force equilibrium. For j pins, there are, therefore, $3j$ equations of equilibrium so if the number, m, of truss members plus the number, n, of reaction forces satisfies

$$m + n = 3j \qquad (5.13)$$

the truss will be statically determinate. Eq. (5.13) is the space truss equivalent of Eq. (5.1a).

Example 5.3

Consider the space truss shown in Figure 5.11 consisting of the five members AB, AD, DC, DB, and BC. The truss is supported by ball and socket joints at A and C and by a short, horizontal weightless link (two-force member) parallel to the y-direction at D. Members AB and BC lie in the x-y plane. Using the method of pins, determine all the forces in the members and the reactions at A, C, and D.

There are five forces in the members of the truss and seven reaction forces so there are a total of 12 unknowns. Likewise, there are 12 equilibrium equations at the four pins, so the truss is statically determinate. Space trusses inherently involve many calculations, so the use of MATLAB is the preferred method of solution.

Free Body Diagrams

Free body diagrams of the forces at the four pins are shown in Figure 5.12. To make these free body diagrams useful, we need to define the unit vectors in Figure 5.12. Then, we can form the force equilibrium

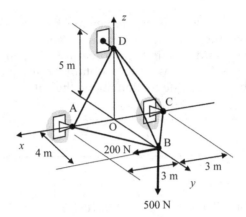

FIGURE 5.11
A space truss. There are ball and socket joints at A and C and a short horizontal link at D.

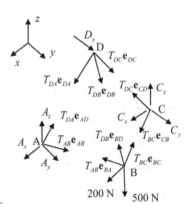

FIGURE 5.12
The forces at the pins of the space truss of Figure 5.10.

equations at each pin and collect and solve those equations. We will use a symbolic MATLAB solution method.

First, we need the unit vectors that define the directions of the member forces at the pins and the applied and reaction forces. Note that, like for the two-dimensional trusses, we will show the member forces at the pins as if those members are in tension, so the unit vectors that define those forces will always act away from the pins along the various members (see Figure 5.12). Here are the unit vectors shown in Figure 5.12 obtained in MATLAB:

```
DA = [3  0 -5];
eDA = DA/norm(DA);
DC = [-3  0 -5];
eDC = DC/norm(DC);
DB = [0  4 -5];
eDB = DB/norm(DB);
AB = [-3  4  0];
eAB = AB/norm(AB);
CB = [3  4  0];
eCB = CB/norm(CB);
eCD = -eDC;
eAD = -eDA;
eBD = -eDB;
eBA = -eAB;
eBC = -eCB;
```

Having these unit vectors, we can then define the unknown force magnitudes symbolically, and then compute the resultant forces at each pin (which must, of course, be all zero forces):

```
syms Ax Ay Az Cx Cy Cz Dy Tda Tdb Tdc Tab Tb
FD = Tda*eDA + Tdb*eDB + Tdc*eDC + [0 Dy 0];    % resultant forces
                                                % at the pins
FA = [Ax Ay Az] + Tda*eAD + Tab*eAB;
FB = [200 0 -500] + Tab*eBA + Tdb*eBD + Tbc*eBC;
FC = [Cx Cy Cz] + Tbc*eCB + Tdc*eCD;
```

Equations of Equilibrium
We can form all these forces into a set of vector equilibrium equations and solve those equations, placing the results in a MATLAB structure, S:

```
Eqs = [FA; FB; FC; FD];
S = solve(Eqs);
```

Note that again we did not have to formally set each of these equations equal to zero. We simply placed the left-hand side of all the equilibrium equations in the vector Eqs. We can put the answers in final numerical form with the function double:

Sn = double([S.Ax; S.Ay; S.Az; S.Cx; S.Cy; S.Cz; S.Dy; S.Tab; ...
    S.Tbc; S.Tda; S.Tdb; S.Tdc])

Sn =	% Forces in Newtons
-400.0000	% Ax
333.3333	% Ay
250.0000	% Az
200.0000	% Cx
66.6667	% Cy
250.0000	% Cz
-400.0000	% Dy
-416.6667	% Tab
-83.3333	% Tbc
-291.5476	% Tda
640.3124	% Tdb
-291.5476	% Tdc

There are quite a few forces here. We can ensure that overall equilibrium is satisfied for the truss by writing the reaction forces at A, C, and D, and the applied force at B and computing the net force and net moment about the origin O of all these forces:

```
A = [Sn(1) Sn(2) Sn(3)];     % reaction force at A
C = [Sn(4) Sn(5) Sn(6)];     % reaction force at C
D = [0 Sn(7) 0];             % reaction force at D
PB = [200 0 -500];           % applied force at B
RF = A + C + D + PB          % resultant of all the forces on the truss
RF = 0 0 0
rA = [3 0 0];                % position vectors from the origin O to A, B, C, D
rB = [0 4 0];
rC = [-3 0 0];
rD = [0 0 5];
% resultant moment about O of all the forces on the truss


RM = cross(rA, A) + cross(rB, PB) + cross(rC, C) + cross(rD, D)
RM = 0 0 0
```

Although this overall equilibrium check is not necessary, it is again a demonstration that by satisfying equilibrium at all the pins, the equilibrium of the entire truss is also satisfied.

5.5 Problems

The problems which ask for all the forces in the truss are best done using the method of pins and MATLAB while those that ask for one or several forces in specific members are best done by hand (i.e., with a calculator) using the method of sections.

P5.1 The three-bar truss of Fig. P5.1 is supported by a pin at A and rollers at C. The force at B has a magnitude F = 1000 lb. Determine the forces in the members and the reactions at A and C. The angle $\theta = 30°$.

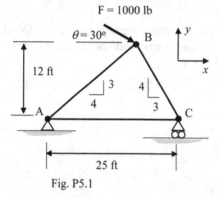

Fig. P5.1

P5.2 The three-bar truss of Fig. P5.2 is supported by a pin at B and rollers at A. The magnitude of the force at C is P = 1500 lb. Determine the forces in the members and the reactions at A and B. The angle $\theta = 25°$.

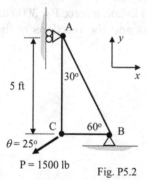

Fig. P5.2

P5.3 The truss in Fig. P5.3 is pinned at A and on rollers at B. The force F = 1000 lb acts vertically down at D. Determine the forces in the members and the reactions at A and B.

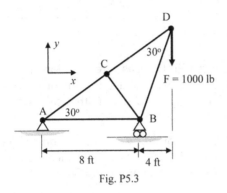

Fig. P5.3

P5.4 The truss in Fig. P5.4 carries a force F = 2000 lb and a force P = 5000 lb, as shown. Determine the forces in the members and the reactions at the supports. The distances a = 8 ft and b = 6 ft.

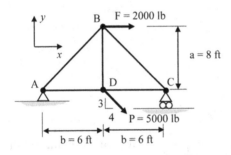

Fig. P5.4

P5.5 The truss in Fig. P5.5 carries a force F = 1000 lb and a force P = 3000 lb, as shown. Determine the forces in the members and the reactions at the supports. The distances a = 9 ft and b = 6 ft.

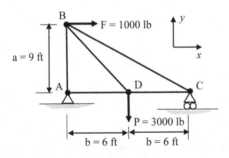

Fig. P5.5

Trusses

P5.6 The truss in Fig. P5.6 carries a force P = 5000 lb as shown. Determine the forces in the members and the reactions at the supports. The distances a = 3 ft, b = 5 ft, and c = 5 ft.

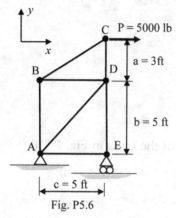

Fig. P5.6

P5.7 The truss in Fig. P5.7 carries a uniform sign which has a weight W = 350 lb. Determine the forces in the members and the reactions at the supports. The distances a = 10 ft and b = 12 ft.

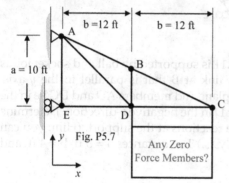

Fig. P5.7

P5.8 The truss in Fig. P5.8 carries a force P = 400 N as shown in Fig. P5.8. Determine the forces in the members and the reactions at the supports. The distances a = 3 m and b = 4 m.

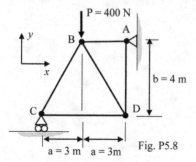

Fig. P5.8

P5.9 Identify all the zero-force members of the truss in Fig. P5.9.

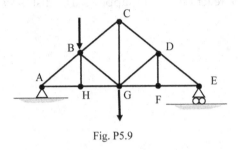

Fig. P5.9

P5.10 Identify all the zero-force members of the truss in Fig. P5.10.

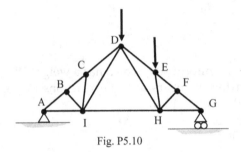

Fig. P5.10

P5.11 The five-bar truss in Fig. P5.11 is supported by ball and socket joints at A and C and by a weightless link at B that is parallel to the y-axis. Members AB and BC lie in the x-z plane and members AD and DC lie in the x-y plane. A force P = 600 lb acts at D in the negative z-direction. Determine the forces in the members and the reactions at the support. (Hint: you can use symmetry to simplify the analysis.) The distances a = 5 ft, b = 4 ft, and c = 3 ft.

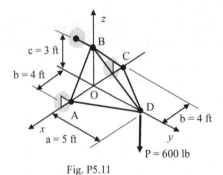

Fig. P5.11

P5.12 Members AB, BC, and AC of the six-bar truss in Fig. P5.12 all lie in plane parallel to the *x-y* plane and 1 ft above it. Members BC, BD, and CD all lie in the *y-z* plane. The truss is supported by a ball and socket joint at A and by weightless links at B and C where the link at B is parallel to the *z*-axis and the links at C are parallel to the *y*- and *z*-axes. Forces F = 300 lb and P = 200 lb act at D in the *x*- and *y*-directions, respectively. Determine the forces in the members and the reactions at the supports. The distances a = 3 ft, b = 4 ft, c = 5 ft, and d = 5 ft.

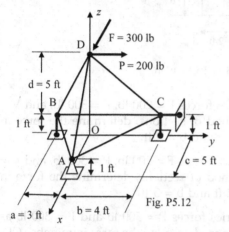

Fig. P5.12

P5.13 Members AE and ED of the seven-bar truss of Fig. P5.13 lie in the *x-y* plane while members AB, BC, and CD lie in the *x-z* plane. A force P = 600 lb is applied to the truss at E in the negative *z*-direction. The truss is supported by ball and socket joints at A and D and by weightless links at B and C that are parallel to the *y*-axis. Determine the forces in the members and the reactions at the supports. (Hint: you can use symmetry to simplify the analysis.) The distances a = 6 ft, b = 4 ft, and c = 3 ft.

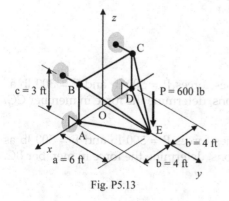

Fig. P5.13

P5.14 The truss in Fig. P5.14 carries forces F = 200 lb, P = 300 lb, and V = 150 lb, as shown. Using the method of sections, determine the force in member CD. The distances a = 2 ft and b = 3 ft.

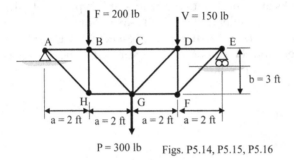

Figs. P5.14, P5.15, P5.16

P5.15 The truss in Fig. P5.15 carries forces F = 200 lb, P = 300 lb, and V = 150 lb, as shown. Using the method of sections, determine the force in member BG. The distances a = 2 ft and b = 3 ft.

P5.16 The truss in Fig. P5.16 carries forces F = 200 lb, P = 300 lb, and V = 150 lb, as shown. Using the method of sections, determine the force in member GH. The distances a = 2 ft and b = 3 ft.

P5.17 The truss in Fig. P5.17 carries forces P = 500 lb and F = 300 lb as shown. Using the method of sections, determine the force in member CD. The distance a = 12 ft.

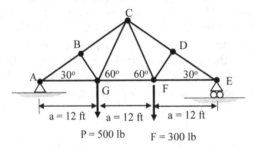

Figs. P5.17, P5.18, P5.19

P5.18 The truss in Fig. P5.18 carries forces P = 500 lb and F = 300 lb as shown. Using the method of sections, determine the force in member CG. The distance a = 12 ft.

P5.19 The truss in Fig. P5.19 carries forces P = 500 lb and F = 300 lb as shown. Using the method of sections, determine the force in member BC. The distance a = 12 ft.

5.5.1 Review Problems

These problems typically have the level of difficulty found on exams. They should be done by hand (i.e., with a calculator).

R5.1 The truss in Fig. R5.1 carries the forces $P_1 = 200$ lb, $P_2 = 300$ lb, and $P_3 = 400$ lb. Using the method of sections, determine the magnitude of the force in member EF and indicate if it is in tension (T) or compression (C). The distances a = 6 ft, b = 5 ft.

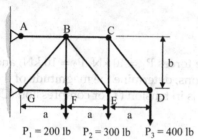

Figs. R5.1, R5.2, R5.3

Choices (in lb)

1. 675 T
2. 895 T
3. 1125 C
4. 1230 C
5. 1320 C

R5.2 The truss in Fig. R5.2 carries the forces $P_1 = 200$ lb, $P_2 = 300$ lb, and $P_3 = 400$ lb. Using the method of sections, determine the magnitude of the force in member BG and indicate if it is tension (T) or compression (C). The distances a = 6 ft and b = 5 ft.

Choices (in lb)

1. 1250 T
2. 1406 C
3. 1502 T
4. 758 C
5. 945 T

R5.3 The truss in Fig. R5.3 carries the forces $P_1 = 200$ lb, $P_2 = 300$ lb, and $P_3 = 400$ lb. Using the method of sections, determine the magnitude of the force in member BE and indicate if it is tension (T) or compression (C). The distances a = 6 ft and b = 5 ft.

Choices (in lb)

1. 255 T

2. 1093 T

3. 350 T

4. 1160 C

5. 750 C

R5.4 The truss of Fig. R5.4 carries the forces P_1 = 40 kN, P_2 = 10 kN, and P_3 = 50 kN. Using the method of sections, determine the magnitude of the force in member BC and indicate if it is in tension (T) or compression (C). The distances a = 3 m and b = 2 m.

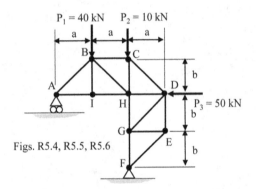

Figs. R5.4, R5.5, R5.6

R5.5 The truss of Fig. R5.5 carries the forces P_1 = 40 kN, P_2 = 10 kN, and P_3 = 50 kN. Using the method of sections, determine the magnitude of the force in member DG and indicate if it is tension (T) or compression (C). The distances a = 3 m and b = 2 m.

R5.6 The truss of Fig. R5.6 carries the forces P_1 = 40 kN, P_2 = 10 kN, and P_3 = 50 kN. Using the method of sections, determine the magnitude of the force in member CD and indicate if it is in tension (T) or compression (C). The distances a = 3 m and b = 2 m.

6

Frames and Machines

OBJECTIVES

- To solve equilibrium problems for a variety of structures and devices.

- To demonstrate solutions with and without the use of MATLAB®.

In Chapter 4, we introduced the concept of free body diagrams and examined the equilibrium equations in various problems. In this chapter, we will continue to explore the use of equilibrium in more complex structures that are often classified as *frames and machines*. The frames we will consider are generally pin-connected rigid members where the members carry more complex internal forces than do trusses. The term "machine" will generally refer to a device, such as pliers, where the applied forces are amplified in the device. Here, we will consider the types of problems found in other statics texts where direct solutions (i.e., those not requiring the solution of simultaneous equations) are obtained for specific unknowns, but we will also determine all the unknown forces, acting either externally or internally on one or more rigid bodies with the aid of MATLAB.

6.1 Solving Equilibrium Problems – II

Example 6.1

Consider a 400 lb load (which acts at the center of gravity, G) that is held on one side of a dock by a hydraulic cylinder CF, which is pinned to member ACD and to the dock as shown in Figure 6.1. The platform is held by two other pinned members DCA and EB. There is an identical hydraulic cylinder and supports on the other side of the dock, but we will analyze only this side as if the other side were absent. All the unknown

DOI: 10.1201/9781003372592-6

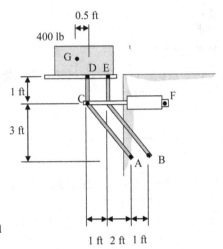

FIGURE 6.1
A 400 lb load held by a hydraulic cylinder and
supporting structure (only one side shown).

forces obtained can then be simply divided by two due to the symmetry
of the geometry. Determine (1) the force in the hydraulic cylinder, and
(2) all the forces at the pinned supports, which are assumed to be smooth.

Free Body Diagrams
Part (1): Usually, it is worthwhile to first examine a free body diagram of
the entire structure (Figure 6.2), particularly if the force we are trying to
find is present in that free body diagram. If we recognize that member EB is
a two-force member, then the direction of the force at pin B is known but

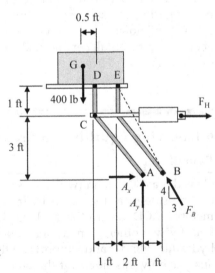

FIGURE 6.2
A free body diagram of the entire supporting
structures (only one side shown).

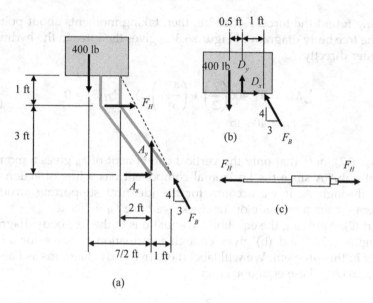

(a)

FIGURE 6.3
Free body diagrams of (a) the load and supporting frame (FBD I), (b) the load and supporting platform (FBD II), and (c) the hydraulic cylinder.

the magnitude of that force is unknown. There are four unknown forces but only three equations of equilibrium for this free body diagram so we cannot find all these forces, including the force in the hydraulic cylinder, F_H, and we need to examine parts of the system. Figure 6.3 shows three free body diagrams, one for the load and the supporting structure, one for the load and the platform it sits on, and one for the hydraulic cylinder. The free body diagram for the hydraulic cylinder is shown simply for completeness. There are a total of six unknown forces and six equilibrium equations for the two free body diagrams in Figure 6.3(a) and (b) so we can find all the forces, including the force in the hydraulic cylinder.

Equations of Equilibrium
We do not have to go through all six equilibrium equations if all that is wanted is the force in the hydraulic cylinder. Taking moments about point D for the free body diagram of Figure 6.3(b):

$$\sum M_D = 0 \quad \frac{4}{5}F_B(1) + 400\left(\frac{1}{2}\right) = 0$$

$$\rightarrow F_B = -250 \text{ lb}$$

(6.1)

Having found the force, F_B, for BE, then taking moments about point A for the free body diagram of Figure 6.3(a) gives the force in the hydraulic cylinder directly:

$$\sum M_A = 0 \quad 400\left(\frac{7}{2}\right) + \left(\frac{4}{5}F_B\right)(1) - F_H(3) = 0 \tag{6.2}$$
$$\rightarrow F_H = 400 \text{ lb}$$

In Eq. (6.2), note that only the vertical component of F_B gives a moment about point A since the horizontal component has a line of action that goes through A. If we account for the identical supporting structure present on the other side of the dock, we have $F_H = 200$ lb.

Part (2): Applying the equilibrium equations to the free body diagrams of Figure 6.3(a) and (b) gives enough information to solve for all the forces in this problem. We will label these free body diagrams as I and II, respectively. These equations are:

$$_I\sum F_x = 0 \quad F_H - \frac{3}{5}F_B + A_x = 0$$

$$_I\sum F_y = 0 \quad A_y + \frac{4}{5}F_B = 400$$

$$_I\sum M_A = 0 \quad \left(\frac{4}{5}F_B\right)(1) - F_H(3) = -400\left(\frac{7}{2}\right)$$

$$_{II}\sum F_x = 0 \quad D_x - \frac{3}{5}F_B = 0 \tag{6.3}$$

$$_{II}\sum F_y = 0 \quad D_y + \frac{4}{5}F_B = 400$$

$$_{II}\sum M_D = 0 \quad \frac{4}{5}F_B(1) = -400\left(\frac{1}{2}\right)$$

We can go through these equations sequentially, solving for all the unknowns without solving simultaneous equations but MATLAB makes the process very simple if we define these equations in MATLAB symbolically:

```
syms Ax Ay Dx Dy Fh Fb
Eq(1) = Fh - 0.6*Fb + Ax == 0;
Eq(2) = Ay + 0.8*Fb - 400 == 0;
Eq(3) = 0.8*Fb - 3*Fh + 1400 == 0;
Eq(4) = Dx - 0.6*Fb == 0;
Eq(5) = Dy + 0.8*Fb - 400 == 0;
Eq(6) = 0.8*Fb + 200 == 0;
```

then we can solve them and place them in a MATLAB structure S:

```
S = solve(Eq)
S =
struct with fields: Ax: -550
                    Ay: 600
                    Dx: -150
                    Dy: 600
                    Fb: -250
                    Fh: 400
```

and we can convert these symbolic values to numeric values

```
double([S.Ax S.Ay S.Dx S.Dy S.Fb S.Fh])

 ans = -550 600 -150 600 -250 400
```

Because of the supports on the other side, we can divide all these values by two. To obtain a purely numerical solution, we need to define the matrix of coefficients for a set of simultaneous equations $[E]\{x\} = \{P\}$ from Eq. (6.3). Letting

$$\{x\}^T = [F_H \ F_B \ A_x \ A_y \ D_x \ D_y] \tag{6.4}$$

the equilibrium matrix and known force vector are

```
E = [1.0000  -0.6000  1.0000  0       0       0;
     0        0.8000  0       1.0000  0       0;
    -3.0000   0.8000  0       0       0       0;
     0       -0.6000  0       0       1.0000  0;
     0        0.8000  0       0       0       1.0000;
     0        0.8000  0       0       0       0];

P = [0; 400; -1400; 0; 400; -200];
```

Then, the solution can be obtained directly as

```
x = E\P
x = 400     % Fh
   -250     % Fb
   -550     % Ax
    600     % Ay
   -150     % Dx
    600     % Dy
```

and again, we can divide all these values by two. Free body diagrams of the entire system and all the parts of the system with the known forces placed on them are given in Figure 6.4.

When this system was analyzed, it was separated into parts at the pinned connections. This was deliberate because if instead an imaginary cut was made, for example, through members DCA or member EB at a 0.5 ft distance below ED, as shown in Figure 6.5(a), then a much more complex set of internal forces and couples would be present in those members, as shown in Figure 6.5(b) and (c). This is true for EB because while it is a two-force member where the forces at pins E and B are simply along the line EB, the member is not straight so the internal forces will be more complex. For a straight two-force member like the hydraulic cylinder, the internal force at a cut as shown in Figure 6.5(d) is simply a 400 lb tension. Once all the forces at the pins are obtained, we can determine the internal forces at these cuts from the free body diagrams of Figure 6.5(b)–(d). We will not show those results here.

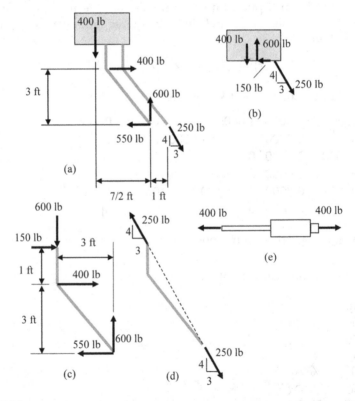

FIGURE 6.4
(a) The free body diagram of the entire system, and (b)–(e) the free body diagrams of the system parts.

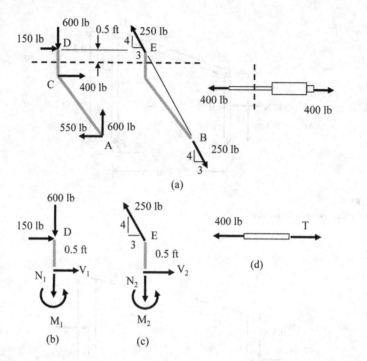

FIGURE 6.5
(a) The case when imaginary cuts are made through the system members, as shown by the dashed lines. (b)–(d) The internal forces and couples exposed at those cuts.

Example 6.2

Consider the structure shown in Figure 6.6 which is pressed against a wall at point E where there is a smooth roller. Determine all the forces acting on the structure using MATLAB. Neglect the weight of all members.

Free Body Diagrams

We will again obtain a symbolic solution to this problem. We recognize member CD as a two-force member so we will only examine the free body diagrams of the other two members AC and DF. Figure 6.7 shows the free body diagrams of those members. There are six unknown forces and six equilibrium equations, so the problem is statically determinate.

Equations of Equilibrium

From the free body diagrams, the equilibrium equations are:

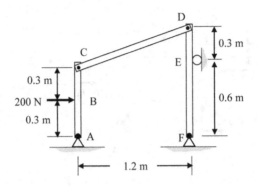

FIGURE 6.6
A two-dimensional structure.

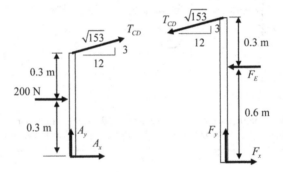

FIGURE 6.7
Free body diagrams of the vertical members AC and DF in Figure 6.6.

$$_{AC}\Sigma F_x = 0 \quad A_x + 200 + (12/\sqrt{153})T_{CD} = 0$$
$$_{AC}\Sigma F_y = 0 \quad A_y + (3/\sqrt{153})T_{CD} = 0$$
$$_{AC}\Sigma M_A = 0 \quad (0.6)(12/\sqrt{153})T_{CD} + (0.3)(200) = 0$$
$$_{DF}\Sigma F_x = 0 \quad F_x - F_E - (12/\sqrt{153})T_{CD} = 0$$
$$_{DF}\Sigma F_y = 0 \quad F_y - (3/\sqrt{153})T_{CD} = 0$$
$$_{DF}\Sigma M_F = 0 \quad (0.9)(12/\sqrt{153})T_{CD} + (0.6)F_E = 0$$

(6.5)

Placing these equations into MATLAB, we can again use the solve function. In setting up the equations it is not necessary to include the == 0. The function solve assumes this implicitly. Here is an example:

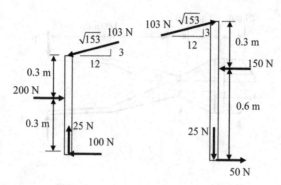

FIGURE 6.8
The forces acting on the vertical members of the structure in Figure 6.6.

```
syms Ax Ay Tcd Fe Fx Fy
Eq(1) = Ax + 200 + (12/sqrt(153))*Tcd;
Eq(2) = Ay + (3/sqrt(153))*Tcd;
Eq(3) = 0.6*(12/sqrt(153))*Tcd + 0.3*200;
Eq(4) = Fx - Fe - (12/sqrt(153))*Tcd;
Eq(5) = Fy - (3/sqrt(153))*Tcd;
Eq(6) = 0.9*(12/sqrt(153))*Tcd + 0.6*Fe;
S = solve(Eq)
S =
  struct with fields:
    Ax: -100
    Ay: 25
    Fe: 150
    Fx: 50
    Fy: -25
    Tcd: -25*17^(1/2)


double([S.Ax S.Ay S.Fe S.Fx S.Fy S.Tcd])
ans =


-100.0000  25.0000  150.0000  50.0000  -25.0000  -103.0776
```

All these forces acting on the two members are shown in Figure 6.8.

Example 6.3

A bolt cutter operated by hand is shown in Figure 6.9. The parts of the cutter are attached to each other by smooth pins at points A, B, C, D, and E.

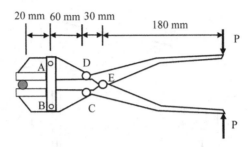

FIGURE 6.9
A bolt cutter.

For a force P = 100 N acting on the handle, determine the force F developed by each jaw on the rod to be cut as well as the other forces at the pins using MATLAB. All the dimensions are in mm.

Free Body Diagrams
Figure 6.10 shows three free body diagrams for the upper half of the cutter, the bolt itself, and the cutter head. There are six unknown forces and six equations of equilibrium in the free body diagrams of Figure 6.10(a) and (c), so the problem is statically determinate.

Equations of Equilibrium
By inspection the forces $E_x = 0$ and $D_x = 0$, so we only have four equations for four unknowns. For the upper half of the cutter, which we will call free body diagram I, and for the cutter head, which we will call free body diagram II, we have:

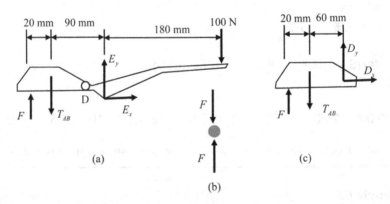

FIGURE 6.10
Free body diagrams for (a) the upper half of the cutter, (b) the bolt, and (c) the cutter head.

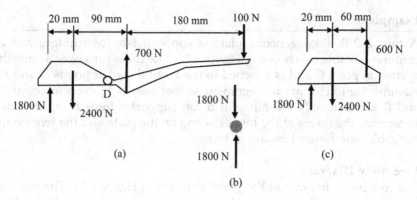

FIGURE 6.11
The forces present in the free body diagrams of Figure 6.10.

$$_I\sum F_y = 0 \quad F - T_{AB} + E_y - 100 = 0$$
$$_I\sum M_E = 0 \quad 90T_{AB} - (180)(100) - 110F = 0$$
$$_{II}\sum F_y = 0 \quad F - T_{AB} + D_y = 0 \qquad\qquad (6.6)$$
$$_{II}\sum M_D = 0 \quad 60T_{AB} - 80F = 0$$

and the solution in MATLAB is:

```
syms F Tab Ey Dy
Eq(1) = F - Tab + Ey - 100;
Eq(2) = 90*Tab - 18000 - 110*F;
Eq(3) = F - Tab + Dy;
Eq(4) = 60*Tab - 80*F;
S = solve(Eq)
S = struct with fields: Dy: 600
                        Ey: 700
                        F: 1800
                        Tab: 2400
```

which we will leave in these symbolic forms. The solution is shown in Figure 6.11 for the free body diagrams of Figure 6.10. You can draw the consistent free body diagrams for the upper handle by itself and the entire lower assembly from these results.

All the examples so far have been two-dimensional problems. Now, consider a three-dimensional geometry.

Example 6.4

A thin 200 lb homogeneous plate is supported by two hinges and a continuous cable, as shown in Figure 6.12. The cable runs over a smooth eyebolt at point C and is attached to the vertical wall at points D and E. Assume the hinges are in alignment so that they only exert forces at A and B and that only the hinge at B can support a force along its axis. Determine the forces at the hinges acting on the plate and the tension in the cable. All dimensions are in inches.

Free Body Diagram

The free body diagram of the plate is shown in Figure 6.13. The plate is homogeneous so that the center of gravity, G, where the weight of plate acts is at the plate center. There are six unknowns and six equations of equilibrium, so the problem is statically determinate.

Equations of Equilibrium

Let's first do the calculations by hand. We need to express the force in the cable acting from C to D and C to E in terms of components. We have:

$$\overrightarrow{CD} = 4i - 28j + 19k$$

$$|\overrightarrow{CD}| = \sqrt{(4)^2 + (-28)^2 + (19)^2} = 34.073 \text{ in}$$

$$\overrightarrow{CE} = -13i - 28j + 19k$$

$$|\overrightarrow{CE}| = \sqrt{(-13)^3 + (-28)^2 + (19)^2} = 36.249 \text{ in}$$

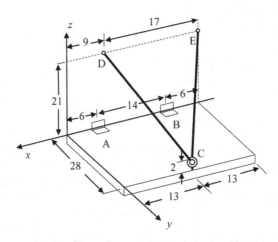

FIGURE 6.12

A plate supported by hinges and a cable. All dimensions are in inches.

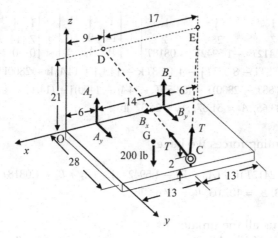

FIGURE 6.13
A free body diagram of the plate in Figure 6.12. All dimensions are in inches.

so that

$$\mathbf{T}_{CD} = T\frac{\overrightarrow{CD}}{|\overrightarrow{CD}|} = T(0.1174\mathbf{i} - 0.8218\mathbf{j} + 0.5576\mathbf{k}) \text{ lb}$$

$$\mathbf{T}_{CE} = T\frac{\overrightarrow{CE}}{|\overrightarrow{CE}|} = T(-0.3586\mathbf{i} - 0.7724\mathbf{j} + 0.5242\mathbf{k}) \text{ lb}$$

where T is the tension in the cable. The total force at C is the sum of these forces:

$$\mathbf{T}_C = \mathbf{T}_{CD} + \mathbf{T}_{CE} = T(-0.2412\mathbf{i} - 1.5942\mathbf{j} + 1.0818\mathbf{k}) \text{ lb}$$

In three-dimensional problems like this, we need to use all six equations to obtain the six unknowns, so it is best to start with the vector moment equation and take moments about a convenient point that eliminates as many unknowns as possible. Summing moments about point B, we have the position vectors:

$$\mathbf{r}_{BC} = 7\mathbf{i} + 28\mathbf{j} + 2\mathbf{k}$$
$$\mathbf{r}_{BA} = 14\mathbf{i}$$
$$\mathbf{r}_{BG} = 7\mathbf{i} + 14\mathbf{j}$$

so the moment about B is

$$\sum M_B = \begin{vmatrix} \mathbf{i} & \mathbf{j} & \mathbf{k} \\ 7 & 28 & 2 \\ -0.2412T & -1.5942T & 1.0818T \end{vmatrix} + \begin{vmatrix} \mathbf{i} & \mathbf{j} & \mathbf{k} \\ 14 & 0 & 0 \\ 0 & A_y & A_z \end{vmatrix} + \begin{vmatrix} \mathbf{i} & \mathbf{j} & \mathbf{k} \\ 7 & 14 & 0 \\ 0 & 0 & -200 \end{vmatrix}$$

$$= 33.4788T\mathbf{i} - 8.055T\mathbf{j} - 4.4058T\mathbf{k} - 14A_z\mathbf{j} + 14A_y\mathbf{k} - 2800\mathbf{i} + 1400\mathbf{j}$$

$$= (33.4788T - 2800)\mathbf{i} - (8.055T + 14A_z - 1400)\mathbf{j} + (14A_y - 4.4058T)\mathbf{k} = 0$$

$$\rightarrow T = 83.65, \ A_z = 51.87, \ A_y = 26.32 \ \text{lb}$$

Finally, summing forces we have

$$\sum \mathbf{F} = (B_x - 0.2412T)\mathbf{i} + (A_y + B_y - 1.5942T)\mathbf{j} + (A_z + B_z + 1.0818T - 200)\mathbf{k} = 0$$

$$\rightarrow \ B_x = 20.18, \ B_y = 107.03, \ B_z = 57.63 \ \text{lb}$$

which gives us all the unknowns.

When using MATLAB, it is not necessary to choose a "good" point to take moments about that eliminates the maximum number of unknowns so that we can more easily solve the problem by hand. Instead, it makes more sense to choose a point where calculating moments is easiest, such as the origin O of the coordinate system, which is the choice we will make.

We begin by defining the unknown components of the forces symbolically and then expressing the weight vector and the vector forces at the hinges in terms of those unknowns and the given weight:

```
syms T Ay Az Bx By Bz
W = [0 0 -200];
B = [Bx By Bz];
A = [0 Ay Az];
```

Next, we must express the cable forces acting at point C along CD and CE so, as in the solution done by hand, we write the position vectors from C to D and C to E and then get unit vectors along those directions. The vector forces are then just the unknown tension T times those unit vectors.

```
CD = [4 -28 19];
eCD = CD/norm(CD);      % unit vector along CD
Td = T*eCD;             % cable force along CD
CE = [-13 -28 19];
eCE = CE/norm(CE);      % unit vector along CE
Te = T*eCE;             % cable force along CE
```

To compute moments about the origin O, we need the position vectors from the origin to all the forces

```
ra = [-6 0 0];        % position vectors from O to A, B, C
rb = [-20 0 0];
rc = [-13 28 2];
rw = [-13 14 0];      % position vector from O to G for the weight
```

Now, we are in the position to compute the net vector force, F, and net vector moment, M, about point O:

```
F = A + B + W + Td + Te;
M = cross(ra, A) + cross(rb, B) + cross(rc, Td) + cross(rc, Te)...
    + cross(rw, W);
```

The left side of each vector equation of equilibrium $\mathbf{F} = 0$ and $\mathbf{M} = 0$ will then be placed in a symbolic equilibrium vector, Eq. Again, we do not have to place == 0 in those equations to solve them with the MATLAB function solve.

```
Eq = [F M]
S = solve(Eq);
```

The symbolic results are in a MATLAB structure S whose values are changed to numerical values, so they are easier to read. All the answers are in pounds.

```
double([S.Ay S.Az S.Bx S.By S.Bz S.T])
```

```
ans = 26.3141 51.8796 20.1763 107.0192 57.6442 83.6372
```

There are many steps to three-dimensional problems like the one just solved, but this MATLAB solution does not require any intermediate calculations, as was the case when we did the problem by hand, so if one sets up the position and force vectors carefully, MATLAB performs all the remaining calculations. This significantly reduces the possibility for errors. Thus, it is highly recommended that you use MATLAB for such three-dimensional problems whenever possible.

Now, consider some two-dimensional structures that carry distributed loads/unit length on some of their members.

Example 6.5

For the two-dimensional frame shown in Figure 6.14, determine the reaction forces acting on the pin at A and on the roller at E. All pins and surfaces are smooth. Neglect the weights of all the members.

Free Body Diagram

The free body diagram of the frame is shown in Figure 6.15 where there are three unknowns. We have three equations of equilibrium so we can solve directly for all the reactions.

Equations of Equilibrium

We can use the principle of superposition and break the distributed load into a uniform and triangular part, as shown in Figure 6.15. The resultant forces from these parts are

$$R_1 = (400)(0.6) = 240 \text{ N}$$
$$R_2 = (0.5)(400)(0.6) = 120 \text{ N} \tag{6.7}$$

and their locations at the centroids of these parts are shown in Figure 6.15. From that figure if we first take moments about point A, we can find the reaction at E

$$\sum M_A = 0$$
$$(E \cos 30°)(0.6) + (E \sin 30°)(0.6 + 0.6 \tan 15°)$$
$$-(240)(0.3 + 0.6 \tan 15°) - (120)(0.4 + 0.6 \tan 15°) = 0 \tag{6.8}$$
$$\rightarrow E = 197.63 \text{ N}$$

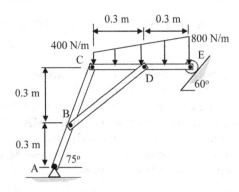

FIGURE 6.14
A structure with a distributed force.

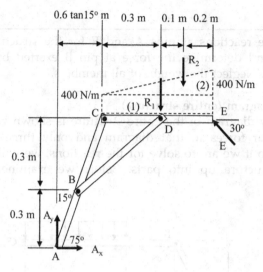

FIGURE 6.15
A free body diagram of the structure in Figure 6.14.

and then summing forces in the x- and y-directions gives us the pin reactions at A:

$$\Sigma F_x = 0 \quad A_x - E \cos 30° = 0$$
$$\rightarrow \quad A_x = 171.15 \text{ N}$$
$$\Sigma F_y = 0 \quad A_y + E \sin 30° - 240 - 120 = 0 \tag{6.9}$$
$$\rightarrow \quad A_y = 261.19 \text{ N}$$

Using the moment equation about point A first gives us the reaction force at E directly. Thus, it is often useful to calculate moments first to eliminate as many unknowns as possible before applying the force equations. This approach often lets us avoid having to solve simultaneous equations, so it is a good strategy to follow, particularly on tests. Note that member BD is a two-force member so that there are a total of six unknowns forces in this structure (two forces at pins A and C, the reaction force magnitude at E, and the tension or compression in BD). We have six equations of equilibrium for members ABC and CDE so we can solve for all the forces present. Obtain the complete solution yourself for this problem with MATLAB.

Now, consider an example where we must consider the structure and its parts to obtain a solution.

Example 6.6

Determine the reactions at pins A and E for the structure shown in Figure 6.16 and determine the force at pin B exerted by the pin on member ABC. Neglect the weight of all members.

Free Body Diagram (entire structure)
The free body diagram of the entire structure is shown in Figure 6.17. There are four forces in that diagram and only three equations of equilibrium so if we are to solve for the reactions, we will be forced to break the structure up into parts. Before we examine other parts,

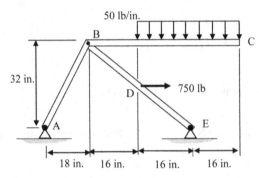

FIGURE 6.16
A pin-connected structural frame.

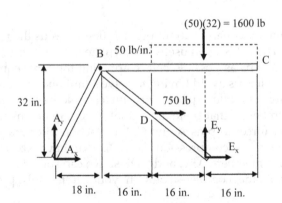

FIGURE 6.17
A free body diagram of the structure in Figure 6.16.

however, we can find the reactions A_y and E_y, and an equation relating A_x and B_x that we will use later. The distributed force can be replaced by its resultant, as shown in Figure 6.17.

Equations of Equilibrium (entire structure)
For later use, we will label the equations for the entire structure as equations (1), (2), and (6), as shown below (the labeling will indicate the order in which the equations are solved):

(1) $\Sigma M_A = 0$ $(E_y)(50) - (750)(16) - (1600)(50) = 0$
$\rightarrow$ $E_y = 1840$ *lb*
(2) $\Sigma F_y = 0$ $A_y + E_y - 1600 = 0$ (6.10)
$\rightarrow$ $A_y = -240$ *lb*
(6) $\Sigma F_x = 0$ $A_x + E_x + 750 = 0$

Free Body Diagram (ABC)
If we examine the free body diagram of part ABC (Figure 6.18), there are three unknowns and three equations of equilibrium so we can solve for all the forces in this free body diagram and then go back to equation (6) to find the remaining unknown E_x.

Equations of Equilibrium
The three equilibrium equations for ABC and the solution of Eq. (6.6) from the entire structure:

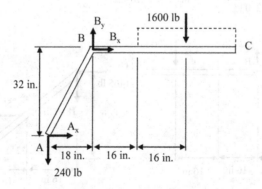

FIGURE 6.18
A free body diagram of member ABC.

(3) $\Sigma M_B = 0$ $(A_x)(32) + (240)(18) - (1600)(32) = 0$
→ $A_x = 1465$ *lb*
(4) $\Sigma F_x = 0$ $A_x + B_x = 0$
→ $B_x = -1465$ *lb*
(5) $\Sigma F_y = 0$ $B_y - 240 - 1600 = 0$ (6.11)
→ $B_y = 1840$ *lb*
(6) $\Sigma F_x = 0$ $A_x + E_x + 750 = 0$
→ $E_x = -2215$ *lb*

All the forces in the structure are shown in Figure 6.19. The entire
structure is shown in Figure 6.19(a), member ABC is shown in
Figure 6.19(b), and member BDE is shown in Figure 6.19(c). The entire
structure in Figure 6.19(a) shows the reaction forces at A and E that were
asked for. Member ABC shows the forces at pin B exerted on that
member, which was the other force asked for. Note that if instead we had
asked for the force at pin B exerted on BDE, we would obtain the forces at
B shown in Figure 6.19(c) instead as the answer.

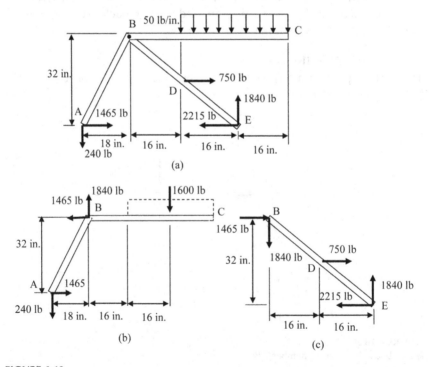

FIGURE 6.19
(a) Forces on the entire structure. (b) Forces on member ABC. (c) Forces on member BDE.

If we do this problem in MATLAB symbolically, we can enter equations (6.1) – (6.6) and then use the MATLAB function solve:

```
syms Ax Ay Bx By Ex Ey
Eq(1) = 50*Ey - 750*16 - 1600*50;
Eq(2) = Ay + Ey - 1600;
Eq(3) = 32*Ax - 18*Ay - 1600*32;
Eq(4) = Ax + Bx;
Eq(5) = By + Ay - 1600;
Eq(6) = Ax + Ex + 750;
S = solve(Eq)
S = struct with fields:
    Ax: 1465
    Ay: -240
    Bx: -1465
    By: 1840
    Ex: -2215
    Ey: 1840
```

These symbolic results agree with our previous solution. If we want to perform a completely numerical solution in MATLAB, we need to set up the equilibrium equation in matrix-vector form as [E]{x} = {P} where [E] is the equilibrium matrix, {x} = [Ax;Ay; Ex; Ey; Bx; By], and {P} = [92000; 1600; 51200; 0; 1600; –750]. In MATLAB, we have:

```
E = [ 0    0  0 50 0 0;
      0    1  0  1 0 0;
     32  -18  0  0 0 0;
      1    0  0  0 1 0;
      0    1  0  0 0 1;
      1    0  1  0 0 0]

P = [92000; 1600; 51200; 0; 1600; -750]

x = E\P
x = 1465      %Ax
    -240      %Ay
   -2215      %Ex
    1840      %Ey
   -1465      %Bx
    1840      %By
```

which again gives the same solution.

Here is another pinned frame with a distributed load problem.

Example 6.7

Member BD of the pin-connected frame shown in Figure 6.20 is loaded by a linearly varying distributed load whose intensity varies from zero at pin B to 150 lb/ft at pin D. Determine the x- and y-components of the force at pin D acting on member CDE.

Free Body Diagrams

Since the force acting at D on CDE is asked for, drawing a free body diagram of that member is a good starting point. From Figure 6.21, there are a total of five forces acting on that member, so we need to find two forces acting on this member before we can find the remaining three with our three equilibrium equations for CDE. From a free body diagram of the entire structure (Figure 6.22(a)), we can find the force at E directly by taking moments about A. Consider also a free body diagram of BD (Figure 6.22(b)), where we have replaced the distributed load by its

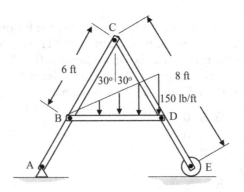

FIGURE 6.20
A frame with a distributed force.

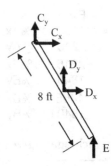

FIGURE 6.21
A free body diagram of member CDE.

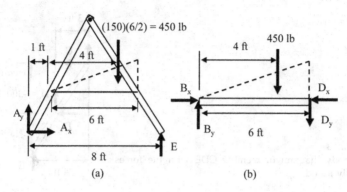

FIGURE 6.22
(a) A free body diagram of the entire structure. (b) A free body diagram of BD.

resultant. A moment equation about B for BD can be used to get D_x. This will give us enough information to find the remaining forces on CDE.

Equations of Equilibrium
From the free body diagram of the entire structure (Figure 6.22(a), the moment equation about A gives

$$\Sigma M_A = 0 \quad (8)(E) - (5)(450) = 0$$
$$\rightarrow E = 281.25 \text{ lb} \tag{6.12}$$

Similarly, taking moments about B for the free body diagram of BD in Figure 6.22(b), we have

$$\Sigma M_B = 0 \quad (6)(D_y) - (4)(450) = 0$$
$$\rightarrow D_y = -300 \text{ lb} \tag{6.13}$$

Note that the forces at D are the forces that CDE exerts on BD, so they are equal and opposite to the forces that BD exerts on CDE, which are the forces asked for in this problem and the ones shown in Figure 6.21. *It is very important that we draw all our free body diagrams consistently* so that internal forces such as the forces at D always occur in equal and opposite pairs. Now that we have found two of the forces that appeared on CDE, we can return to that free body diagram (Figure 6.23) and find the other component of the force at D by summing moments about point C:

$$\Sigma M_C = 0 \quad (4)(281.25) + (3)(-300) + (6\cos(30°))(D_x) = 0$$
$$\rightarrow D_x = -43.3 \text{ lb} \tag{6.14}$$

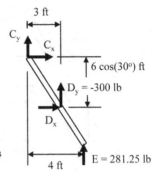

FIGURE 6.23
A free body diagram of member CDE with the forces previously found.

We can, of course, also obtain the forces at C but those forces were not required. It follows that on CDE, the vector components at D are

$$
\begin{aligned}
\mathbf{D}_x &= 43.3 \ \text{lb} \leftarrow \\
\mathbf{D}_y &= 300 \ \text{lb} \downarrow
\end{aligned}
\tag{6.15}
$$

Note that we found these results using only three equations of equilibrium, so our solution was very efficient. This is a typical problem found in many statics texts where we are asked to only find a particular set of forces in a structure. Implicitly, such problems are usually posed so that they can be solved without ever solving any simultaneous equations. They are designed so you examine various free body diagrams of the structure and find a short sequence of steps that can lead to the solution, solving for as few unknowns as possible. Acquiring the skill needed to find a solution in this fashion is useful but it is also important to also know how to determine all the unknown forces in a structure. In this problem, there are a total of nine unknowns. If we use three free body diagrams, we have nine equations of equilibrium, so all the unknown forces can be found. Try obtaining the complete solution with MATLAB.

6.2 Problems

P6.1 The pin-connected L-shaped frame AB in Fig. P6.1 is acted upon by a uniform load w = 15 lb/ft and is connected to member BC. Determine the magnitude of the force acting in member BC at C. The distances a = 8 ft and b = 6 ft.

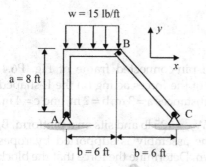

Figs. P6.1, P6.2

Choices (approximately, in lb):

1. 14
2. 17
3. 20
4. 28
5. 38

P6.2 The pin-connected L-shaped frame in Fig. P6.2 is acted upon by a uniform load w = 15 lb/ft. Determine the Cartesian components of the forces at pins A and B acting on the frame AB. The distances a = 8 ft and b = 6 ft.

P6.3 A force P = 350 N acts on the pin-connected frame of Fig. P6.3. Determine the vertical component of the force acting at pin C on member BCE. The distances a = 3 m, b = 2 m, and c = 4 m.

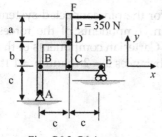

Figs. P6.3, P6.4

Choices (in N):

1. −311.1
2. −393.8
3. 393.8

4. 787.5

5. −787.5

P6.4 A force P = 350 N acts on the pin-connected frame of Fig. P6.4. Determine the Cartesian components of the forces acting on the L-shaped member ABD at pins A, B, and D. The distances a = 3 m, b = 2 m, and c = 4 m.

P6.5 Block A in Fig. P6.5 has a weight W_A = 125 lb and sits on a platform, B, which has a weight W_B = 35 lb. The assembly is supported by ropes running over smooth pulleys, as shown. Determine the force that the block exerts on the platform. The distances a = 1 ft and b = 3 ft.

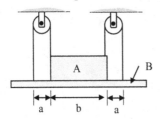

Fig. P6.5

Choices (in lb):

1. 35

2. 40

3. 45

4. 80

5. 90

P6.6 A force P = 250 lb acts on the pin at F for the pin-connected system shown in Fig. P6.6. Determine the Cartesian components of the forces acting at pins A and B on the system, and the Cartesian components of the force at pin C acting on member ACD. The distances a = 2 ft and b = 1 ft.

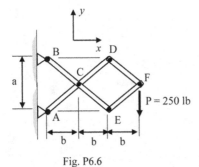

Fig. P6.6

P6.7 A mechanism that is used to control the bucket of a bulldozer is shown in Fig. P6.7. A horizontal force P is acting on the mechanism at C and the arm ABCD has a weight $W_1 = 400$ lb, with its center of gravity (cg) at pin B, while arm DEFG has a weight of $W_2 = 200$ lb with its center of gravity at pin E. The bucket and its load have a combined weight $W_3 = 2200$ lb, acting at a center of gravity H. The weight of the hydraulic cylinders can be neglected. The angles of arms ABCD and DEFG are $\theta = 30°$, as shown. The distances a = 25 in., b = 15 in., and c = 10 in. Determine the force P, the force in the hydraulic cylinders BF and EI, and the Cartesian components of the forces at pin A (acting on ABCD), at pin D (acting on DEFG), and at pin G (acting on the bucket). Cylinder EI is horizontal in the position shown and the angle of cylinder BF can be obtained from the given geometry (for the given parameters, you should show the angle β of BF with respect to the x-axis is $\beta = 23.41°$). (There are many distances needed in this rather complex geometry, so you should draw the free body diagrams with care.)

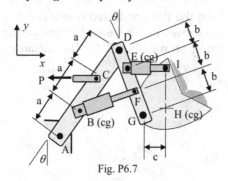

Fig. P6.7

P6.8 A hoist mechanism is shown in Fig. P6.8, where a weight W = 400 lb is supported at A. There are smooth rollers at G and E. Determine the magnitude of the force in the hydraulic cylinder BF. The distances a = 30 in., b = 60 in, and c = 20 in.

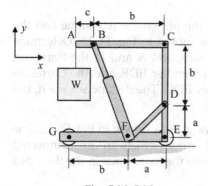

Figs. P6.8, P6.9

Choices (in lb):

1. 562
2. 324
3. 612
4. 495
5. 283

P6.9 A hoist mechanism is shown in Fig. P6.9, where a weight W = 400 lb is supported at A. There are smooth rollers at G and E. Determine the reactions on the hoist at G and E from the floor, the force in the link DF and in the hydraulic cylinder BF, and the Cartesian force components at pin C acting on member ABC. The distances a = 30 in., b = 60 in, and c = 20 in.

P6.10 The force P = 300 lb acts at F on the pin-connected frame and the force F = 125 lb acts on the pin at C, as shown in Fig. P6.10. Determine the Cartesian force components at B acting on member BDF. The distances a = 6 ft, b = 5 ft, and c = 2 ft.

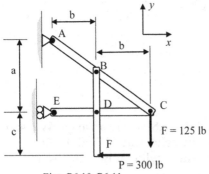

Figs. P6.10, P6.11

P6.11 The force P = 300 lb acts at F on the pin-connected frame and the force F = 125 lb acts on the pin at C, as shown in Fig. P6.11. Determine the Cartesian force components at the supports A and E, the Cartesian force components at B and D acting on member BDF, and the Cartesian force components at C acting on member EDC. The distances a = 6 ft, b = 5 ft, and c = 2 ft.

P6.12 A force P = 500 lb acts on a scissors jack, as shown in Fig. P6.12. The lengths, L = 12 in., of all the inclined arms (which are not connected where they cross) are the same. Determine the axial force in the threaded screw that is used for raising and lowering the jack if the angle $\theta = 20°$. The distance a = 3 in.

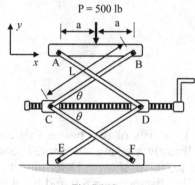

Fig. P6.12

6.2.1 Review Problems

These problems typically have the level of difficulty found on exams. They should be done by hand (i.e., with a calculator).

R6.1 A force P = 2 kN acts on the horizontal bar ABCD in Fig. R6.1 while a couple M = 4 kN-m acts on the inclined bar EBF. The two bars are connected by a pin at B attached to ABCD that rides in a smooth slot in bar EBF. Determine the reaction at C on ABCD. The distances a = 4 m and b = 3 m. Neglect the weight of the bars.

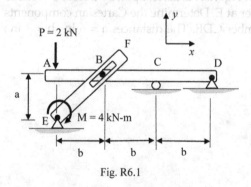

Fig. R6.1

R6.2 The frame shown in Fig. R6.2 carries forces P = 350 lb and F = 100 lb and a uniform distributed load w = 25 lb/ft. Determine the horizontal and vertical force components at pin B acting on member ABD. The distances a = 5 ft, b = 6 ft, and c = 3 ft.

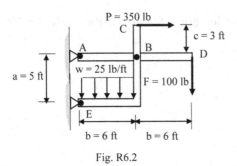

Fig. R6.2

R6.3 The compound beam of Fig. R6.3 consists of two beams that are pinned together at C and support a linearly varying distributed load whose maximum intensity is w = 50 lb/ft, as shown. Determine the reactions at the supports A, B, and D. The distances a = 8 ft and b = 4 ft.

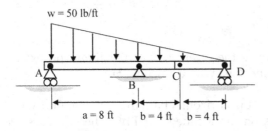

Fig. R6.3

R6.4 The pin-connected frame in Fig. R6.4 is acted upon by a force P = 130 lb and is attached to a smooth roller at E. Determine the Cartesian components of the force at C acting on member CDE. The distances a = 6 in., b = 3 in., and c = 5 in.

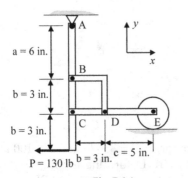

Fig. R6.4

R6.5 The pin-connected frame of Fig. R6.5 carries a uniform distributed
load w = 12 lb/ft and a force P = 125 lb. Determine the Cartesian
components of the reaction force at D on the frame. The distances a = 3 ft,
b = 5 ft, and c = 4 ft.

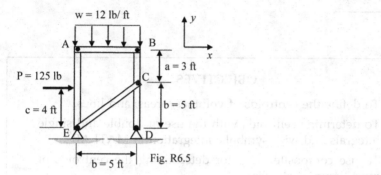

Fig. R6.5

7

Centroids

OBJECTIVES

- To define the centroids of volumes, areas, and lines.
- To determine centroids with the use of double and single integrals and with symbolic integration in MATLAB®.
- To use composite areas for determining the centroids of more complex shapes.
- To use the concept of centroids for obtaining the resultants of distributed forces.

Centroids are properties that engineers encounter in many areas of engineering beyond statics, including dynamics and strength of materials. Centroids are also closely related to concepts such as the center of mass and center of gravity. Resultants of forces distributed along lines and their locations can also be found in terms of areas and centroids. In this chapter, we will examine methods for determining centroids including integration and the use of *composite body* formulae.

7.1 Centroids of Volumes, Areas, and Lines

In Chapter 3, we examined the force of gravity and the concept of the center of gravity. For a body which occupies a volume V and whose weight per unit volume is a constant (called a homogeneous body), we found that the center of gravity was located at a point (x_c, y_c, z_c) called the *centroid* of the volume V, where

DOI: 10.1201/9781003372592-7

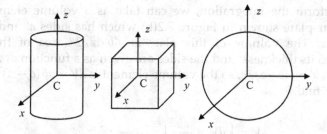

FIGURE 7.1
The centroids of some simple shapes are at their geometrical center.

$$x_c = \frac{\int_V x\,dV}{V} = \frac{\int_V x\,dV}{\int_V dV}$$

$$y_c = \frac{\int_V y\,dV}{V} = \frac{\int_V y\,dV}{\int_V dV} \tag{7.1}$$

$$z_c = \frac{\int_V z\,dV}{V} = \frac{\int_V z\,dV}{\int_V dV}$$

The centroid is a function of the geometry of the body only and does not depend on its other properties. If a geometry has a line or plane of symmetry, the centroid must lie on that line of symmetry or in that plane of symmetry, so for simple shapes like cylinders, blocks, and spheres, the centroid C will be at the geometrical center (Figure 7.1). For other shapes, we can find the centroid location by integration. Here is an example.

Example 7.1

Consider the tetrahedron shown in Figure 7.2(a) whose sides are a, b, and c. Determine the location of the centroid.

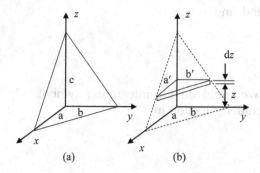

FIGURE 7.2
(a) A tetrahedron. (b) A volumetric element consisting of the plate of width dz as shown.

To perform the integration, we can take as a volume element the triangular plate shown in Figure 7.2(b) which has sides a' and b' and width dz. The volume of this plate is $a'b'dz/2$ (area of the plate face times its thickness) and the sides are given as a function of z as $a' = a(c - z)/c$, $b' = b(c - z)/c$, so the volume element is $dV = ab(c - z)^2dz/2c^2$. Then we find

$$V = \int dV = \frac{ab}{2c^2} \int_{z=0}^{z=c} (c - z)^2 dz$$

$$= \frac{ab}{2c^2} \int_{z=0}^{z=c} (c^2 - 2cz + z^2) dz \tag{7.2}$$

$$= \frac{ab}{2c^2} [c^2z - cz^2 + z^3/3]_0^c = \frac{abc}{6}$$

and

$$\int z dV = \frac{ab}{2c^2} \int_0^c z(c - z)^2 dz$$

$$= \frac{ab}{2c^2} \int_0^c (c^2z - 2cz^2 + z^3) dz \tag{7.3}$$

$$= \frac{ab}{2c^2} [c^2z^2/2 - 2cz^3/3 + z^4/4]_0^c = \frac{abc^2}{24}$$

This last integral is called a *first moment of the volume*. Then, the centroid is at

$$z_c = \frac{\int z dV}{V} = \frac{c}{4} \tag{7.4a}$$

In an entirely similar fashion, we can choose plate elements normal to the x- and y-axes and find

$$x_c = \frac{a}{4}, \quad y_c = \frac{b}{4} \tag{7.4b}$$

Alternatively, we could do the integration symbolically in MATLAB, using the integration function int:

```
syms a b c z                          % define all variables symbolically
V = (a*b/(2*c^2))*int ((c − z)^2, z, [0 c]);   % compute the volume
zV = (a*b/(2*c^2))*int (z*(c − z)^2, z, [0 c]); % compute the first moment
                                      % of the volume
zc = zV/V                             % z-coordinate of the centroid
zc = c/4
```

Note that we could have omitted the common factor outside these integrals since they cancel out. In using the integration function int, the first argument of that function is the integrand for the integral being performed, the second argument is the variable of integration, and the third argument is a vector containing the limits of integration. The arguments are all separated by commas.

In the same way that we define centroids for volumes, we can define centroids for areas and lines. Figure 7.3(a) shows a plane area and an area element, while Figure 7.3(b) shows a line in the x-y plane and an element of that line.

For the area, the location of the centroid (x_c, y_c) is given as

$$
x_c = \frac{\int_A x\,dA}{\int_A dA} = \frac{\int_A x\,dA}{A}
$$

$$
y_c = \frac{\int_A y\,dA}{\int_A dA} = \frac{\int_A y\,dA}{A}
$$

(7.5)

while for the centroid of the line we have

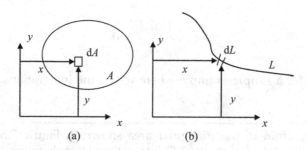

(a)   (b)

FIGURE 7.3
(a) A plane area A and an area element dA. (b) A line in two dimensions of length L and an elemental length dL.

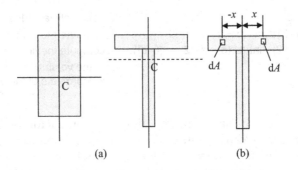

FIGURE 7.4
(a) Areas and their axes of symmetry. (b) A T-section, showing canceling first moments of
the area elements about the axis of symmetry.

$$x_c = \frac{\int_L xdL}{\int_L dL} = \frac{\int_L xdL}{L}$$

$$y_c = \frac{\int_L ydL}{\int_L dL} = \frac{\int_L ydL}{L}$$

(7.6)

The centroid of an area with an axis of symmetry must lie along that
axis of symmetry. For a rectangle which has two axes of symmetry, the
centroid must, therefore, be at the geometric center, as shown in
Figure 7.4(a). The T-section in that figure has one axis of symmetry,
so the centroid must lie somewhere along that axis. We can see that this
must be the case, since for the T-section there are always canceling first
moments xdA and $-xdA$ for every point in the cross section, so when
we compute the integral as the sum of all these canceling elements we
must find

$$x_c = \frac{\int_A xdA}{A} = 0$$

Let's consider a simple example where we do the integrations.

Example 7.2

Find the centroid of the triangular area shown in Figure 7.5. We can
always do the calculations by 2-D integrations, but there are also other
ways, as we will see shortly. First, do the 2-D integrations by hand. For
the area calculation, we have

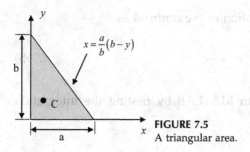

FIGURE 7.5
A triangular area.

$$A = \iint dA$$
$$= \int_{y=0}^{y=b} \int_{x=0}^{x=a(b-y)/b} dxdy = \int_{y=0}^{y=b} [x]_{x=0}^{x=a(b-y)/b} dy$$
$$= \int_{y=0}^{y=b} \frac{a}{b}(b-y)dy = -\int_{u=b}^{0} \frac{a}{b}udu \tag{7.7}$$
$$= \frac{ab}{2}$$

which is the well-known area for a triangle. Here, we did the x-integration first and integrated from $x = 0$ to the line $x = a(b - y)/b$ (see Figure 7.5). We did the y-integration by letting $u = b - y$, $du = - dy$ to make that integration easier. Similarly, for the first moment integrations, using similar steps

$$x_c A = \iint xdA$$
$$= \int_{y=0}^{y=b} \int_{x=0}^{x=a(b-y)/b} xdxdy = \int_{y=0}^{y=b} \frac{x^2}{2}\Big|_{x=0}^{x=a(b-y)/b} dy$$
$$= \int_{y=0}^{y=b} \frac{a^2}{2b^2}(b-y)^2dy = -\int_{u=b}^{0} \frac{a^2}{2b^2}u^2du \tag{7.8}$$
$$= \frac{a^2b}{6}$$

and the x-location of the centroid, then is

$$x_c = \frac{a^2b/6}{ab/2} = \frac{a}{3} \tag{7.9a}$$

Similarly, we can find the y-location of the centroid as

$$y_c = \frac{b}{3} \tag{7.9b}$$

We can do these integrations in MATLAB by nesting the integration function, int:

```
syms a b x y
A = int(int(1, x, [0 a*(b - y)/b]), y, [0 b])
A = (a*b)/2
xA = int(int(x, x, [0 a*(b - y)/b]), y, [0 b])
xA = (a^2*b)/6
xc = xA/A
xc = a/3
```

Here, we do the x-integration first and then use that integral, which is a function of y, as the integrand of the y-integration. Again, we can find the y-location in the same fashion.

7.1.1 Integration with Strips

Areas and area moments can always be done with two-dimensional integrals, as just illustrated. Usually, however, if we are doing the integrations by hand, it is easier to choose strip area elements and reduce the problem to single integrations. We can see how this works by examining the 2-D integration results of the last example here. Consider the area integration. When we calculated the x-integral first, we found, after that x-integration

$$A = \int_{y=0}^{y=b} \frac{a}{b}(b - y)dy = \int_{y=0}^{y=b} dA_s \tag{7.10}$$

which we can interpret as just summing up strips with area $dA_s = [a(b - y)/b]dy$ (Figure 7.6(a)). The first moment calculations similarly gave after the x-integration

$$x_c A = \int_{y=0}^{y=b} \frac{a^2}{2b^2}(b - y)^2 dy = \int_{y=0}^{y=b} \bar{x}_s dA_s \tag{7.11}$$

where $\bar{x}_s = a(b - y)/2b$ is the *distance to the centroid of the strip* in the x-direction (note the factor of ½). Thus, we can calculate the x-location of the centroid with single integrals as

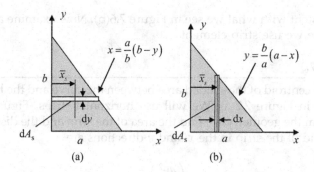

FIGURE 7.6
(a) Use of a horizontal strip for integration. (b) Use of a vertical strip for integration.

$$x_c = \frac{\int_A \bar{x}_s dA_s}{\int_A dA_s} \tag{7.12}$$

This was using horizontal strips, but we could use vertical strips instead, as shown in Figure 7.6(b). This corresponds to doing the y-integration first so let's examine those 2-D integrals. For the area integral

$$A = \int_{x=0}^{x=a} \left\{ \int_{y=0}^{y=b(a-x)/a} dy \right\} dx$$
$$= \int_{x=0}^{x=a} \frac{b}{a}(a-x)dx = \int_{x=0}^{x=a} dA_s \tag{7.13}$$

and for the first moment integral

$$x_c A = \int_{x=0}^{x=a} x \left\{ \int_{y=0}^{y=b(a-x)/a} dy \right\} dx$$
$$= \int_{x=0}^{x=a} x \frac{b}{a}(a-x)dx = \int_{x=0}^{x=a} \bar{x}_s dA_s \tag{7.14}$$

where now the area of the strip and the distance to the centroid of the strip are

$$dA_s = \frac{b}{a}(a-x)dx$$
$$\bar{x}_s = x \tag{7.15}$$

so we are again led to single integrals and can use the form of Eq. (7.12). Note that the distance to the centroid of the vertical strip has no factor of

1/2 consistent with what we see in Figure 7.6(b). Now examine a specific case where we use strip elements.

Example 7.3

Locate the centroid of the shaded area between the curve and the line $y = b$, as shown in Figure 7.7(a). We will use horizontal strips (Figure 7.7(b)) where from the geometry we find the area of the strip and the distances to the centroid of the strip in the x- and y-directions are

$$dA_s = \left(\frac{ay^2}{b^2}\right)dy$$

$$\bar{x}_s = \frac{1}{2}\frac{ay^2}{b^2}, \quad \bar{y}_s = y \tag{7.16}$$

We find the area of the shaded area as

$$A = \int dA_s = \int_{y=0}^{y=b}\left(\frac{ay^2}{b^2}\right)dy = \frac{ay^3}{3b^2}\bigg|_{y=0}^{y=b} = \frac{ab}{3} \tag{7.17}$$

and the first moments as

$$x_c A = \int \bar{x}_s dA_s = \int_{y=0}^{y=b}\left(\frac{ay^2}{2b^2}\right)\left(\frac{ay^2}{b^2}\right)dy = \frac{a^2y^5}{10b^4}\bigg|_{y=0}^{y=b} = \frac{a^2b}{10}$$

$$y_c A = \int \bar{y}_s dA_s = \int_{y=0}^{y=b} y\left(\frac{ay^2}{b^2}\right)dy = \frac{ay^4}{4b^2}\bigg|_{y=0}^{y=b} = \frac{ab^2}{4} \tag{7.18}$$

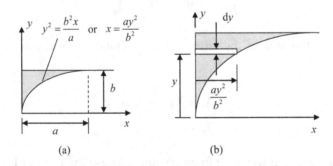

(a)   (b)

FIGURE 7.7
(a) A shaded area whose centroid is to be determined. (b) Choice of a strip element.

giving

$$x_c = \frac{x_c A}{A} = \frac{3a}{10}, \quad y_c = \frac{y_c A}{A} = \frac{3b}{4} \tag{7.19}$$

Doing these calculations in MATLAB gives

```
syms a b y
hs = a*y^2/b^2;          %  hs = dAs/dy is the length of the strip
xs = a*y^2/(2*b^2);      %  define x-and y-centroids of the strip
ys = y;
A = int(hs, y, [0 b])    %  do area integral
A = (a*b)/3
xA = int(xs*hs, y, [0 b]) %  do first moment integrals
xA = (a^2*b)/10
yA = int(ys*hs, y, [0, b])
yA = (a*b^2)/4
xc = xA/A                %  calculate x-and y-centroid locations
xc = (3*a)/10
yc = yA/A
yc = (3*b)/4
```

where the variable hs is the length of the strip, which is the integrand we need for the int function when integrating on y.

Now let's consider the location of the centroid of a line.

Example 7.4

A slender curved rod is shown in Figure 7.8(a). Determine the x- and y-locations of its centroid.

The centroids are given by Eq. (7.6) which we repeat here

$$x_c = \frac{\int_L x\,dL}{\int_L dL} = \frac{\int_L x\,dL}{L}$$

$$y_c = \frac{\int_L y\,dL}{\int_L dL} = \frac{\int_L y\,dL}{L} \tag{7.20}$$

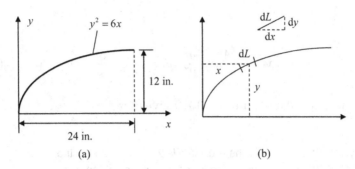

FIGURE 7.8
(a) A slender rod, and (b) a small element along its length.

The length of a small element of the rod is

$$dL = \sqrt{dx^2 + dy^2} = dy\sqrt{\left(\frac{dx}{dy}\right)^2 + 1}$$

$$= dy\sqrt{\left(\frac{y}{3}\right)^2 + 1} = \frac{1}{3}\sqrt{y^2 + 9}\ dy \tag{7.21}$$

so the length is given by the integral

$$L = \int_{y=0}^{y=12} \frac{1}{3}\sqrt{y^2 + 9}\ dy = \frac{1}{3}\left[\frac{y}{2}\sqrt{y^2 + 9} + \frac{9}{2}\ln\left(y + \sqrt{y^2 + 9}\right)\right]_{y=0}^{y=12} \tag{7.22}$$

$$= 27.88\ \text{in.}$$

where we have written out the integral explicitly from a table of integrals. The first moments, again from tables, are

$$x_c L = \int_{y=0}^{y=12} \frac{y^2}{6}\frac{1}{3}\sqrt{y^2 + 9}\ dy = \frac{1}{18}\left[\frac{y}{4}\sqrt{(y^2 + 9)^3} - \frac{9y}{8}\sqrt{y^2 + 9}\right.$$

$$\left. - \frac{81}{8}\ln\left(y + \sqrt{y^2 + 9}\right)\right] = 304.96\ \text{in}^2 \tag{7.23}$$

$$y_c L = \int_{y=0}^{y=12} y\frac{1}{3}\sqrt{y^2 + 9}\ dy = \frac{1}{3}\left[\frac{1}{3}\sqrt{(y^2 + 9)^3}\right]_{y=0}^{y=12} = 207.28\ \text{in}^2$$

so the centroid coordinates are

$$x_c = \frac{304.96}{27.88} = 10.94\ \text{in}$$

$$y_c = \frac{207.28}{27.88} = 7.43\ \text{in} \tag{7.24}$$

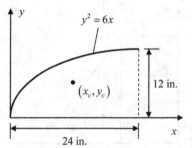

FIGURE 7.9
Location of the centroid for the slender rod.

The results are shown in Figure 7.9, where it is obvious that the centroid does not lie along the line.

If we do the integrations in MATLAB, we have

```
syms y
Ls = sqrt(y^2 + 9)/3;        % Ls = dL/dy which is the integrand
xb = y^2/6;                  % x- and y-locations to the element
yb = y;
L = int(Ls, y, [0 12]);      % total length of the line
xL = int(xb*Ls, y, [0 12]);  % first moments
yL = int(yb*Ls, y, [0 12]);
xc = xL/L;                   % centroid locations
yc = yL/L;
double([xc yc])              % convert to numerical values
ans = 10.9381   7.4345
```

The answer is the same as before. All the integration details are handled automatically.

7.2 Composite Areas

Now, consider more complicated areas that can be decomposed into simpler parts whose results are much easier to obtain. These are called *composite areas*. By using the definition of the centroid locations, we can show that these centroids can be defined as a weighted sum of the areas of the parts divided by the total area, where the weighting factors are just the centroid locations for the individual parts. For example, consider the area in Figure 7.10, which is composed of rectangular and triangular parts whose centroid locations are known. Call those areas (A_1, A_2) and

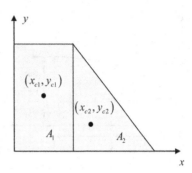

FIGURE 7.10
A composite area.

their centroid locations (x_{c1}, y_{c1}) and (x_{c2}, y_{c2}), respectively. Then by the definition of the centroid location in the x-direction, for example, we have

$$x_c = \frac{\iint xdA}{A} = \frac{\iint_{A_1} xdA + \iint_{A_2} xdA}{A_1 + A_2}$$

$$= \frac{x_{c1}A_1 + x_{c2}A_2}{A_1 + A_2} \tag{7.25}$$

and similarly for the y-coordinate

$$y_c = \frac{y_{c1}A_1 + y_{c2}A_2}{A_1 + A_2} \tag{7.26}$$

We can use the same idea to calculate centroids of composite lines and volumes. The center of gravity and the center of mass for composite bodies follow similar formulae:

$$x_c = \frac{x_{c1}L_1 + x_{c2}L_2 + \dots}{L_1 + L_2 + \dots} \quad \text{etc.} \quad \text{lines}$$

$$x_c = \frac{x_{c1}V_1 + x_{c2}V_2 + \dots}{V_1 + V_2 + \dots} \quad \text{etc.} \quad \text{volumes}$$

$$x_G = \frac{x_{G1}W_1 + x_{G2}W_2 + \dots}{W_1 + W_2 + \dots} \quad \text{etc.} \quad \text{weights} \tag{7.27}$$

$$x_m = \frac{x_{m1}M_1 + x_{m2}M_2 + \dots}{M_1 + M_2 + \dots} \quad \text{etc.} \quad \text{masses}$$

Example 7.5

Locate the centroid of the shaded area in Figure 7.11, where area A_2 is the area of a quarter circle of radius 150 mm. The area A_3 is a square hole, so

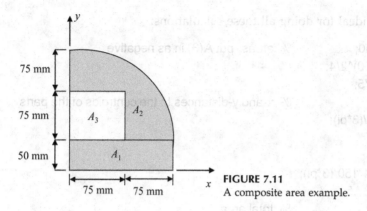

FIGURE 7.11
A composite area example.

we want to subtract its area and first moment values in the composite area formulae. For a quarter circle of radius r, we can find its centroid location easily in tables so from Figure 7.15 we have

$$A_2 = \frac{\pi r^2}{4} = \frac{(\pi)(150)^2}{4} = 17{,}671 \text{ mm}^2$$

$$x_{c2} = \frac{4r}{3\pi} = \frac{4(150)}{3\pi} = 63.66 \text{ mm} \tag{7.28}$$

$$y_{c2} = 50 + \frac{4r}{3\pi} = 113.66 \text{ mm}$$

The other centroid locations are easily found from Figure 7.11, giving

$$
\begin{aligned}
x_c &= \frac{x_{c1}A_1 + x_{c2}A_2 - x_{c3}A_3}{A_1 + A_2 - A_3} \\
&= \frac{(75)(7500) + (63.66)(17{,}671) - (37.5)(5625)}{7500 + 17{,}671 - 5625} = 75.5 \text{ mm}
\end{aligned} \tag{7.29}
$$

and

$$
\begin{aligned}
y_c &= \frac{y_{c1}A_1 + y_{c2}A_2 - y_{c3}A_3}{A_1 + A_2 - A_3} \\
&= \frac{(25)(7500) + (113.66)(17{,}671) - (87.5)(5625)}{7500 + 17{,}671 - 5625} = 87.2 \text{ mm}
\end{aligned} \tag{7.30}
$$

MATLAB is ideal for doing all these calculations:

```
A(1) = 50*150;              % areas, put A(3) in as negative
A(2) = pi*(150)^2/4;
A(3) = -75*75;
x(1) = 75;                  % x-and y-distances to the centroids of the parts
x(2) = 4*150/(3*pi);
x(3) = 75/2;
y(1) = 25;
y(2) = 50 + 4*150/(3*pi);
y(3) = 50 + 37.5;
At = sum(A);                % total area
xc = sum(x.*A)/At           % centroid locations
xc = 75.5412
yc = sum(y.*A)/At
yc = 87.1711
```

In MATLAB, we enter the areas and the coordinates of the centroids of the components into vectors and then do sums or weighted sums. Note the use of the MATLAB operation .*, which represents element-by-element multiplication of the components of the two vectors to yield a vector of the same size containing the products. This is a very easy way to get the weighted terms we need. The MATLAB function sum then sums up these products.

7.3 Distributed Line Loads

In Chapter 3, we briefly discussed the resultants of forces that are distributed along a line (Figure 7.12(a)). We found that the resultant force, R, of a force distribution described by its intensity $w(x)$ (force/unit length) was given by (Eq. (3.16))

$$R = \int_{x=x_1}^{x=x_2} w(x)dx \tag{7.31}$$

and the location of the resultant along the x-axis was (Eq. (3.18))

$$x_R = \frac{\int_{x=x_1}^{x=x_2} xw(x)dx}{R} \tag{7.32}$$

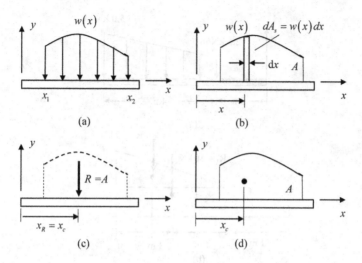

FIGURE 7.12
(a) A distributed line load. (b) An equivalent area and area element. (c) The resultant of the line load. (d) The centroid of the equivalent area.

These results can be viewed in terms of areas and their centroids by letting $dA_s = w(x)dx$ be a strip element of area, dA_s, for the planar area A under the intensity curve (Figure 7.12(b)). Then, the resultant, R, is just the total area under the curve. The first moment appearing in Eq. (7.32) can then also be interpreted as the first moment of that area, so the centroid location in the x-direction is just the centroid of the total area (see Figures 7.12(c) and (d)) and this is also the location of the resultant.

In Chapter 6, we used the principle of superposition to obtain the resultant in terms of the sum of simpler distributions. By viewing distributed loads in the context of centroids, we can also use the concept of composite areas to find the resultant and its location for more complex distributions. Here is an example.

Example 7.6

Determine the resultant force of the distributed force distribution acting on the beam shown in Figure 7.13(a) and locate it with respect to the origin, O.

Breaking the load distribution into rectangular and triangular distributions, whose centroids are located as shown in Figure 7.13(b) and whose areas are the resultant forces, we find that the resultant force is given by

$$R = R_1 + R_2 = (2.5)(4) + (2.5)(4)/2$$
$$= 10 + 5 = 15 \text{ kN}$$

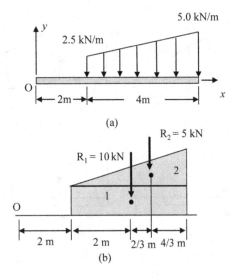

FIGURE 7.13
(a) A distributed force acting on a beam. (b) The composite areas and the forces they produce, which are located at the centroids of the component areas.

and the location of the force from O is

$$x_R = \frac{x_{c1}R_1 + x_{c2}R_2}{R_1 + R_2} = \frac{(4)(10) + \left(4 + \frac{2}{3}\right)(5)}{10 + 5} = 4.22 \text{ m}$$

7.4 Problems

P7.1 Determine by integration the location of the x-coordinate of the shaded area in Fig. P7.1 where $h = 40$ mm.

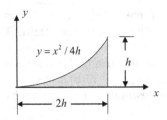

Figs. P7.1,P7.2

Choices (in mm):

1. 50
2. 55
3. 60
4. 65

P7.2 Determine by integration the location of the *y*-coordinate of the shaded area in Fig. P7.2 where $h = 40$ mm.

Choices (in mm):

1. 10
2. 12
3. 14
4. 16

P7.3 Determine by integration the location of the *x*-coordinate of the shaded area in Fig. P7.3. Let $a = 20$ in. and $b = 36$ in.

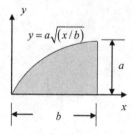

Figs. P7.3,P7.4

Choices (in inches):

1. 21.6
2. 23.3
3. 24.6
4. 25.3

P7.4 Determine by integration the location of the *y*-coordinate of the shaded area in Fig. P7.4. Let $a = 20$ in. and $b = 36$ in.

Choices (in inches):

   1. 5.5

   2. 6.5

   3. 7.5

   4. 8.5

P7.5 Determine by integration the location of the x-coordinate of the shaded area in Fig. P7.5. Let $a = 12$ mm and $b = 6$ mm.

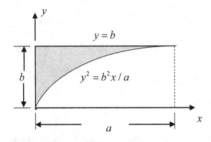

Figs. P7.5, P7.6

Choices (in mm):

   1. 4.6

   2. 2.8

   3. 3.2

   4. 3.6

P7.6 Determine by integration the location of the y-coordinate of the shaded area in Fig. P7.6. Let $a = 12$ mm and $b = 6$ mm.

Choices (in mm):

   1. 3.7

   2. 4.5

   3. 3.8

   4. 4.8

P7.7 Determine by integration the location of the coordinates of the centroid of the shaded area shown in Fig. P7.7. Let $a = 25$ in. and $b = 50$ in.

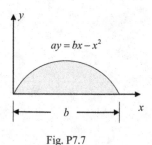

Fig. P7.7

P7.8 Determine by integration the location of the coordinates of the centroid of the thin wire shown in Fig. P7.8. Let $b = 50$ in.

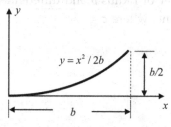

Fig. P7.8

P7.9 Determine by integration the volume and the location of the centroid of a hemisphere of radius $r = 10$ in. as shown in Fig. P7.9. Use a 1-D integral by considering as a volume element a cylinder of radius ρ and differential thickness dz at a height z above the x-y plane, where $\rho = \sqrt{x^2 + y^2}$.

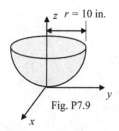

Fig. P7.9

P7.10 Determine by integration the volume and the location of the centroid of the paraboloid shown in Fig. P7.10. Use a 1-D integral by considering as a volume element a cylinder of radius ρ and differential thickness dy at a distance y from the x-z plane, where $\rho = \sqrt{x^2 + z^2}$.

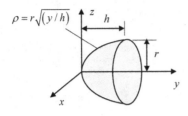

Fig. P7.10

P7.11 Determine by integration the volume and the location of the centroid of the circular cone shown in Fig. P7.11. Use a 1-D integral by considering as a volume element a cylinder of radius ρ and differential thickness dz at a height z from the x-y plane, where $\rho = \sqrt{x^2 + y^2}$.

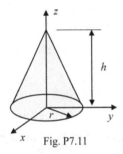

Fig. P7.11

P7.12 Determine the location of the centroid of the area shown in Fig. P7.12 for a = 12 in., b = 10 in., and t = 2 in.

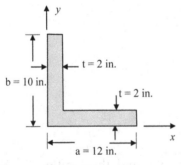

Fig. P7.12

P7.13 Determine the location of the centroid of the area shown in Fig. P7.13 for a = 40 mm, b = 50 mm, c = 30 mm, and d = 25 mm.

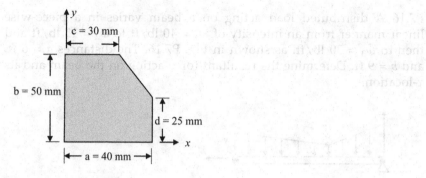

Fig. P7.13

P7.14 Determine the location of the centroid of the shaded cross-sectional area of the I-beam shown in Fig. P7.14 for $a = 100$ mm, $b = 200$ mm, $c = 150$ mm, and $t = 12$ mm.

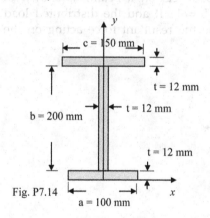

Fig. P7.14

P7.15 Determine the location of the centroid of the shaded area shown in Fig. P7.15 for $a = 60$ mm and $r = 30$ mm.

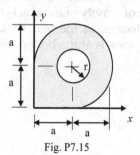

Fig. P7.15

P7.16 A distributed load acting on a beam varies in a piece-wise linear manner from an intensity of w_1 = 40 lb/ft to w_2 = 15 lb/ft and then to w_3 = 30 lb/ft. as shown in Fig. P7.16. The distances a = 6 ft. and b = 9 ft. Determine the resultant force acting on the beam and its x-location.

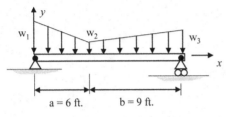

Fig. P7.16

P7.17 A non-uniform distributed load acts on a beam, as shown in Fig. P7.17. The length of the beam L = 12 ft and the distributed load parameter w_0 = 100 lb/ft. Determine the resultant force acting on the beam and its x-location.

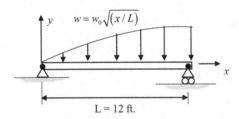

Fig. P7.17

P7.18 An 8×10 in. and a 5×10 in. plate, both made of aluminum (γ = 0.1 lb/in³), are joined to a 11×10 in. steel plate (γ = 0.284 lb/in³) to form a bracket, as shown in Fig. P7.18. The thickness of all the plates is t = 1.0 in. Determine the location of the centroid and center of gravity of the bracket in the (x, y, z) coordinates. Note: the z-axis is located at the rear back edges of the steel and aluminum plates and the y-axis is at the bottom of the lower Al plate.

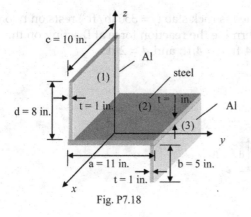

Fig. P7.18

7.4.1 Review Problems

These problems typically have the level of difficulty found on exams. They should be done by hand (i.e., with a calculator).

R7.1 Determine the location of the centroid of the bent slender rod shown in Fig. R7.1. Neglect the small gap at the ends of the rod.

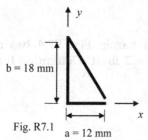

Fig. R7.1

$a = 12$ mm

R7.2 Determine the location of the centroid of the shaded area shown in Fig. R7.2. Let $a = 2$ in. and $b = 3$ in.

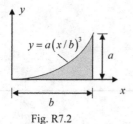

Fig. R7.2

R7.3 A one-foot-thick homogeneous rock slab ($\gamma = 350$ lb/ft^3) rests on two smooth rollers at A and B. Determine the reaction force at B acting on the slab. The lengths $a = 6$ ft, $b = 4$ ft, $c = 4$ ft, and $d = 2$ ft.

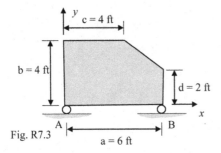

Fig. R7.3

Choices (in lb):

1. 2642
2. 2936
3. 3581
4. 4119
5. 5673

R7.4 The thin homogeneous circular-shaped bar in Fig. R7.4 has a radius $r = 4$ ft and weight/unit length of $\gamma = 2$ lb/ft. Determine the reaction at A.

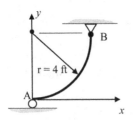

Fig. R7.4

7.5 Tables of Centroids

Centroids of some simple volumetric shapes, areas, and lines are tabulated in Figures 7.14, 7.15, and 7.16, respectively.

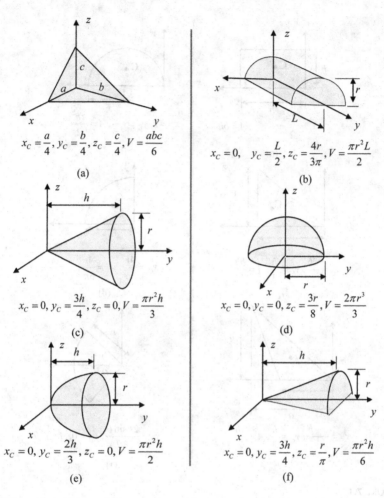

$$x_C = \frac{a}{4}, y_C = \frac{b}{4}, z_C = \frac{c}{4}, V = \frac{abc}{6}$$

(a)

$$x_C = 0, \quad y_C = \frac{L}{2}, z_C = \frac{4r}{3\pi}, V = \frac{\pi r^2 L}{2}$$

(b)

$$x_C = 0, y_C = \frac{3h}{4}, z_C = 0, V = \frac{\pi r^2 h}{3}$$

(c)

$$x_C = 0, y_C = 0, z_C = \frac{3r}{8}, V = \frac{2\pi r^3}{3}$$

(d)

$$x_C = 0, y_C = \frac{2h}{3}, z_C = 0, V = \frac{\pi r^2 h}{2}$$

(e)

$$x_C = 0, y_C = \frac{3h}{4}, z_C = \frac{r}{\pi}, V = \frac{\pi r^2 h}{6}$$

(f)

FIGURE 7.14
The centroids and volumes of some simple volumes. (a) Rectangular tetrahedron. (b) Semi-cylinder. (c) Right circular cone. (d) Hemisphere. (e) Paraboloid. (f) Half-cone.

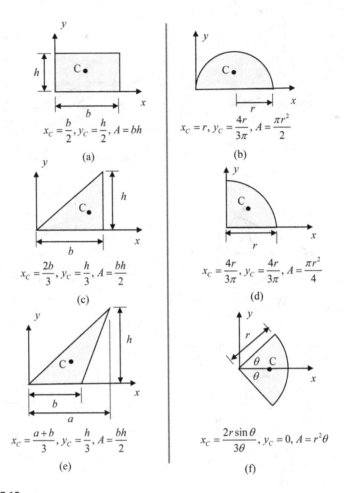

FIGURE 7.15
The centroids and areas of some simple shapes. (a) A rectangular area. (b) A semicircular area. (c) A right triangular area. (d) A quadrant of a circular area. (e) An oblique triangular area. (f) An area of a circular sector.

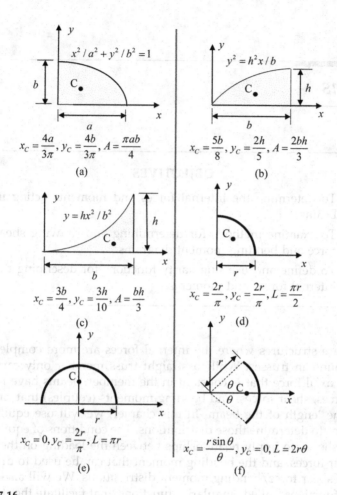

FIGURE 7.16
Centroids of simple areas and lines. (a) Quadrant of an ellipsoidal area. (b) Quadrant of a parabolic area. (c) A parabolic spandrel area. (d) Quadrant of a circular arc. (e) A semicircular arc. (f) A circular arc.

8

Beams

OBJECTIVES

- To determine the internal forces and moments acting in beams.
- To examine methods for determining and drawing shear force and bending moment diagrams.
- To define and use singularity functions for describing the internal forces and moments.

Beams are structures where the internal forces are more complex than those found in trusses. While a straight truss member only carries an internal axial force that is constant in the member, beams have internal axial forces, shear forces, and bending moments (couples) that can vary along the length of the beam. In this chapter, we will use equilibrium equations to determine those distributions. The conditions of equilibrium will be shown to imply relationships between the loading on the beam, the shear forces, and the bending moment that can be used to draw the internal shear force/bending moment distributions. We will also define special functions, called singularity functions, that facilitate the determination of the internal shear forces and bending moments, even under relatively complex loading conditions.

8.1 Internal Forces and Moments

We will center our initial discussion of the internal forces and moments around the example shown in Figure 8.1(a). The beam is loaded by an inclined force, is pinned at A, and sits on a smooth roller at B. If we imagine taking a cut through the beam at point P, a free body diagram of the section of the beam to the left of the cut is shown in Figure 8.1(b), where the internal axial force is N, the internal shear force is V, and the

DOI: 10.1201/9781003372592-8

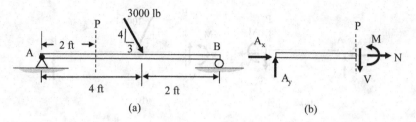

FIGURE 8.1

(a) A simply supported beam, and (b) the external reactions at A and the internal forces and moment at a cut taken at point P.

internal moment (called a bending moment) is M. By convention, the directions for these internal forces and moment seen in Figure 8.1(b) are the ones usually taken as positive for a cut on the right side of the section from A to P. *This means that the shear force is shown acting down on a right-side cut and the bending moment is acting counterclockwise. It is important to follow this convention so that one obtains consistent results throughout the beam.* The axial force, as in trusses, is shown as if the cut section is in tension. The internal forces and moment are the resultants of distributions of internal forces, as shown in Figure 8.2. The normal force is produced by a uniform distribution of axial forces acting over the cross section in the x-direction, while the shear force is due to a non-uniform distribution of forces acting in the negative y-direction. The bending moment (couple) is produced by a linearly varying axial force distribution where there are pairs of equal and opposite forces in the distribution. In a strength of materials class, one examines those distributions in more detail but in statics we will only deal with the resultants N, V, and M. As with all internal forces, these occur in equal and opposite pairs. Thus, if we examine the cut on the left side of the section of the beam from P to B, as shown in Figure 8.3(b), we will see the

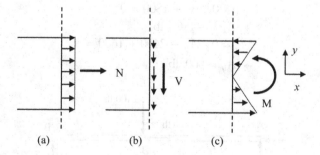

FIGURE 8.2

(a) A uniform force distribution acting in the x-direction, producing the normal force, N.
(b) A varying force distribution in the negative y-direction, producing the shear force, V.
(c) The linearly varying force distribution, producing the bending moment, M.

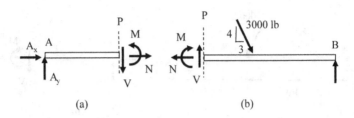

(a)   (b)

FIGURE 8.3
(a) The internal forces on the right side of the section from A to P, and (b) the internal forces on the left side of the section from P to B.

internal forces and moment that cancel those on the right side of the section taken from A to P in Figure 8.3(a).

Example 8.1

First, consider the steps needed to obtain the internal forces/moments at any specific location in a beam. We will continue to use the beam of Figure 8.1 for this example. To determine the internal forces, we must know all the external forces acting on the beam, including the reaction forces at the supports. Thus, we must first find those unknown reactions.

Free Body Diagram
The free body diagram of the entire beam is as shown in Figure 8.4.

Equations of Equilibrium
Summing moments about A and using the force equilibrioum equations, we find

$$
\begin{aligned}
\Sigma M_A = 0 \quad & 6B - (2400)(4) = 0 \\
\rightarrow \quad & B = 1600 \ \text{lb} \\
\Sigma F_x = 0 \quad & A_x + 1800 = 0 \\
\rightarrow \quad & A_x = -1800 \ \text{lb} \\
\Sigma F_y = 0 \quad & A_y - 2400 + 1600 = 0 \\
\rightarrow \quad & A_y = 800 \ \text{lb}
\end{aligned}
\tag{8.1}
$$

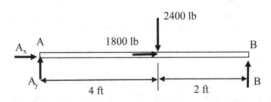

FIGURE 8.4
A free body diagram of the beam.

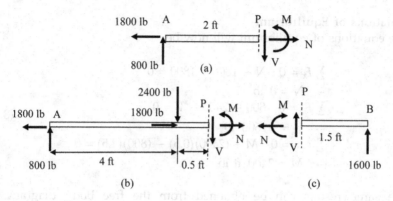

FIGURE 8.5
(a) A free body diagram of a section of the beam for a cut taken 2 ft to the right of A. (b) A free body diagram of a section of the beam for a cut taken 0.5 ft to the right of the appled force. (c) A free body diagram for a cut taken at the same location as in (b), but where the section of the beam to the right of the cut is shown instead.

With the reaction forces known, the internal forces and moment can be obtained at any location in the beam.

Free Body Diagram
For example, consider a cut taken 2 ft to the right of A. The free body diagram is shown in Figure 8.5(a).

Equations of Equilibrium
The equations of equilibrium for this piece are

$$\Sigma F_x = 0 \quad N - 1800 = 0$$
$$\rightarrow \quad N = 1800 \text{ lb}$$
$$\Sigma F_y = 0 \quad 800 - V = 0$$
$$\rightarrow \quad V = 800 \text{ lb} \tag{8.2}$$
$$\Sigma M_P = 0 \quad M - (800)(2) = 0$$
$$\rightarrow \quad M = 1600 \text{ ft-lb}$$

where we took moments about the point P of the cut. In general, this is a good point to take moments about since the shear force and normal force then do not contribute to the moment.

Obviously, the values of the internal forces and moment will depend on the location of the cut. The free-body diagram also can change depending on the location of the cut.

Free Body Diagram
For example, for a cut taken 0.5 ft to the right of the applied force, the free body diagram is given in Figure 8.5(b).

Equations of Equilibrium
The equations of equilibrium will now be:

$$\sum F_x = 0 \quad N - 1800 + 1800 = 0$$
$$\rightarrow \quad N = 0 \text{ lb}$$
$$\sum F_y = 0 \quad 800 - 2400 - V = 0$$
$$\rightarrow \quad V = -1600 \text{ lb} \tag{8.3}$$
$$\sum M_P = 0 \quad M + (2400)(0.5) - (800)(4.5) = 0$$
$$\rightarrow \quad M = 2400 \text{ ft-lb}$$

The same results can be obtained from the free body diagram of Figure 8.5(c), where we examined the section of the beam to the right of the cut. This is also a simpler free body diagram to use as there are fewer forces involved. *Note, however, that we must draw the unknown internal forces and moment in* Figure 8.5(c) *consistent with those shown in* Figure 8.5(b).

Example 8.2

Since the values of the internal forces and moment depend on the location of the cut taken, instead of taking cuts at specific locations, it is also possible to let the location of the cut to be at an arbitrary location in the beam. For example, let x be a variable distance to the right of A for the beam of Figure 8.1. Then, we can find the internal forces and moment as a function of x for any cut. Note, however, if the free body changes, as we saw it did when we went from a cut that was to the left of the load to one that was to the right of the load, then the functions of x that describes the internal forces and moment will change. For example, for our present problem, consider a cut anywhere to the left of the applied force, i.e., $0 < x < 4$ ft. Then, the free body diagram is shown in Figure 8.6(a) and from equilibrium we obtain

$$\sum F_x = 0 \quad N(x) - 1800 = 0$$
$$\rightarrow \quad N(x) = 1800 \text{ lb}$$
$$\sum F_y = 0 \quad 800 - V(x) = 0$$
$$\rightarrow \quad V(x) = 800 \text{ lb} \tag{8.4}$$
$$\sum M_P = 0 \quad M(x) - (800)(x) = 0$$
$$\rightarrow \quad M(x) = 800x \text{ ft-lb}$$

However, if $4 < x < 6$ ft, then equilibrium for the free body diagram of Figure 8.6(b) gives

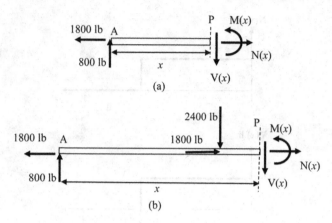

FIGURE 8.6
(a) A free body diagram for a cut taken where $0 < x < 4$ ft, and (b) a free body diagram for a cut taken where $4 < x < 6$ ft.

$$\sum F_x = 0 \quad N(x) - 1800 + 1800 = 0$$
$$\rightarrow \quad N(x) = 0 \text{ lb}$$
$$\sum F_y = 0 \quad 800 - 2400 - V(x) = 0$$
$$\rightarrow \quad V(x) = -1600 \text{ lb}$$
$$\sum M_P = 0 \quad M(x) + (2400)(x - 4) - (800)(x) = 0$$
$$\rightarrow \quad M(x) = -1600x + 9600 \text{ ft-lb}$$

(8.5)

We can collect these results and write the expressions for the internal forces and moment as:

$$N(x) = \begin{cases} 1800 & 0 < x < 4 \\ 0 & 4 < x < 6 \end{cases}$$

$$V(x) = \begin{cases} 800 & 0 < x < 4 \\ -1600 & 4 < x < 6 \end{cases}$$

(8.6)

$$M(x) = \begin{cases} 800x & 0 < x < 4 \\ -1600x + 9600 & 4 < x < 6 \end{cases}$$

The plots of these distributions are shown in Figure 8.7. If one examines those distributions, one notices a connection between the behavior of the applied forces and the shear force distribution and a connection between the behavior of the shear force distribution and the bending moment distribution. Those connections are highlighted in Figure 8.8. Wherever there is a concentrated applied force acting on the beam, the shear force

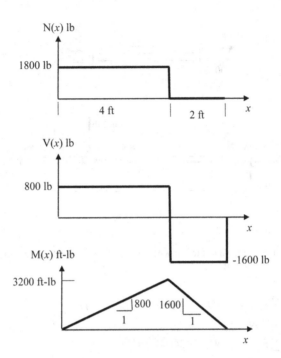

FIGURE 8.7
The internal normal force, shear force, and bending moment for the beam of Figure 8.1 (a).

diagram exhibits a jump at that location, where the jump is equal in value to the magnitude of the force and occurs in the same direction as the applied force. These jumps are indicated by the dashed lines and values seen in Figure 8.8. Similarly, the constant values of the shear force in the sections $0 < x < 4$ and $4 < x < 6$ are seen to correspond to the constant slopes of the bending moment in those sections, as indicated by the arrows connecting those diagrams in Figure 8.8. These relationships seen in this problem are special cases of general relationships we will now obtain and discuss.

As shown in Figure 8.9, a beam may have distributed forces as well as concentrated forces and moments acting along its length. Consider examining a small element of the beam under the distributed load of length dx and at a distance, x. A free body diagram of the element is given in Figure 8.10. Since the element is small, the distributed load is nearly uniform over the element so that the resultant of the distributed load is just the area under the loading, $w(x)dx$, and is located at $dx/2$, as seen in Figure 8.10. The shear force and bending moments have small changes from one side of the element to the other. If we sum forces in the y-direction and moments about the point P, we find two relations:

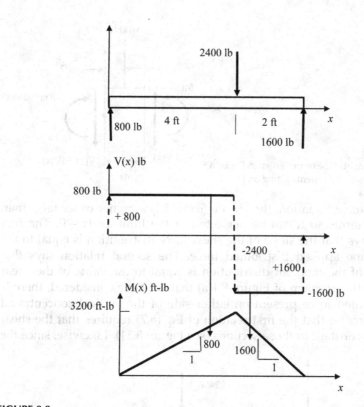

FIGURE 8.8
The applied loads, shear force diagram, and bending moment diagram, showing the connections discussed in the text.

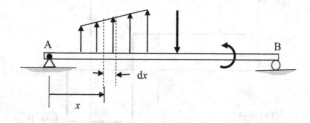

FIGURE 8.9
A beam with a distributed force and a concentrated force and moment.

$$\sum F_y = 0 \quad -[V(x) + dV(x)] + V(x) + w(x)dx = 0$$

$$\rightarrow \quad \frac{dV}{dx} = w(x)$$

$$\sum M_P = 0 \quad [M(x) + dM(x)] - M(x) - V(x)dx - [w(x)dx][dx/2] = 0 \quad (8.7)$$

$$\rightarrow \quad \frac{dM}{dx} = V(x)$$

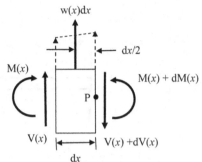

FIGURE 8.10
An small element of the beam shown in Figure 8.8 and the forces and moments acting on it.

In the moment equation, the term $w(dx)^2/2$ is a higher-order term than the other terms so it can be neglected in the limit as $dx \to 0$. The first relation says that the slope of the shear force distribution is equal to the value of the applied distributed force. The second relation says that the slope of the moment distribution is equal to the value of the shear force. For the problem of Figure 8.1(a) that we have considered, there is no distributed force present on either side of the vertical concentrated 2400 lb force so that the first relation of Eq. (8.7) requires that the shear force be a constant in those sections (see Figure 8.11). Likewise, since the

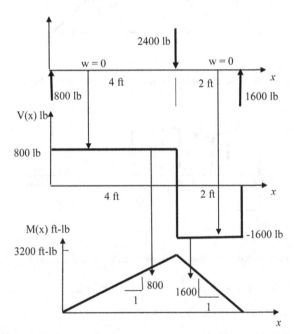

FIGURE 8.11
The differential relations of Eq. (8.7) for the shear force and bending moment diagrams for the beam of Figure 8.1(a).

shear force is a constant in those sections, the bending moment must have a constant slope (Figure 8.11). When there are concentrated forces acting along the beam, then we have jumps in the shear diagram. Similarly, at concentrated couples, the bending moment distribution must have a jump. There are no concentrated couples for the beam of Figure 8.1(a), so the bending moment diagram is continuous in Figure 8.11 but there are jumps in the shear force diagram at both the external end reactions and the applied force. The explicit relationship for these jumps can be obtained by examining equilibrium of a very small section at a concentrated force or couple, as shown in Figure 8.12, where there are different values on either side of the applied load. For the concentrated force of Figure 8.12(a), equilibrium gives

$$\sum F_y = 0 \quad -V_+ + V_- + P = 0$$
$$\rightarrow \quad V_+ - V_- = [[\,V\,]] = P \tag{8.8}$$

while for the concentrated couple of Figure 8.12(b) we find

$$\sum M = 0 \quad M_+ - M_- - C = 0$$
$$\rightarrow \quad M_+ - M_- = [[\,M\,]] = C \tag{8.9}$$

Thus, the shear force exhibits a jump, $[[\,V\,]]$, at a concentrated force, where the jump is equal to the value of the force. An upward force P will cause the shear force diagram to jump up an amount P while a downwards force P will produce a jump down by an amount P. At a concentrated couple, the jump, $[[\,M\,]]$, in the bending moment diagram is equal to the value of the concentrated couple. A clockwise couple C will cause the bending moment distribution to jump up by an amount C while a counterclockwise couple C will cause the bending moment to jump down by an amount C. The jumps seen in Figure 8.11 demonstrate this behavior for the shear force.

In drawing shear force and bending moment diagrams, we need to find the values of the force or moment at various points to accurately draw

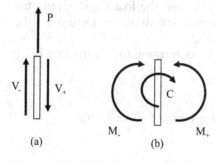

(a) (b)

FIGURE 8.12
(a) The behavior of the shear force at a concentrated force, and (b) the behavior of the bending moment at a concentrated couple.

the behavior of these distributions. We can often get such information by integrating the differential relations obtained in Eq. (8.7) between two locations x_1 and x_2. We find

$$V(x_2) - V(x_1) = \int_{x_1}^{x_2} w(x)dx$$
$$M(x_2) - M(x_1) = \int_{x_1}^{x_2} V(x)dx$$

(8.10)

which shows that the change of the shear force between two locations is equal to the area under the distributed force between those locations and the change in the bending moment between two points is equal to the area under the shear force diagram between those points. There is no distributed load in the problem of Figure 8.11, so Eq. (8.10) just says the shear force must be a constant (except at the concentrated forces, of course). The value of the bending moment 4 ft to the right of $x = 0$ can be found from Eq. (8.10) as

$$M(4) - M(0) = M(4) - 0 = (800)(4)$$
$$\rightarrow \quad M(4) = 3200 \text{ ft-lb}$$

(8.11a)

while the value of the bending moment at 6 ft to the right is

$$M(6) - M(4) = (-1600)(2)$$
$$\rightarrow \quad M(6) = 0$$

(8.11b)

which obviously agree with the distributions shown.

One can often use the relations seen in Eqs. (8.7)–(8.10) to draw shear force and bending moment distributions directly, without writing any equations, as long as the applied loads are not too complex. We will illustrate this process with the next example.

Example 8.3

Plot the shear force and bending moment diagrams for the beam of Figure 8.13(a) using the relationships between the loading diagram, the shear force diagram, and the bending moment diagram. Determine the maximum bending moment in the beam.

First, the external reactions need to be determined from the free body diagram of the entire beam.

Free Body Diagram
The free body diagram of the beam is given in Figure 8.13(b).

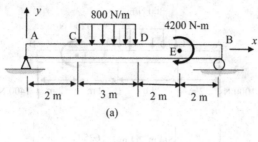

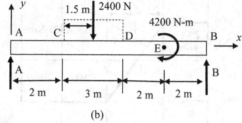

FIGURE 8.13
(a) A simply supported beam, and (b) its free body diagram.

Equations of Equilibrium

Summing moments about A and summing forces in the *y*-direction:

$$\Sigma M_A = 0 \quad 9B - 4200 - (2400)(3.5) = 0$$
$$\rightarrow \quad B = 1400 \text{ N}$$
$$\Sigma F_y = 0 \quad A - 2400 + 1400 = 0 \tag{8.12}$$
$$\rightarrow \quad A = 1000 \text{ N}$$

Plotting the Shear Force and Bending Moment Distributions

Starting from the free body diagram of the applied loads, we can draw the shear force diagram and then the bending moment diagram, as shown in Figure 8.14. The various quantities needed are also shown in Figure 8.14, as we will explain, section-by-section. First, consider the shear force.

AC: At A, there is a 1000 N force acting up so the shear force diagram, which starts at zero, will have a jump upward of 1000 N since $[[V]] = P$ (Eq. (8.8)). There is no distributed load in the first 2 m of the beam so from $\Delta V = \int w dx$ (Eq. (8.10)), the shear force will remain constant in that section.

CD: In section CD, there is a uniform downward distributed load of 800 N/m so from $dV/dx = w$ ((Eq. 8.7)), the shear force will have a constant slope in that section of $-800/1$, which means that over

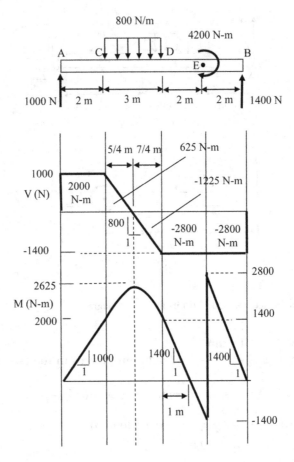

FIGURE 8.14
The loading diagram for the beam and the shear force and bending moment diagrams.

3 m the shear force will drop 2400 N, reaching a value of −1400 N at D.

DB: From D to B, there is no distributed load so from $dV/dx = w$ (Eq. (8.7)), the shear force will remain at a value of −1400 N until B where because of $[[V]] = P$(Eq. (8.8)) the shear jumps up 1400 N to a value of zero because of the 1400 N end reaction there. This completes the shear force diagram.

To draw the bending moment diagram, we will need, however, additional information from the shear force diagram. Specifically, we need the areas under the shear force diagram for various sections so we can evaluate changes in the moment using $\Delta M = \int V dx$ (Eq. (8.10)). In section AC, for example, the area is 2000 N-m. The other areas for the sections are indicated in Figure 8.14. We also need to know where in section CD the shear force goes to zero. Since the slope in section CD is $-800/1$, to get to zero from a value of 1000 N we need to go through a distance of $5/4$ m from C, as shown in Figure 8.14.

Now we can draw the bending moment section-by-section.

AC: The moment starts from zero at A. Since the shear force has a constant value of 1000 N in the first 2 m, from $dM/dx = V$ (Eq. (8.7)), the bending moment will have a constant slope of $1000/1$ in that section, reaching a value of 2000 N-m at C.

CD: In section CD, the bending moment will be a quadratic function since the shear force varies linearly in that section. We do not need to know the exact form of that quadratic, however, to sketch out the bending moment behavior. The area under the shear force diagram from C to the point where the shear force goes to zero is 625 N-m so using $\Delta M = \int V dx$ (Eq. (8.10)), the change of the bending moment from C to where the slope of the bending moment goes to zero is equal to that area. Thus, the value of the maximum bending moment in CD is 2625 N-m. From the location of that maximum to D, the area under the shear force is -1225 N-m, so again using $\Delta M = \int V dx$ (Eq. (8.10)), the bending moment curve must drop down to a value of $2625 -1225 = 1400$ N-m at D.

DE: In section DE, the shear force is a constant -1400 N so from $dM/dx = V$ (Eq. (8.7)), the slope of the bending moment in that section is $-1400/1$, which means the bending moment will go to zero at a distance of 1 m in the section, as shown, and drop to a value of -1400 N-m at E. This change is consistent with using the area under the shear force diagram from D to E.

EB: At E, there is a concentrated clockwise couple of 4200 N-m so because $[[M]] = C$ (Eq. (8.8)), the bending moment diagram will jump up at E the same amount, to a value of 2800 N-m. In section EB, from $dM/dx = V$ (Eq. (8.7)) the slope of the bending moment is again a constant $1400/1$, so the bending moment will decrease to zero at B which is consistent with using $\Delta M = \int V dx$ (Eq. (8.10)) and the area of -2800 N-m under the shear force diagram in section EB.

This completes the bending moment diagram. From this diagram, we see that the maximum bending moment in the entire beam is 2800 N-m.

The internal shear force and bending moment diagrams for many other problems can be obtained directly from the relationships we have outlined, which are summarized here:

Relations between the Loading, Shear Force and Bending Moment Diagrams

slope of the shear force = the value of the distributed force

$$\frac{dV}{dx} = w(x)$$

slope of the bending moment = the value of the shear force

$$\frac{dM}{dx} = V(x)$$

jump of the shear force = the value of a concentrated force at that point

$$[[V]] = V_+ - V_- = P$$

jump of the bending moment = the value a concentrated moment (couple) at that point

$$[[M]] = M_+ - M_- = C$$

change in the shear force = the area under the distributed force diagram

$$\Delta V = V(x_2) - V(x_1) = \int_{x_1}^{x_2} w(x)dx$$

change in the bending moment = the area under the shear force diagram

$$\Delta M = M(x_2) - M(x_1) = \int_{x_1}^{x_2} V(x)dx$$

8.2 Singularity Functions

In the last section, we showed how to sketch out shear force and bending moment diagrams from the relationships that exist between these

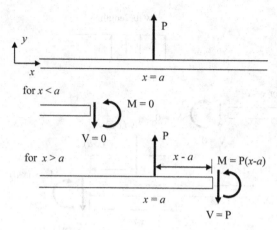

FIGURE 8.15
Behavior of the shear force and bending moment due to a concentrated force P.

functions and the applied loads. While those sketches are important, if we want, for example, to calculate the beam deflections, we will need to know the actual $V(x)$ and $M(x)$ distribution functions. As seen, even in the simple example considered in the last section, the applied loads on the beam cause changes in the free body diagram for the internal force and moment, so we need to define the shear force and bending moment expressions in a piece-wise fashion. But what if we could define the shear force and bending moment explicitly in terms of functions for the entire beam? This is possible with the use of functions called *singularity functions*. For example, at a concentrated force (Figure 8.15), we can write the shear force and bending moment as a function of x. We could express these functions in a piece-wise manner by writing

$$V(x) = \begin{cases} 0 & x < a \\ P & x > a \end{cases}$$

$$M(x) = \begin{cases} 0 & x < a \\ P(x - a) & x > a \end{cases} \tag{8.13}$$

or we can write these expressions as

$$V(x) = P\langle x - a \rangle^0$$
$$M(x) = P\langle x - a \rangle^1 \tag{8.14}$$

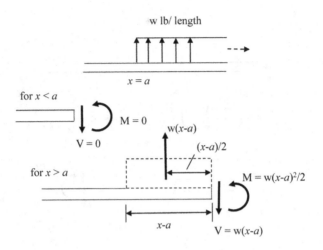

FIGURE 8.16
Behavior of the shear force and bending moment for a uniform force distribution of intensity w lb/unit length that starts at $x = a$.

where the singularity functions are defined as

$$\langle x - a \rangle^n = \begin{cases} 0 & x < a \\ (x - a)^n & x > a \end{cases} \tag{8.15}$$

and where $x = a$ is where the concentrated force P is located. In the same fashion, we can consider a constant distributed force that starts at $x = a$ (Figure 8.16) which we can write in piece-wise fashion as

$$V(x) = \begin{cases} 0 & x < a \\ w(x - a) & x > a \end{cases}$$

$$M(x) = \begin{cases} 0 & x < a \\ \dfrac{w}{2}(x - a)^2 & x > a \end{cases} \tag{8.16}$$

or, in terms of singularity functions as

$$V(x) = w \langle x - a \rangle^1$$
$$M(x) = \frac{w}{2} \langle x - a \rangle^2 \tag{8.17}$$

Singularity functions can be differentiated and integrated like ordinary functions since

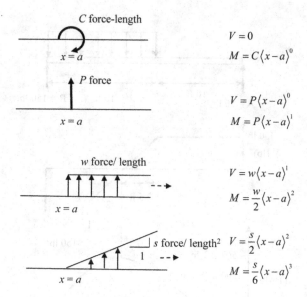

FIGURE 8.17
The singularity functions for a variety of different forces or a moment acting on a beam.

$$\frac{d}{dx}\langle x - a\rangle^n = n\langle x - a\rangle^{n-1} \quad n \geq 1$$

$$\int \langle x - a\rangle^n dx = \frac{\langle x - a\rangle^{n+1}}{n+1} + C \quad n \geq 0 \tag{8.18}$$

which are analogous to the forms seen for an ordinary function x^n. Figure 8.17 gives a summary of the singularity functions for a variety of applied loads. Combinations of these functions can represent different loading conditions.

Figure 8.18 gives an example where we have solved for the end reaction forces on the beam. The singularity functions can then be used to describe the shear force and bending moment for the entire beam as

$$V(x) = 50\langle x - 0\rangle^0 - 20\langle x - 10\rangle^1 + 150\langle x - 20\rangle^0$$
$$M(x) = 50\langle x - 0\rangle^1 - 10\langle x - 10\rangle^2 + 150\langle x - 20\rangle^1 \tag{8.19}$$

The behavior of these functions is shown in Figure 8.19. The function step_sf models a discontinuous function so if the jump in the function occurs at the first point in the plot, the function needs to be zero at that first point where $x = a$ and then jump to one at the next point to be able to properly plot that jump. At all other points in the plot, we can take the

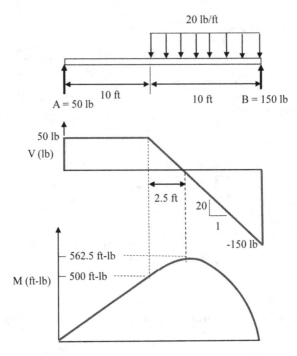

FIGURE 8.18
A beam and its shear force and bending moment diagrams.

function to be one at $x \geq a$ and the jump will be properly plotted, including at the end point of the x-interval. The other singularity functions are continuous and equal to zero at $x = a$ so their definitions are simpler. *The bending moment expression is just the integral of the shear force expression so once we have the shear force it is easy to get the bending moment. If concentrated couples are present, they simply need to be added to the integrated shear force terms.* There is no constant of integration if the bending moment is zero for x values less than the starting point of integration, as it is here and for all other beams. Note that in the bending moment expression in Eq. (8.19), we do not need the last singularity function since that function is zero for all $x \leq 20$, so it does not affect the values anywhere in the beam. Including it, however, is also fine.

Since we have explicit functions of x here, we can easily obtain the values at any location we want. As an example, consider the value of the moment at $x = 12.5$ ft, the location of the maximum moment (see Figure 8.18). We have, using the definitions of the singularity functions,

$$M(12.5) = 50(12.5) - 10(2.5)^2 + 150(0) = 562.5$$

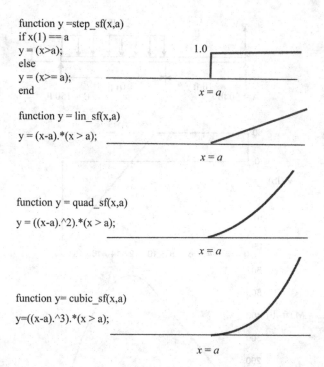

```
function y =step_sf(x,a)
if x(1) == a
y = (x>a);
else
y = (x>= a);
end

function y = lin_sf(x,a)
y = (x-a).*(x > a);

function y = quad_sf(x,a)
y = ((x-a).^2).*(x > a);

function y= cubic_sf(x,a)
y=((x-a).^3).*(x > a);
```

FIGURE 8.19
Singularity functions as defined in MATLAB and sketches of their behavior.

MATLAB can be used to great advantage with singularity functions since we can easily define the singularity functions in MATLAB, as seen in Figure 8.19, and then use those functions to define and plot the shear and bending moment diagrams. These MATLAB functions are also given in Appendix B. The shear force diagram for the problem in Figure 8.18, for example, can be obtained and plotted in MATLAB (see Figure 8.20) in the following manner:

```
V = @(x) 50*step_sf(x, 0) - 20*lin_sf(x,10) + 150*step_sf(x,20);
x = linspace(0, 20, 1000);
plot(x, V(x))
```

and here is how the bending moment diagram in Figure 8.20 is obtained, which is just the integral of the shear force:

```
M = @(x) 50*lin_sf(x,0) - 10*quad_sf(x,10) + 150*lin_sf(x,20);
figure(2)
plot(x, M(x))
```

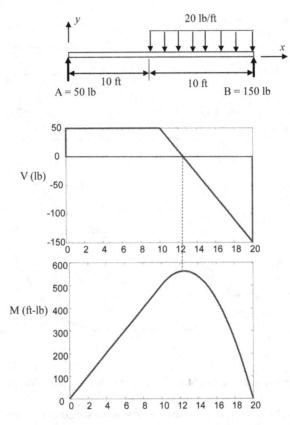

FIGURE 8.20
The shear force and bending moment diagrams for the beam problem shown in the figure, as plotted by using MATLAB singularity functions.

Note that, strictly speaking, we should terminate the distributed load at $x = 20$ ft by writing two terms in V as:

$-20*lin\_sf(x,10) + 20*lin\_sf(x,20)$

but the last term has no contribution for $x \leq 20$ ft so it is omitted. We can also omit the $150*lin\_sf(x, 20)$ term in the bending moment expression for the same reason. Keeping them in, if you like, is fine.

Here we wrote the shear force and bending moment expressions as *anonymous functions* (see Appendix A for more details). This allows us to evaluate those expressions at any specific point or points we want. One such critical point is where the bending moment is a maximum, which is also where the shear force is zero. In this simple example, we easily found that location as $x = 12.5$ ft, but in more complex cases, we may need to numerically find where the shear force is zero and then evaluate the bending moment at that location. Generally, we can find where $V = 0$ with the MATLAB function fzero as follows:

```
xm = fzero(@(x) V(x), 12) % determine the location xm where V = 0
xm = 12.5000
```

The function fzero does an iterative search for where the function $V(x)$ is zero and the value of 12 in the fzero function here is an initial guess of the location of that zero. Once this location is known, we can evaluate the bending moment there:

```
M(xm)
ans = 562.5000
```

which agrees with our previous result. Now, let's give a more complex example that shows the value of using singularity functions.

Example 8.4

Consider the beam shown in Figure 8.21, where the reaction forces at the supports have been calculated from equilibrium. Plot the shear force and bending moment diagrams using the MATLAB singularity functions and identify critical points in the plots. Note that there is an overhang at the left end of the beam, so it starts at $x = -3$ ft. The shear force defined and plotted in MATLAB is:

```
V = @(x) - 1000*lin_sf(x, -3)+8000*step_sf(x, 0) +1000*lin_sf(x, 3)...
-5000*step_sf(x, 6) + 9000*step_sf(x, 9) -4000*lin_sf(x, 9) + (2000/3)*quad_sf(x, 9);
x = linspace(-3, 12, 1000);
plot(x, V(x))
```

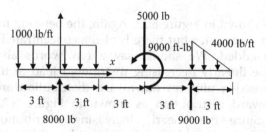

FIGURE 8.21
A beam with complex loads.

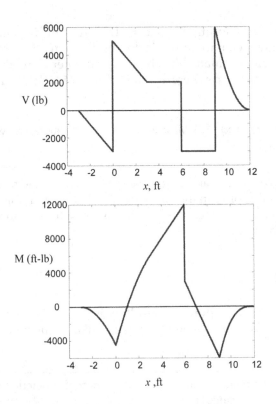

FIGURE 8.22
The shear force and bending moment diagrams for the beam of Figure 8.21, using singularity functions and MATLAB.

and the bending moment is given by:

```
M = @(x) −500*quad_sf(x, −3) +8000*lin_sf(x, 0) +500*quad_sf(x, 3) −5000*lin_sf(x,6)...
−9000*step_sf(x, 6) +9000*lin_sf(x, 9) − 2000*quad_sf(x, 9)+(2000/9)*cubic_sf(x, 9);
figure(2)
plot(x, M(x))
```

Both plots are shown in Figure 8.22. Again, the bending moment is the integral of the shear force but there is also a concentrated couple term that must be added. We should say a few words about how we represented the linearly decreasing distributed load at the right end. We superimposed a uniform downward distribution and a linearly increasing upward distribution as shown in Figure 8.23(b) and (c), respectively. Since the linearly increasing distribution continues beyond $x = 12$ ft, we should also eliminate it by adding in the linearly decreasing distribution shown in Figure 8.23(d) but this distribution is

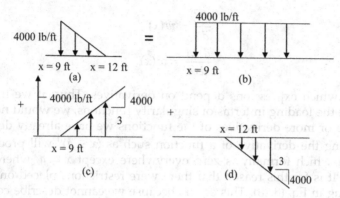

FIGURE 8.23
(a) A linearly decreasing load distribution over the interval 9 < x <12 ft, represented as the
sum of (b) a uniform load, (c) a linearly increasing load, and (d) a linearly decreasing load.

zero for $x \le 12$ ft, so it has no effect on the distribution in the beam and
so was omitted.

We again defined the shear force and bending moment as anonymous
functions that we then plotted. This also lets us evaluate these functions
at any critical points we want. For example:

V(0)
ans = −3000
V(0 + eps)
ans = 5000
V(4)
ans = 2000
M(0)
ans = −4500
M(6)
ans = 12000
M(9)
ans = −6000

and we can see those values in Figure 8.22.

You may wonder why these functions are called singularity functions
since these functions appear well-behaved. However, note that we only
defined these functions for the shear force and bending moment. From
Eq. (8.7), we can write the shear force and bending moment in terms of
the loading, $w(x)$ as

$$\frac{dV}{dx} = w(x)$$

$$\frac{d^2M}{dx^2} = w(x)$$

(8.20)

both of which expressions depend on derivatives. Thus, if we tried to describe the loading in terms of singularity functions, we would need to take one or more derivatives of the functions we have already defined. But taking the derivative of a function such as $\langle x - a \rangle^0$ will produce a function which formally is zero everywhere except $x = a$, where it is infinite! It is for this reason that there were restrictions placed on the n appearing in Eq. (8.18). This occurs because we cannot describe concentrated forces and moments in terms of ordinary functions for $w(x)$ that have non-singular behavior. We can define generalized functions that are singular and do represent concentrated forces and moments, but they are not needed if we only describe the behavior of the shear force and the bending moment. The functions we have been using are also called *Macaulay bracket functions*, which is a more descriptive name. In strength of materials, you will find that the vertical deflection of the beam, $v(x)$, is related to the bending moment through the moment-curvature relation given by

$$EI\frac{d^2v}{dx^2} = M(x)$$

(8.21)

where E is Young's modulus, a material property, and I is a property of the cross-sectional area of the beam called an area moment (which we will discuss in Chapter 11). Thus, to find the deflection we need to integrate the bending moment function twice. Since we know how to integrate the singularity functions, if we write the bending moment in terms of them the integrations can be done for the entire beam rather than performing integrations section by section, as would be needed if we had used ordinary piece-wise functions for each section of the beam. Thus, singularity functions provide a very efficient way for determining beam deflections.

8.3 Problems

P8.1 A cantilever beam is loaded by a force $P = 50$ lb and a couple $M = 2000$ ft-lb. Determine the bending moment at the cross section a-a shown in Fig. P8.1 if that cross-section is located at $x = 9$ ft to the right of the force.

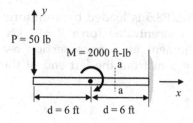

Figs. P8.1, P8.2, P8.3, P8.4

Choices (in ft-lb):

1. 1250
2. 1450
3. 1550
4. 1650
5. 2050

P8.2 A cantilever beam is loaded by a force $P = 50$ lb and a couple $M = 2000$ ft-lb. Determine the bending moment at the cross section a-a shown in Fig. P8.2 if that cross-section is located anywhere in the section between the couple and the fixed wall at a distance x to the right of the force P.

Choices (in ft-lb):

1. $-2000 - 50x$
2. $50 - 2000x$
3. $2000 + 50x$
4. $2000 - 50x$
5. $50 + 20000x$

P8.3 A cantilever beam is loaded by a force $P = 50$ lb and a couple $M = 2000$ ft-lb. Draw the shear force and bending moment diagrams for the beam of Fig. P8.3 using the relations between the loading, the shear force, and the bending moment. Determine values at all critical points in the plots.

P8.4 A cantilever beam is loaded by a force $P = 50$ lb and a couple $M = 2000$ ft-lb. Draw the shear force and bending moment diagrams for the beam of Fig. P8.4 using singularity functions. Determine values at all critical points in the plots.

P8.5 The simply supported beam of Fig. P8.5 is loaded by a uniform distributed load w = 4 kN/m and a concentrated force P = 3 kN, as shown. Determine the bending moment at a cross section a-a in section BC located at a distance x = 5 m from the left end of the beam.

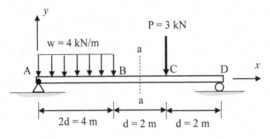

Figs. P8.5, P8.6, P8.7, P8.8

Choices (in kN-m):

1. 12.25
2. 13.50
3. 14.25
4. 15.75
5. 16.00

P8.6 The simply supported beam of Fig. P8.6 is loaded by a uniform distributed load w = 4 kN/m and a concentrated force P = 3 kN, as shown. Determine the bending moment at any location x in section CD between the force P and the right end of the beam, as measured from the left end.

Choices (in kN-m):

1. $12.75 - 4x$
2. $16 - 4x$
3. $12.75 - 6.25x$
4. $16 - 12.75x$
5. $32 - 3.25x$

P8.7 The simply supported beam of Fig. P8.7 is loaded by a uniform distributed load w = 4 kN/m and a concentrated force P = 3 kN, as shown. Draw the shear force and bending moment diagrams for the beam using the relations between the loading, the shear force, and the bending moment. Determine values at all critical points in the plots.

P8.8 The simply supported beam of Fig. P8.8 is loaded by a uniform distributed load $w = 4$ kN/m and a concentrated force $P = 3$ kN, as shown. Draw the shear force and bending moment diagrams for the beam using singularity functions. Determine values at all critical points in the plots.

P8.9 A force $P = 140$ lb acts on a pin at D which is also connected to two members AD and DB, as shown in Fig. P8.9. Those members are in turn connected to the simply supported beam ABC at A and B through the pins at those locations. Draw the shear force and bending moment diagrams in the beam ABC. Determine values at all the critical points in the plots. Is there any axial force in the beam?

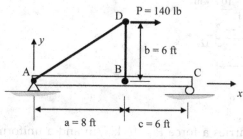

Fig. P8.9

8.3.1 Review Problems

These problems typically have the level of difficulty found on exams. They are best done by hand (i.e., with a calculator).

R8.1 A simply supported beam carries a couple $M = 40$ kN-m and a force $P = 10$ kN, as shown in Fig. R8.1. The distance $d = 2$ m. Determine the internal bending moment at a cross-section a-a in BC located at a distance $x = 3$ m from B.

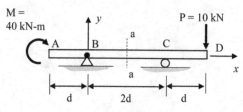

Figs. R8.1, R8.2, R8.3

R8.2 A simply supported beam carries a couple $M = 40$ kN-m and a force $P = 10$ kN, as shown in Fig. R8.2. The distance $d = 2$ m. Determine the internal bending moment at any location x in BC where the distance x is measured from B.

R8.3 A simply supported beam carries a couple $M = 40$ kN-m and a force $P = 10$ kN, as shown in Fig. R8.3. The distance $d = 2$ m. Draw the shear force and bending moment diagrams for the beam AD. Determine the values at all critical values in the plots.

R8.4 The beam in Fig. R8.4 carries a force $P = 16$ kN/m and a uniform distributed load $w = 10$ kN/m. The distance $d = 1.5$ m. Determine the internal bending moment at cross section a-a which is in section CD at a distance $x = 4$ m from A.

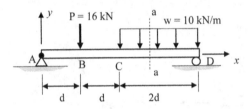

Figs. R8.4, R8.5, R8.6

R8.5 The beam in Fig. R8.5 carries a force $P = 16$ kN/m and a uniform distributed load $w = 10$ kN/m. The distance $d = 1.5$ m. Determine the internal bending moment at any location x in CD where x is measured from A.

R8.6 The beam in Fig. R8.6 carries a force $P = 16$ kN and a uniform distributed load $w = 10$ kN/m. The distance $d = 1.5$ m. Draw the shear force and bending moment diagrams for the beam. Determine the values at all critical points in the beam.

9

Frictional Forces

OBJECTIVES

- To determine the conditions under which structures lose equilibrium by slipping or tipping.
- To solve equilibrium problems where either slipping or tipping is impending.
- To examine the equilibrium of flat belts or ropes in contact with rough surfaces.
- To solve friction problems involving wedges.

Friction is a type of reaction force found at rough surfaces. A characteristic feature of frictional forces is that they are limited in how large they can become before equilibrium is no longer possible. If the maximum frictional force is exceeded, a body will slide along the rough surface. Another way a body can lose equilibrium at a rough surface is through tipping. In this chapter, we will examine the conditions under which both types of loss of equilibrium are possible.

9.1 Slipping and Tipping

A block on an inclined rough plane, as shown in Figure 9.1(a), is the classic problem used to describe friction. The plane will exert both a normal force N and a tangential frictional force F on the block. From the free body diagram of Figure 9.1(b), equilibrium of the block requires

$$\sum F_x = 0 \quad F - W \sin \theta = 0$$
$$\rightarrow \quad F = W \sin \theta$$
$$\sum F_y = 0 \quad N - W \cos \theta = 0 \tag{9.1}$$
$$\rightarrow \quad N = W \cos \theta$$

DOI: 10.1201/9781003372592-9

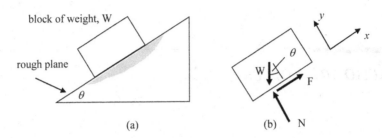

FIGURE 9.1

(a) A block on an inclined rough surface. (b) The free body diagram of the block.

where W is the weight of the block. It follows that we must have

$$\frac{F}{N} = \tan \theta \tag{9.2}$$

If the angle of the block is increased, there will be a maximum angle θ_{max} at which the block will begin to slide, and a corresponding maximum frictional force, where we have

$$\frac{F_{max}}{N} = \tan \theta_{max} \tag{9.3}$$

If we define the tangent of that maximum angle as the *static coefficient of friction*, μ_s, then it follows that

$$F_{max} = \mu_s N \tag{9.4}$$

is the maximum frictional force when slipping of the block impends. Thus, for equilibrium to be possible, in a problem involving friction, we must have the magnitude of the frictional force less than this maximum value:

$$|F| < \mu_s N \tag{9.5}$$

When the magnitude of the friction force has reached its maximum value and slipping is imminent (slipping impending), then

$$|F| = |F_{max}| = \mu_s N \tag{9.6}$$

and *the direction of the frictional force must be opposite to the direction of impending motion* (since the frictional force always opposes the tendency of the body to slide – it never aids in that tendency). Once sliding occurs, a frictional force still exists. Consider, for example, the block on the horizontal rough surface in Figure 9.2(a), where the free body diagram is shown in Figure 9.2(b). As the applied force P increases, the frictional force (which must equal P from equilibrium) will increase until it reaches the

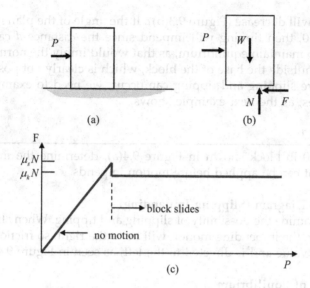

FIGURE 9.2
(a) A block of weight W on a rough surface acted upon by a horizontal force P. (b) The free body diagram. (c) The friction force as a function of the applied force P.

maximum value of $\mu_s N$. If the force P is increased beyond that value, the block moves and the frictional force drops to a slightly smaller value $F = \mu_k N$, where μ_k is called the *kinetic coefficient of friction*. This frictional force remains constant during motion of the block, independent of the size of the force P (Figure 9.2(c)).

Having the maximum possible frictional force exceeded so the body slips is not the only way bodies in frictional contact can lose equilibrium. It is also possible for the body to tip over. We can analyze this tipping tendency by again examining a block on an inclined surface (Figure 9.3(a)). As the angle θ of the plane increases, the distance d, where the normal force N acts along the plane underneath

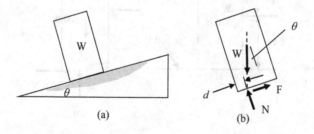

FIGURE 9.3
(a) A block of weight W in equilibrium on an inclined rough surface. (b) The free body diagram for examining the possibility of tipping.

the block, will decrease (Figure 9.3(b)). If the angle of the plane is $\theta = \theta_t$ when $d = 0$, then tipping will impend since the distance d cannot be negative to maintain equilibrium, as that would imply the normal force would lie outside the base of the block, which is clearly not possible. In cases where slipping and tipping can occur, we need to examine both possibilities, as the next example shows.

Example 9.1

For the 300 lb block shown in Figure 9.4(a), determine the maximum force P that can be applied before motion impends.

Free Body Diagram (Slipping Impending)
We will examine the possibility of slipping and tipping. When slipping is impending, the impending motion will be to the right, so friction has its maximum value and is directed to the left, as seen in Figure 9.4(b).

Equations of Equilibrium
Summing forces in the y-direction, we can find the normal force and from summing forces in the x-direction, we obtain the force P for slipping impending:

$$
\begin{aligned}
\sum F_y = 0 \quad & N - 300 = 0 \\
\rightarrow \quad & N = 300 \text{ lb} \\
\sum F_x = 0 \quad & P - (0.5)(300) = 0 \\
\rightarrow \quad & P = 150 \text{ lb}
\end{aligned}
\tag{9.7}
$$

Free Body Diagram 1 (Tipping Impending)
We can examine tipping in two ways, as will be shown. The force P and the frictional force F must form a couple because $P = F$ always from

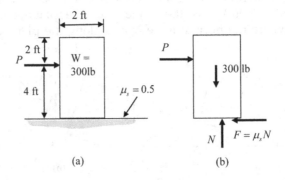

(a)   (b)

FIGURE 9.4
(a) A block on a rough surface. (b) The free body diagram of the block when slipping is impending.

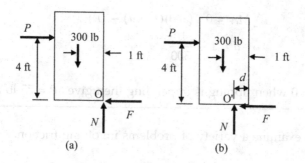

FIGURE 9.5
(a) The forces on the block when tipping is impending. (b) The free body diagram for the general case where the normal force, N, is at a distance, d, from the edge.

equilibrium. This couple must be balanced by a couple formed by the 300 lb weight and the normal force, N, which also must be equal. However, the distance between P and F is fixed, but not the distance between the 300 lb weight and N, so as P increases the distance of N from the line of action of the 300 lb weight must increase. The largest this distance can be is when the normal force is at the right edge of the block, as seen in Figure 9.5(a). This is the condition of tipping impending, where at larger values of P the block will rotate clockwise as it tips. The first way to analyze tipping, therefore, is to draw the free body diagram when N is at the edge of the block (Figure 9.5(a)) and solve for the force P.

Equations of Equilibrium
Summing moments about point O at the block edge gives:

$$\sum M_O = 0 \quad (300)(1) - (P)(4) = 0$$
$$\rightarrow \quad P = 75 \text{ lb} \tag{9.8}$$

This is smaller than the force P when slipping impends, so tipping will occur first and $P_{max} = 75$ lb.

Free Body Diagram 2 (tipping impending)
Alternatively, we could also analyze the problem by drawing the free body diagram for the block for any value of the force P where the normal force is under the block at some distance d from the edge (Figure 9.5(b)).

Equations of Equilibrium
In this case, we can sum moments about point O' where the normal force acts, to find the distance d as a function of the force, P:

$$\sum M_{O'} = 0 \quad (300)(1 - d) - (P)(4) = 0$$
$$\rightarrow \ d = 1 - \frac{4P}{300} \tag{9.9}$$

Setting $d = 0$ when tipping is impending then gives $P = 75$ lb again.

Now, let's examine a variety of problems involving friction.

Example 9.2

Consider the three blocks in Figure 9.6(a), which are in frictional contact with each other and the horizontal plane. The upper block is held by an inclined cable. Determine the largest force P that can be applied to block B before motion impends.

Free Body Diagram 1

Here, tipping is not an issue, but slipping can occur in one of two ways. First, block B could tend to move by itself, slipping on both blocks A and C. But it is also possible that blocks B and C could slip as a unit together, slipping on block A and the plane. We need to examine both possibilities to determine which case of slipping occurs first. For either possibility, slipping must occur between A and B, so for either case, the free body diagram of A will be as shown in Figure 9.6(b), where the maximum frictional force will be developed between A and B and there is some tension, T, in the cable. Motion will be impending to the left so why is the maximum frictional force drawn in Figure 9.6(b) also to the left? It is because it is the maximum frictional force that A exerts on B that must be opposite to the direction of impending motion of B, while in Figure 9.6(b)

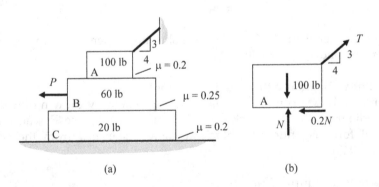

(a)   (b)

FIGURE 9.6
(a) Three blocks in frictional contact and held by a cable. (b) The free body diagram of block A when slipping is impending.

we are showing the equal and opposite force that B exerts on A. Another way to look at this is to recognize when B starts to slip, it will tend to drag A to left also, which is why the frictional force on A should be drawn to the left.

Equations of Equilibrium

From the free body diagram of Figure 9.6(b), we can find the normal force N (and the tension T, but it is not needed):

$$\sum F_x = 0 \quad -0.2N + 4T/5 = 0$$
$$\sum F_y = 0 \quad N - 100 + 3T/5 = 0 \qquad (9.10)$$
$$\rightarrow N = 87 \text{ lb}, \quad T = 21.7 \text{ lb}$$

Free Body Diagram 2

Now consider the first possibility where B moves alone as shown in the free body diagram of Figure 9.7(a). In this case, we must have the maximum frictional forces developed between B and A and between B and C and they both must be opposite to the impending motion, which is to the left.

Equations of Equilibrium

When B moves alone, the equilibrium equations for slipping impending are

$$\sum F_y = 0 \quad N_1 - 60 - 87 = 0$$
$$\rightarrow N_1 = 147 \text{ lb}$$
$$\sum F_x = 0 \quad -P + (87)(0.2) + 0.25N_1 = 0 \qquad (9.11)$$
$$\rightarrow P = 54.2 \text{ lb}$$

so the load $P = 54.2$ lb when this motion impending occurs.

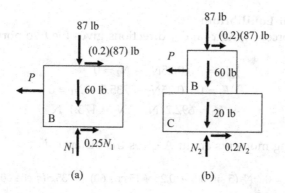

(a)   (b)

FIGURE 9.7

(a) The case when B slips alone. (b) The case where B and C slip together.

Free Body Diagram 3
Now, consider the possibility that B and C slide together, as shown in the free body diagram of Figure 9.7(b).

Equations of Equilibrium
For B and C sliding together, when slipping impends we have

$$\sum F_x = 0 \quad N_2 - 20 - 60 - 87 = 0$$
$$\rightarrow \quad N_2 = 167 \text{ lb}$$
$$\sum F_x = 0 \quad -P + (0.2)(87) + 0.2N_2 = 0 \tag{9.12}$$
$$\rightarrow \quad P = 50.8 \text{ lb}$$

Here, the force $P = 50.8$ lb. This value is slightly smaller than the previous case, so it is the maximum possible force. Note that the values of the force P in this problem are not too different, so there is no way we could come to the right conclusion without examining both possibilities.

Example 9.3

Determine the maximum distance d that a 75 kg person can climb before the ladder shown in Figure 9.8(a) slips if (1) $\mu = 0.25$ at all surfaces, (2) $\mu = 0.25$ at the wall and $\mu = 0.4$ at the ground. Neglect the weight of the ladder.

Free Body Diagram 1
For the first case, the free body diagram is shown in Figure 9.8(b) for slipping impending.

Equations of Equilibrium
Summing forces in the x- and y-directions gives the two normal forces:

$$\sum F_x = 0 \quad 0.25N_1 - N_2 = 0$$
$$\sum F_y = 0 \quad 0.25N_2 - 735.5 + N_1 = 0 \tag{9.13}$$
$$\rightarrow \quad N_1 = 692.2 \text{ N}, \quad N_2 = 173.1 \text{ N}$$

and summing moments about A gives the distance d:

$$\sum M_A = 0 \quad N_2(5 \sin 60) + 0.25N_2(5 \cos 60) - 735.5(d \cos 60) = 0 \tag{9.14}$$
$$\rightarrow \quad d = 2.33 \text{ m}$$

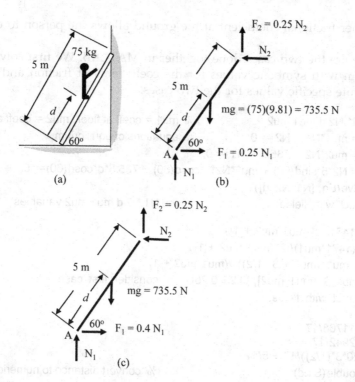

FIGURE 9.8
A person standing on a ladder which is in frictional contact with the wall and ground. (b)
Free body diagram when $\mu = 0.25$ at both surfaces. (c) Free body diagram when $\mu = 0.25$ at
the wall and $\mu = 0.4$ at the ground.

Free Body Diagram 2
For the second case, the free body diagram is shown in Figure 9.8(c) for
slipping impending.

Equations of Equilibrium
From force equilibrium:

$$\sum F_x = 0 \quad 0.4N_1 - N_2 = 0$$
$$\sum F_y = 0 \quad 0.25N_2 - 735.5 + N_1 = 0 \tag{9.15}$$
$$\rightarrow N_1 = 668.6 \text{ N}, \quad N_2 = 267.5 \text{ N}$$

and from moment equilibrium:

$$\sum M_A = 0 \quad N_2(5\sin 60) + 0.25N_2(5\cos 60) - 735.5(d\cos 60) = 0$$
$$\rightarrow d = 3.6 \text{ m} \tag{9.16}$$

A higher frictional coefficient at the ground allows the person to climb higher.

Here are the two cases done together in MATLAB. We first solve the problem with symbolic values for the coefficients of friction and then substitute specific values for the two cases:

```
syms N1 N2 d mu1 mu2              % mu1 = coeff at floor, mu2 = coeff at wall
Eq(1) = mu1*N1 - N2 == 0;         % equations of equilibrium
Eq(2) = mu2*N2 - 735.5 + N1 == 0;
Eq(3) = N2*5*sind(60) + mu2*N2*5*cosd(60) - 735.5*d*cosd(60)== 0;
S = solve(Eq, [N1 N2 d])
S = struct with fields:           % the N1 N2 d mu1 mu2 variables

    N1: 1471/(2*(mu1*mu2 + 1))
    N2: (1471*mu1)/(2*(mu1*mu2 + 1))
     d: (5*mu1*(mu2 + 3^(1/2)))/(mu1*mu2 + 1)
S1 = subs(S, [mu1 mu2], [0.25 0.25])   % consider first case
S1 = struct with fields:

    N1: 11768/17
    N2: 2942/17
     d: (20*3^(1/2))/17 + 5/17
d1 = double(S1.d)                 % convert distance to numerical
d1 = 2.3318
S2 = subs(S, [mu1 mu2], [0.4 0.25]);   %second case
d2 = double(S2.d)                 % convert distance to numerical
d2 = 3.6037
```

Example 9.4

A force P pushes on a block of weight W that lies on an inclined rough plane (Figure 9.9(a)). The coefficient of static friction between the block and the plane is μ. Slipping motion is assumed to be impending up the plane:

1. Using MATLAB, solve the equilibrium equations symbolically and obtain the force, P, as a function of W, μ, and θ for motion impending up the plane

2. Let $W = 50$ lb, $\mu = 0.4$. Plot the force P required to have motion impending as a function of the angle θ from 0 to 45°

3. What is the maximum angle that θ can have and still produce impending motion up the plane with a finite pushing force, P? (take $\mu = 0.4$ again)

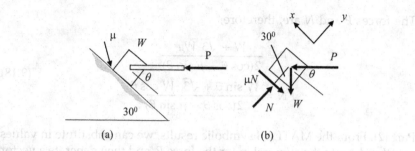

(a)   (b)

FIGURE 9.9
(a) A block being pushed up an inclined plane. (b) The free body diagram for sliding impending up the plane.

Free Body Diagram
The free body diagram for sliding impending up the plane is shown in Figure 9.9(b).

Equations of Equilibrium
Part (1): From summing forces normal and tangent to the plane:

$$\sum F_y = 0 \quad N - W\cos(30°) - P\sin(\theta) = 0$$
$$\sum F_x = 0 \quad P\cos(\theta) - \mu N - W\sin(30°) = 0 \tag{9.17}$$

The MATLAB code for solving these equations is:

```
syms N P W mu t                    % declare symbolic variables, t = theta
Eq(1) = N - cosd(30)*W - P*sin(t);  % equations of equilibrium in
                                    % x- and y-directions
Eq(2) = P*cos(t) - mu*N - W*sind(30);
S = solve(Eq, N, P)                % solve equilibrium equations for N, P
S = struct with fields:
N: (W*sin(t) + 3^(1/2)*W*cos(t))/(2*(cos(t) - mu*sin(t)))
P: (W + 3^(1/2)*W*mu)/(2*(cos(t) - mu*sin(t)))
pretty(S.P)                        % show results in more readable form
 W + sqrt(3)W mu
---------------
2(cos(t) - mu  sin(t))


pretty(S.N)
W  sin(t) + sqrt(3) W  cos(t)
--------------------
   2(cos(t) - mu  sin(t))
```

The forces P and N are, therefore,

$$P = \frac{W + \sqrt{3}\, W\mu}{2(\cos\theta - \mu\sin\theta)}$$

$$N = \frac{W\sin\theta + \sqrt{3}\ W\cos\theta}{2(\cos\theta - \mu\sin\theta)}$$

(9.18)

Part (2): From the MATLAB symbolic results, we can substitute in values for W and μ into the expression for the force P, and then generate a vector of θ values and substitute them in as well. We then can plot the results. Note that the symbolic expressions are in terms of the sin and cos functions, so we need to substitute values in radians and then convert radians to degrees for plotting.

```
Pt = subs(S.P, {W, mu}, [50, 0.4])          % substitute values
Pt = (20*3^(1/2) + 50)/(2*(cos(t) - (2*sin(t))/5))
angr = linspace(0, pi/4, 90);               % generate angles from zero
                                            % to 45 degrees in radians
Pn = subs(Pt, angr);                        % substitute angles
plot(angr*180/pi, Pn)                       % plot P versus angle in degrees
xlabel('Angle, degrees')
ylabel('Force, P(lb)')
```

Alternatively, we can do all the substitutions at once. In this case, we must put all the values in a cell vector {50, 0.4, angr} rather than the ordinary vector used previously. Cell vectors allow us to have components of different types. Here, 50 and 0.4 are scalars but angr is a vector:

```
angr = linspace(0, pi/4, 90);
Pn2 = subs(S.P, {W, mu, t}, {50, 0.4, angr});
plot(angr*180/pi, Pn2)
xlabel('Angle, degrees')
ylabel('Force, P(lb)')
```

The same plot is obtained in either case, as shown in Figure 9.10.

Part (3): Examining the expression for P in Eq. (9.18), P goes to infinity when $\cos\theta - \mu\sin\theta = 0$, which also gives $\tan\theta = 1/\mu = 2.5$, whose solution is $\theta = 68.2°$.

There are conditions under which slipping at a rough surface is not possible (called *self-locking* conditions) as the next example shows.

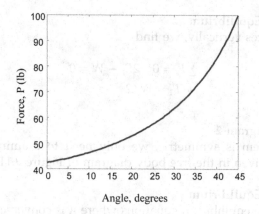

FIGURE 9.10
The required force P to push the block up the plane.

Example 9.5

A broom is held by a mechanism shown in Figure 9.11(a) where the broom can be pushed up between two cylinders and then supported by the friction forces between the broom and the cylinders. The coefficient of friction is μ at all surfaces. Design this broom holder so that no matter what the weight, W, of the broom, the broom cannot drop (self-locking). Neglect the weight of the small cylinders.

Free Body Diagram 1

A free body diagram of the broom is given in Figure 9.11(b), where we have used symmetry to show the friction and normal forces on both sides of the broom as the same.

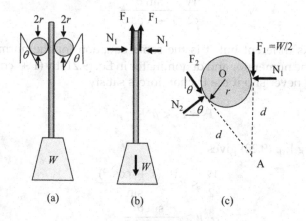

FIGURE 9.11
(a) A broom-holding mechanism. (b) A free body diagram of the broom. (c) A free body diagram of one of the cylinders.

Equations of Equilibrium
If we sum forces vertically, we find

$$\sum F_y = 0 \quad 2F_1 - W = 0$$
$$\rightarrow F_1 = W/2 \tag{9.19}$$

Free Body Diagram 2
Since the system is symmetric, we only need to examine one of the cylinders, as given in the free body diagram of Figure 9.11(c).

Equations of Equilibrium
There are three equilibrium equations where it is convenient to use two moment equations and a force equation to solve for the normal forces on the cylinder:

$$\sum M_O = 0 \quad rF_2 - r(W/2) = 0$$
$$\rightarrow \quad F_2 = W/2$$
$$\sum M_A = 0 \quad N_1 d - N_2 d = 0$$
$$\rightarrow \quad N_2 = N_1 \tag{9.20}$$
$$\sum F_y = 0 \quad W/2 + (W/2)\cos\theta - N_2 \sin\theta = 0$$
$$\rightarrow \quad N_1 = N_2 = \frac{W}{2}\frac{(1 + \cos\theta)}{\sin\theta}$$

If we had used the force equation in the x-direction instead, we would find

$$\sum F_x = 0 \quad -N_1 + N_2 \cos\theta + (W/2)\sin\theta = 0$$
$$\rightarrow \quad N_1 = N_2 = \frac{W}{2}\frac{\sin\theta}{1 - \cos\theta} \tag{9.21}$$

which looks different but it is the same equation for the normal forces (multiply the numerator and denominator in Eq. (9.21) by $(1 + \cos\theta)$). The broom will never slip if the friction forces satisfy

$$F_1 = F_2 < \mu N_2 = \mu N_1 \tag{9.22}$$

which using Eq. (9.20) gives

$$\frac{W}{2} < \mu\frac{W}{2}\frac{(1 + \cos\theta)}{\sin\theta}$$
$$\rightarrow \mu > \frac{\sin\theta}{1 + \cos\theta} \tag{9.23}$$

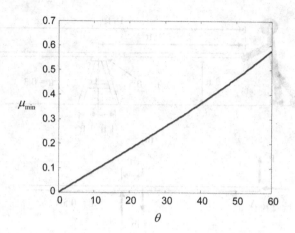

FIGURE 9.12

A plot of the minimum coefficient of friction needed to support a broom of any weight for an angle θ in the mechanism.

If the broom placed in the holder has a coefficient of friction larger than a minimum value given by $\mu_{min} = \sin\theta/(1 + \cos\theta)$, the broom cannot slip, regardless of its weight. This minimum value is plotted versus θ in Figure 9.12.

Example 9.6

A carpenter pushes horizontally on a long board and holds it up to keep it from tipping (Figure 9.13(a)). The board has a uniform weight of 3 lb/ft and the sawhorse has a weight of 15 lb and a center of gravity at G. Neglect the width at the top of the saw horse. As the carpenter increases the horizontal force that she (he) applies, determine how the system loses equilibrium and the value of the horizontal force.

The first thing we need to do in this problem is to examine the possible ways that equilibrium can be lost. First, the board could slip on the top of the sawhorse (the solution likely the carpenter is looking for). Second, it is possible the board and sawhorse may slide together on the floor. Third, the force applied to the top of the sawhorse could cause it to tip. We need to examine all three possibilities.

Free Body Diagram 1

First, consider that the board slips on the sawhorse. The free body diagram is shown in Figure 9.13(b).

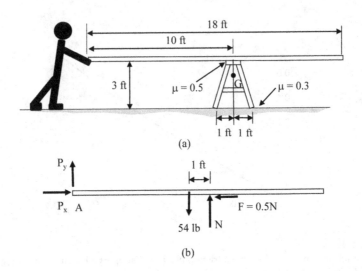

FIGURE 9.13
(a) A carpenter pushes on a long board and holds it up at the left end. (b) The case where slipping is impending for the board on the sawhorse.

Equations of Equilibrium
From equilibrium we have, when slipping is impending:

$$\sum M_A = 0 \quad (N)(10) - (54)(9) = 0$$
$$\rightarrow N = 48.6 \text{ lb}$$
$$\sum F_y = 0 \quad P_y - 54 + 48.6 = 0$$
$$\rightarrow P_y = 5.4 \text{ lb} \tag{9.24}$$
$$\sum F_x = 0 \quad P_x - 0.5N = 0$$
$$\rightarrow P_x = 0.5N = (0.5)(48.6) = 24.3 \text{ lb}$$

Free Body Diagram 2
Next, assume the sawhorse slips on the floor. The free body diagram for this case is shown in Figure 9.14(a).

Equations of Equilibrium
When slipping of the sawhorse on the floor is impending, we have

$$\sum F_y = 0 \quad N_1 + N_2 - 15 - 48.6 = 0$$
$$\rightarrow N_1 + N_2 = 63.6 \text{ lb}$$
$$\sum F_x = 0 \quad -0.3N_1 - 0.3N_2 + P_x = 0 \tag{9.25}$$
$$\rightarrow P_x = (0.3)(63.6) = 19.08 \text{ lb}$$

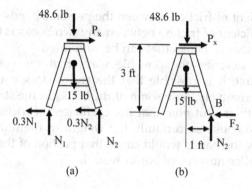

FIGURE 9.14
(a) A free body diagram of the sawhorse when slipping at the floor for the sawhorse is impending. (b) The case where tipping of the sawhorse about B is impending.

Free Body Diagram 3

Finally, consider the possibility of tipping. From Figure 9.14(b), this case is where the normal force at the left leg of the sawhorse goes to zero so the sawhorse is about to tip around B.

Equations of Equilibrium

Summing moments about B, we find

$$\sum M_B = 0 \quad (48.6 + 15)(1) - (P_x)(3) = 0$$
$$\rightarrow P_x = 21.2 \text{ lb} \tag{9.26}$$

Thus, the sawhorse will slide on the floor first, at a force $P \cong 19.1$ lb.

Example 9.7

A person holds a stack of books by applying compressive forces of 80 N to the ends of the stack (Figure 9.15). The mass of each book is 0.95 kg

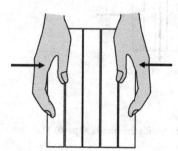

FIGURE 9.15
A person holding a stack of books.

and the coefficient of friction between the person's hands and the books is 0.6. The coefficient of friction between any two books is 0.4. Determine the largest number of books that can be supported.

We need to describe the possible ways that the system can lose equilibrium. First, it is possible that the entire stack could slide from the hands. Alternatively, only some of the books in the stack could slide out from the others. But note that the coefficient of friction is the same between all the books so certainly the weight of the total stack minus the two held by the hands would cause that portion of the stack to slide before any smaller number of books would.

Free Body Diagram 1
Consider first the possibility that all the books except the two in contact with the hands slip. Then, the free body diagram is as shown in Figure 9.16(a).

Equations of Equilibrium
When slipping of the two books less than the total number in the stack impends:

$$\sum F_y = 0 \quad 32 + 32 - (0.95)(9.81)(n - 2) = 0$$
$$\rightarrow n = 8.87$$

(9.27)

In this case, we can only support $n = 8.87$ books, but n must be an integer, so $n = 8$ is the maximum number that could be supported.

Free Body Diagram 2
Now, assume all the books slide from the hands. The free body diagram is seen in Figure 9.16(b).

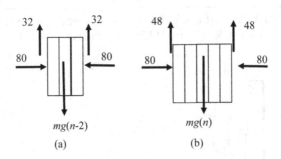

FIGURE 9.16
(a) The case when all but the two of the n books in contact with the hands slip. (b) The case where all the n books between the hands slip.

Equations of Equilibrium
For slipping of the total stack impending:

$$\sum F_y = 0 \quad 48 + 48 - (0.95)(9.81)(n) = 0$$
$$\rightarrow n = 10.3$$

(9.28)

which yields $n = 10$ books. Thus, the maximum number of books that can be supported is eight.

Example 9.8

A 9-foot-long chain that weighs 0.5 lb/ft is draped a distance, d, over the edge of a rough inclined plane (Figure 9.17). The coefficient of friction between the plane and the chain is 0.3. Assume the upper edge of the plane is rounded and smooth so that the tension in the chain is fully transmitted around the corner. Determine the largest and smallest values that d can have for the chain to be in equilibrium.

Free Body Diagrams 1
If we assume slipping of the chain up the plane is impending, Figures 9.18(a) shows the free body diagram of the portion of the chain on the plane and Figure 9.18(b) shows the free body diagram of the hanging portion.

Equilibrium Equations
Consider Figure 9.18(b). Summing forces vertically gives the tension in the chain in terms of the weight of the portion hanging over the edge:

$$\sum F_y = 0 \quad T - (0.5)(d) = 0$$
$$\rightarrow T = 0.5d$$

(9.29)

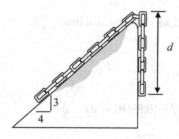

FIGURE 9.17
A chain in contact with a rough inclined surface.

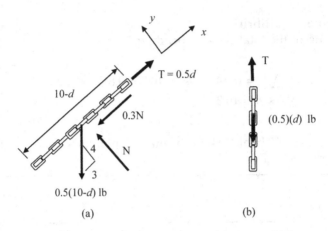

FIGURE 9.18
(a) Free body diagram of the chain on the plane when slipping is impending up the plane.
(b) Free body diagram of the hanging portion of the chain.

while from Figure 9.18(a) we have

$$\sum F_y = 0 \quad N - (4/5)(0.5)(10 - d) = 0$$
$$\rightarrow \quad N = 4 - 0.4d$$
$$\sum F_x = 0 \quad 0.5d - (0.3)(4 - 0.4d) - (3/5)(0.5)(10 - d) = 0 \qquad (9.30)$$
$$\rightarrow \quad d = 4.57 \text{ ft}$$

Free Body Diagrams 2
If the slipping of the chain down the plane is impending, then the free body diagram of the chain on the plane is shown in Figure 9.19(a) and the free body diagram of the hanging portion is shown in Figure 9.19(b).

Equilibrium Equations
Since the free body diagram of the hanging portion remains the same for both cases, we have again

$$\sum F_y = 0 \quad T - (0.5)(d) = 0$$
$$\rightarrow \quad T = 0.5d \qquad (9.31)$$

and the free body diagram of the portion on the inclined plane (Figure 9.19(a)) gives

$$\sum F_y = 0 \quad N - (4/5)(0.5)(10 - d) = 0$$
$$\rightarrow \quad N = 4 - 0.4d$$
$$\sum F_x = 0 \quad 0.5d + (0.3)(4 - 0.4d) - (3/5)(0.5)(10 - d) = 0 \qquad (9.32)$$
$$\rightarrow \quad d = 2.65 \text{ ft}$$

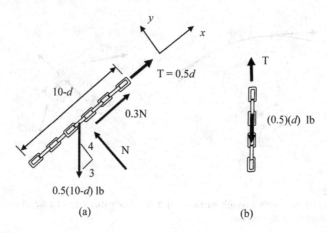

FIGURE 9.19
(a) Free body diagram of the chain when slipping is impending down the plane. (b) Free body diagram of the hanging portion of the chain.

To keep the chain from slipping either up or down the plane, we therefore must have $2.65 < d < 4.57$ ft.

9.2 Flat Belts or Ropes in Contact with Rough Surfaces

When a flat belt or rope is in contact with a smooth surface over some portion of its length, the magnitude of the tension remains the same in the belt/rope on either side of the contact region but the direction of the tensile force can change. For contact with a rough surface, however, the magnitude of the tension itself can change since friction can help support some the load. Suppose, for example, we have a rope in contact with a rough surface over a range of angles from θ_1 to θ_2, where the total angular range of contact is $\beta = \theta_2 - \theta_1$ (Figure 9.20(a)). Let the tensions on either side of the contact region be (T_S, T_L), where T_S is the smaller tension and T_L is the larger tension. To analyze how these tensions are related, consider a very small length of the contact region, ds, which is shown in more detail in Figure 9.20(b). If we sum forces tangential and normal to this segment

$$\sum F_x = 0 \quad dF + T\cos\left(\frac{d\theta}{2}\right) - (T + dT)\cos\left(\frac{d\theta}{2}\right) = 0$$

$$\sum F_y = 0 \quad dN - T\sin\left(\frac{d\theta}{2}\right) - (T + dT)\sin\left(\frac{d\theta}{2}\right) = 0$$

(9.33)

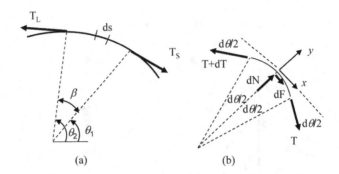

FIGURE 9.20
(a) A rope in contact with a rough surface. (b) A free body diagram of a small portion of the rope in contact with the surface.

where

$$\sin\left(\frac{d\theta}{2}\right) \simeq \frac{d\theta}{2} \quad \cos\left(\frac{d\theta}{2}\right) \simeq 1 \tag{9.34}$$

so that the equations in Eq. (9.33) become simply:

$$\begin{aligned} dF &= dT \\ dN &= Td\theta \end{aligned} \tag{9.35}$$

Equation (9.35) shows that the change in tension is equal to the frictional force, which indicates that indeed the friction is taking up some of the tension in the belt/rope. When slipping is impending, then $dF = \mu dN$ so that:

$$\begin{aligned} \mu Td\theta &= dT \\ \rightarrow \frac{dT}{T} &= \mu d\theta \end{aligned} \tag{9.36}$$

If we integrate both sides in Eq. (9.36), we obtain a relationship

$$\int_{T_S}^{T_L} \frac{dT}{T} = \int_{\theta=\theta_1}^{\theta=\theta_2} \mu d\theta$$

$$\ln\left(\frac{T_L}{T_S}\right) = \mu(\theta_2 - \theta_1) = \mu\beta \tag{9.37}$$

$$\rightarrow \frac{T_L}{T_S} = e^{\mu\beta}$$

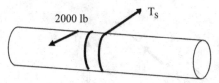

2000 lb T_S

FIGURE 9.21
A rope wrapped twice around a post, where a 2000 lb tension is being carried by a smaller force.

where T_L is the larger tension and T_S is the smaller tension on either side of the contact region. Thus, we now have an expression that relates the two different tensions in the belt/rope when slipping is impending.

Example 9.9

Consider the rope shown in Figure 9.21, which is wrapped twice around a post for which the coefficient of friction, $\mu = 0.5$. If $T_L = 2000$ *lb*, determine T_S.

From Eq. (9.37), we have

$$\frac{T_L}{T_S} = \exp(\mu\beta) = \exp[(0.5)(4\pi)] = \exp(2\pi)$$

$$\rightarrow T_S = 2000\exp(-2\pi) = 3.8 \ lb$$

(9.38)

Thus, a very large force can be held in this manner by letting friction carry most of the tension.

As a final example, consider how friction can affect the results of the chain problem examined in Example 9.8.

Example 9.10

Let's revisit the problem of a chain draped over the edge of an inclined plane (Figure 9.17) but now assume the chain is in full contact with a rough rounded corner with the same static coefficient as the plane ($\mu = 0.3$). The angular extent of the contact, $\beta = 126.87°$, is shown in Figure 9.22(a). In this case, the tension changes around the corner.

Free Body Diagrams 1
If we assume slipping occurs up the plane (see the free body diagrams of Figure 9.22(b) and (c)), then $T > T_1$ and we can use Eq. (9.37) in conjunction with equilibrium to determine d.

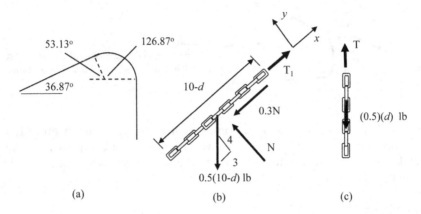

FIGURE 9.22

(a) The case where the rounded corner over which the chain hangs in Figure 9.17 is rough. The angular extent of the contact region is shown. (b) The case where slipping of the chain up the plane is impending, and (c) the free body diagram of the hanging segment of the chain.

Equilibrium Equations

First, from the hanging portion of the chain

$$\sum F_y = 0 \quad T - (0.5)(d) = 0$$
$$\rightarrow T = 0.5d$$

(9.39)

which, when placed into Eq. (9.37), gives

$$\frac{0.5d}{T_1} = \exp\left[(0.3)\frac{\pi}{180}(126.87)\right] = 1.94$$
$$\rightarrow T_1 = 0.257d$$

(9.40)

Using this result in the equilibrium equations for the chain in contact with the inclined plane then gives d:

$$\sum F_y = 0 \quad N - (4/5)(0.5)(10 - d) = 0$$
$$\rightarrow N = 4 - 0.4d$$
$$\sum F_x = 0 \quad 0.257d - (0.3)(4 - 0.4d) - (3/5)(0.5)(10 - d) = 0$$
$$\rightarrow d = 6.2 \text{ ft}$$

(9.41)

Free Body Diagrams 2

We can follow the same steps when slipping of the chain down the plane is impending. The free body diagrams are given in Figure 9.23(a) and (b).

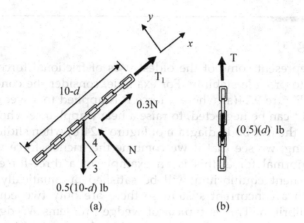

FIGURE 9.23
(a) The case where slipping of the chain down the plane is impending, and (b) the free body diagram of the hanging segment of the chain.

Equilibrium Equations

In this case, $T_1 > T$ and the free body diagram of the hanging portion has not changed, so we have Eq. (9.39) again, and from Eq. (9.37):

$$\frac{T_1}{0.5d} = \exp\left[(0.3)\frac{\pi}{180}(126.87)\right] = 1.94 \qquad (9.42)$$
$$\rightarrow T_1 = .971d$$

Equilibrium on the plane (Figure 9.23(a)) gives d again:

$$\sum F_y = 0 \quad N - (4/5)(0.5)(10 - d) = 0$$
$$\rightarrow N = 4 - 0.4d$$
$$\sum F_x = 0 \quad 0.971d + (0.3)(4 - 0.4d) - (3/5)(0.5)(10 - d) = 0 \qquad (9.43)$$
$$\rightarrow d = 1.56 \text{ ft}$$

Thus, the range of allowable lengths is $1.56 < d < 6.2$ ft. Compare this to our previous result of $2.65 < d < 4.57$ ft for the smooth case. This makes sense. We can drape more of the chain over the edge (in comparison to the smooth case) before the chain is pulled up the plane since the friction in the corner is working with the forces pulling the chain down on the inclined surface to oppose the amount of weight draped over the edge. On the other hand, the weight of the draped portion of chain works in conjunction with the friction in the corner to prevent the remainder of the chain sliding down the plane so we need less chain draped over the edge for equilibrium in this case.

9.3 Wedges

Wedges represent some of the oldest uses of frictional forces to raise bodies or to provide stability. For example, consider the configuration shown in Figure 9.24(a) where a force P is applied to a wedge, whose own weight can be neglected, to raise a heavy appliance whose weight is W. From the free body diagram of Figure 9.24(b) when sliding motion is impending, we see that if we combine the friction force vectorially with the normal force, this is an example of a three-force problem where moment equilibrium will be satisfied automatically, and the forces form a concurrent system so there are only two equations of force equilibrium. This is typical of wedge problems. Wedges can be self-locking where the wedge can be held in place by the frictional force F without an external force, as shown in Figure 9.24(c). For this to be the case, the angle θ and/or the coefficient of friction must satisfy certain conditions. Wedge problems of these types are considered in the next example.

Example 9.11

The weight $W = 200$ lb for the appliance shown in Figure 9.24(a) and the coefficient of friction $\mu = 0.2$. for a wedge where the angle $\theta = 15°$. (a) Determine the force P needed to slide the appliance to the right. (b) If $P = 0$, determine the largest angle θ at which the wedge will be in equilibrium (self-locking) for $\mu = 0.2$.

Free Body Diagram

The free body diagram of the wedge, whose own weight is negligible, for the problem of part (a) is shown in Figure 9.24(b).

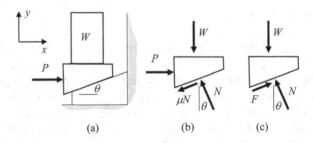

(a)   (b)   (c)

FIGURE 9.24
(a) A wedge being used to lift a weight, W. (b) Free body diagram of the wedge when motion is impending to the right. (c) The case where the weight is held in place by the wedge with no applied force (self-locking).

Equilibrium
In this case, the equilibrium equations are:

$$\sum F_x = 0 \quad P - N\sin(15°) - 0.2N\cos(15°) = 0$$
$$\sum F_y = 0 \quad N\cos(15°) - 0.2N\sin(15°) - 200 = 0 \tag{9.44}$$

From the second equation, we find $N = 218.78$ lb and from the first equation $P = 98.9$ lb.

If we solved this problem in MATLAB, we would have

```
syms P N                                    % symbolic applied force and
                                            % normal force
Eq(1) = P - N*sind(15) - 0.2*N*cosd(15);    % equations of equilibrium
Eq(2) = N*cosd(15) - 0.2*N*sind(15) - 200;
S = solve(Eq);                              % solve equations
double([S. P S. N])                         % convert to numerical values
                                            % for P and N (in lb)
ans=
    98.8893 218.7796
```

Now, consider part (b).

Free Body Diagram
When the force $P = 0$, the free body diagram is shown in Figure 9.24(c).

Equilibrium
The equilibrium equations are:

$$\sum F_x = 0 \quad F\cos\theta - N\sin\theta = 0$$
$$\sum F_y = 0 \quad N\cos\theta + F\sin\theta - W = 0 \tag{9.45}$$

These equations are easy to solve for F and N, where $F = W\sin\theta$ and $N = W\cos\theta$. Thus, to keep $F \leq \mu N$, we must have $\tan\theta \leq \mu = 0.2$, which gives $\theta \leq 11.31°$. Under these conditions, we see the wedge will not slip regardless of the weight, W, so indeed the wedge is self-locking. Since this angle is smaller than the angle specified in part (a), we see that in case (a) if we remove the force P, the wedge will slip out.

9.4 Problems

P9.1 A block A weighing $W_A = 100$ lb is supported by another block B of weight W_B through a cable that runs over a smooth pulley (Fig. P9.1). If the coefficient of friction between A and the inclined plane is $\mu = 0.5$ and the angle of the plane is $\theta = 45°$, determine the minimum weight of B needed to keep A from sliding down the plane.

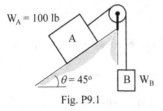

Fig. P9.1

Choices (in lb):

1. 26.3
2. 29.4
3. 32.6
4. 35.4
5. 42.2

P9.2 A bracket supports a weight and slips over a pipe with a smooth collar (at B), as shown in Fig. P9.2. If the coefficient of friction is $\mu = 0.4$ at A, determine the minimum distance d that the bracket can have so that it will support any weight, W, without slipping (self-locking). The distances $a = 200$ mm and $b = 100$ mm.

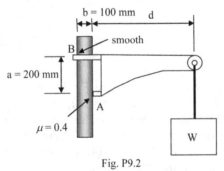

Fig. P9.2

P9.3 A lever arm is attached to a smooth pin at A and is in contact with a block of weight $W = 100$ lb at B. The arm is pressed down by a force $F = 75$ lb, as shown in Fig. P9.3. The distances $a = 2$ ft, $b = 3$ ft, and $c = 5$ ft. If the coefficient of friction $\mu = 0.5$ at all contact surfaces, determine the magnitude of the force P that will cause sliding of the block to impend. Neglect the weight of the arm.

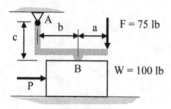

Fig. P9.3

P9.4 A uniform plank whose weight $W = 50$ lb is resting on supports at A and B. A person applies a force P at an angle of $\theta = 50°$, as shown in Fig. P9.4. If the coefficient of friction $\mu = 0.6$ at the supports, determine the force P when motion of the plank is impending. Assume there is no slipping between the person's shoes and the floor.

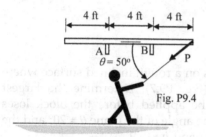

Fig. P9.4

P9.5 A force P acts at the tip of a homogeneous triangular block of weight $W = 3$ lb, as shown in Fig. P9.5. The coefficient of friction between the block and the floor is $\mu = 0.4$.(a) If $b = 4$ in. and $h = 4$ in., determine the largest force P that can be applied before equilibrium is lost. Does the block slip or tip? (b) How does your answer change if the direction of P is reversed?

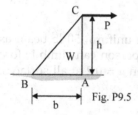

Fig. P9.5

P9.6 A force P holds a homogeneous bar of weight $W = 45$ lb against a rough wall where the coefficient of friction is $\mu = 0.3$ (Fig. P9.6). The floor is assumed to be smooth. When the distances $a = 8$ ft and $b = 6$ ft, determine the minimum force P needed to keep the bar from slipping.

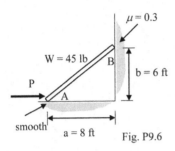

Fig. P9.6

Choices (in lb)

1. 11.2

2. 16.1

3. 21.4

4. 25.8

5. 30.2

P9.7 A block weighing W = 300 N lies on a rough inclined surface where the coefficient of friction μ = 0.5 (Fig. P9.7). Determine the largest magnitude of the force P that can be applied before the block loses equilibrium by slipping or tipping. The angle of the plane θ = 20° and the angle α = 60°. The distances a = 0.6 m and b = 0.9 m.

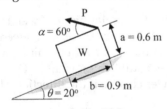

Fig. P9.7

P9.8 A person weighing W = 120 lb walks up a uniform 50 lb beam as seen in Fig. P9.8. Determine the distance d the person can climb before the beam starts to slip if the coefficient of friction μ = 0.2 at all surfaces. The distances a = 6 ft, b = 8 ft, and c = 2 ft.

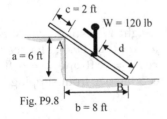

Fig. P9.8

P9.9 Block A in Fig. P9.9 has a weight W_A = 980 N while block B has a weight W_B = 490 N. The blocks are pinned together by a thin rod which is parallel to the inclined plane and whose weight can be neglected. If the coefficient of friction between A and the plane is μ_A = 0.15 and between B and the plane is μ_B = 0.25, determine the maximum angle θ possible for the blocks to remain in equilibrium.

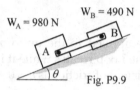

Fig. P9.9

P9.10. A pipe of weight W and radius R is supported by two identical wedges as shown in Fig. P9.10. Determine the maximum angle θ for the wedges so that the pipe can never slip if the coefficient of friction between the pipe and a wedge is μ_1 = 0.2 and between a wedge and the ground is μ_2 = 0.45. Neglect the weights of the wedges. Because of the symmetry of the problem, you need consider only one wedge. Note that your answer is independent of W and R.

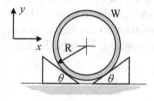

Fig. P9.10

9.4.1 Review Problems

These problems typically have the level of difficulty found on exams. They should be done by hand (i.e., with a calculator).

R9.1 A torque M = 10 ft-lb is applied to a disk that can rotate about O as shown in Fig. R9.1. If a brake arm that is pinned at A and is in frictional contact with the disk at B is used to prevent that rotation, determine the minimum force P that is required to hold the disk stationary if the coefficient of friction between the arm and the disk is μ = 0.25. The distances a = 6 in, b = 5 in., c = 3 in., and r = 2.5 in. Neglect the weight of the arm and the disk.

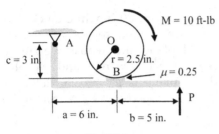

Fig. R9.1

R9.2 Determine the largest force P that can be applied to the uniform hoop in Fig. R9.2 that weighs $W = 20$ lb before the hoop rotates while it is contact with the rough surfaces, where the coefficient of friction is $\mu = 0.2$.

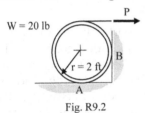

Fig. R9.2

R9.3 The pin connected frame in Fig. R9.3 supports a block of weight W and is in frictional contact with the floor at A and C where the coefficient of friction is $\mu = 0.6$. Determine the maximum angle θ so the frame always remains in equilibrium for any value of the weight W (self-locking). Neglect the weight of the frame.

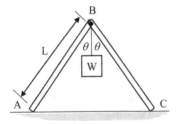

Fig. R9.3

R9.4 A beam carries a force $F = 400$ lb and is pinned at A and supported by a short post at B (whose dimensions are greatly exaggerated in Fig. R9.4). If the coefficient of friction between the post and the beam and the post and the floor are both $\mu = 0.6$, determine the force P required to pull the post out. Neglect the weights of the beam and the post. The distances $a = 12$ ft and $b = 4$ ft.

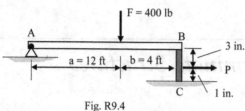

Fig. R9.4

R9.5 A bracket that is designed to hold a poster board of weight W on a wall is shown in Fig. R9.5. The friction coefficient between the cylinder and the board at A and between the bracket and the cylinder at B are the same and the friction coefficient between the back of the poster board and the bracket at A can be neglected. Determine the smallest value for the coefficient of friction so that the poster board will not slip, regardless of its weight (self-locking) if the angle $\theta = 30°$. The weight of the cylinder can be neglected. Note that the poster board can be easily removed by raising it and sliding it sideways.

Fig. R9.5

10

Statically Indeterminate Structures

OBJECTIVES

- To define the flexibility of deformable bodies and the compatibility of deformations.
- To solve statically indeterminate problems for deformable bodies with small deformations with a force-based method involving equilibrium, compatibility, and flexibility.
- To obtain the displacements of a deformable body with small deformations.

The problems we have solved in previous chapters have all been *statically determinate* problems where the number of equations of equilibrium is equal to the number of unknown forces or moments. If this is not the case and we have fewer equations than unknowns, the problem is said to be *statically indeterminate*. In Figure 10.1(a), for example, the beam has four supports. From equilibrium we can solve at most for the reaction forces at two supports since we have one force equation (in the vertical direction) and a moment equation and so the problem is statically indeterminate. If we cannot find the forces at the supports, then we cannot also determine the internal shear force and bending moment in the beam from equilibrium. In Figure 10.1(b), the truss is likewise over-constrained at the supports, so it is statically indeterminate (count the number of unknowns and equilibrium equations for the truss and convince yourself that this is true). In Figure 10.1(c), the external reactions of the truss can be determined from the equations of equilibrium for the entire truss but we cannot solve for all the internal forces in the members (there are 13 unknowns total (three reaction forces and ten member forces) and only 12 equilibrium equations at the six pins). The frame in Figure 10.1(d) is also over-constrained externally, so we cannot solve for all the external reactions and subsequently for the internal axial force, shear force, and bending moment.

DOI: 10.1201/9781003372592-10

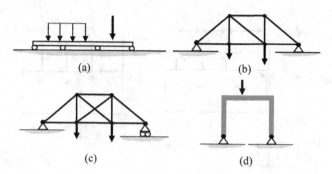

FIGURE 10.1
(a) A beam with multiple supports. (b) A truss that is over-constrained at the supports. (c) A truss that has too many internal members. (d) A frame that is over-constrained at the supports.

The structures in Figure 10.1 are simple types of statically indeterminate structures. More complex structures may be even more over-constrained or have additional internal components that render them highly statically indeterminate. So how do we solve such problems for the forces and moments? To fully answer that question, we normally need to go beyond the topics of statics and deal with such problems in the context of later courses such as strength of materials and the theory of elasticity. However, a recently developed force-based approach to solving statically indeterminate problems allows us to consider them in a statics course, which is what we will do in this chapter. Our discussion will use MATLAB matrix algebra extensively, so it is advisable to first read Appendix A.

10.1 Solving a Statically Indeterminate Problem

To illustrate what is needed in addition to the equations of equilibrium to solve statically indeterminate problems, we will use an example that is typical of the types of problems we have already been analyzing. Consider, for example, the hanging sign shown in Figure 10.2(a). The sign is pinned to a post at A and is supported by three wires.

Free Body Diagram
The free body diagram of the sign is shown in Figure 10.2(b). The center of gravity is at G, where the weight of the sign is $W = 50$ lb. There are five unknown forces and only three equations of equilibrium, so the problem is statically indeterminate.

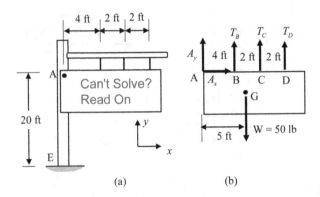

FIGURE 10.2
(a) A hanging sign problem which is statically indeterminate. (b) The free body diagram of the sign.

Equations of Equilibrium

From equilibrium in the x-direction, the horizontal force $A_x = 0$ so we will not include it in any further discussion. Summing forces in the y-direction and moments about point A, we have the two equilibrium equations:

$$\sum F_y = 0 \quad A_y + T_B + T_C + T_D - 50 = 0$$
$$\sum M_{zA} = 0 \quad 4T_B + 6T_C + 8T_D - (5)(50) = 0 \tag{10.1}$$

These are two equilibrium equations for four unknowns. Even if we only examine the moment equation where the reaction force A_y is absent, we have one equation for the three unknown tensions. Thus, the problem is statically indeterminate to the second degree, i.e., we need two more equations for the forces in order to obtain a solution. Examining a free body of the post (Figure 10.3) does not change the static indeterminancy since there are three additional equations of equilibrium but also three additional reactions at the fixed support E. Thus, we will continue by examining only the equilibrium equations for the free body diagram of the sign. We will find it useful to write the equilibrium equations in matrix-vector form. If we use only the moment equation in Eq. (10.1), we will write that equation as $[E_1]\{F_1\} = \{P_1\}$, which here is:

$$[4 \quad 6 \quad 8] \begin{Bmatrix} T_B \\ T_C \\ T_D \end{Bmatrix} = \{250\} \tag{10.2}$$

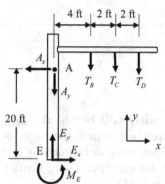

FIGURE 10.3
Free body diagram of the support post.

Compatibility Equations

Equation (10.2) is one equation for three unknowns. To solve for the forces, we need to provide additional information about those forces. To obtain that information, we will remove the assumption that all the components of the system are rigid. In particular, we will assume the supporting wires can deform and that the force in a wire, T, is proportional to its deformation (stretch or elongation), Δ (Figure 10.4), i.e., $T = k\Delta$, where k is called the *elastic spring constant* (it is also called the spring *stiffness*). Thus, each wire acts like a linear elastic spring, although in most cases, it will be a very stiff spring. For a wire of cross-sectional area A and length L and made from a material whose Young's modulus is E (where E is a material property that you will learn more about in a course on strength of materials), the spring constant is given by $k = AE/L$. For a steel wire, for example, $E = 30 \times 10^6\ \text{lb/in}^2$ so if the cross-sectional area was 0.01 in.^{2} and the wire was 24 inches long, the spring constant would be $k = 12,500\ \text{lb/in}$. Such a very large constant would mean that under applied forces such as the weight of the hanging sign, the deformations (elongations) will be very small.

If we assume the other components of the system such as the post and the sign are rigid, then while the wires stretch, the rigid sign can at most rotate through a very small angle about the pin at A so that the deformations of the wires are not independent. They must satisfy relations called *compatibility equations*. In this case, we have the situation shown in Figure 10.5, where if the small angle of rotation of the rigid sign

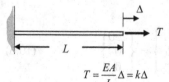

$$T = \frac{EA}{L}\Delta = k\Delta$$

FIGURE 10.4
A wire that deforms in a linear elastic manner, where the force in the wire is proportional to its deformation (stretch).

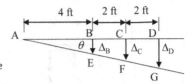

FIGURE 10.5
The deformations of the wires (deformations are greatly exaggerated).

about A is θ, then the deformations of the wires will be $(\Delta_B, \Delta_C, \Delta_D)$, respectively, assuming the wires are unstretched when the sign is horizontal. It can be seen from Figure 10.5 that these deformations are related through similar triangles, so we have, for example, using similar triangle pairs ABE and ACF and pairs ACF and ADG:

$$\frac{\Delta_B}{4} = \frac{\Delta_C}{6}, \quad \frac{\Delta_C}{6} = \frac{\Delta_D}{8} \tag{10.3}$$

or, equivalently

$$\Delta_B - \frac{2}{3}\Delta_C = 0$$
$$\Delta_C - \frac{3}{4}\Delta_D = 0 \tag{10.4}$$

which are two compatibility equations for the deformations.

Equilibrium and Compatibility
If we let $\Delta_B = (1/k_B)T_B$, $\Delta_C = (1/k_C)T_C$, and $\Delta_D = (1/k_D)T_D$, where $(1/k_B, 1/k_C, 1/k_D)$ are called the *flexibilities* of the wires, then our two compatibility equations become two additional equations for the forces given by

$$T_B/k_B - \frac{2}{3}(T_C/k_C) = 0$$
$$T_C/k_C - \frac{3}{4}(T_D/k_D) = 0 \tag{10.5}$$

and we can append these equations to our one equilibrium equation to obtain a set of "system" equations we can solve for the forces. In matrix-vector form, these are

$$\begin{bmatrix} 4 & 6 & 8 \\ 1/k_B & -2/3k_C & 0 \\ 0 & 1/k_C & -3/4k_D \end{bmatrix} \begin{Bmatrix} T_B \\ T_C \\ T_D \end{Bmatrix} = \begin{Bmatrix} 250 \\ 0 \\ 0 \end{Bmatrix} \tag{10.6}$$

Now, let's solve this system symbolically in MATLAB, letting all the flexibilities of the wires have the same value $1/k$:

```
syms Tb Tc Td k                          % symbolic variable
Esys = [4  6  8; 1/k −2/(3*k) 0; 0 1/k −3/(4*k)];   % matrix of system equations
Psys = [250; 0; 0];                      % column vector of system equations
T = Esys\Psys;                           % solve for tensions
double(T)                                % tensions Tb, Tc, Td(in lb)
ans = 8.6207
     12.9310
     17.2414
Ay = 50 − sum(T);                        % determine Ay from force
                                         % equation in Eq. (10.1)
double(Ay)                               % reaction force Ay (in lb)
ans = 11.2069
```

We see that the solution is independent of k, which makes sense since a common flexibility can be eliminated from the compatibility equations, as seen in Eq. (10.6), before we solve for the forces. We also determined the reaction force at A once the tensions were found.

10.2 A Matrix-Vector Force-Based Method

The solution we have described is the common way in which statically indeterminate problems of this type are obtained. However, this approach is not easily extended to more complex problems. If we followed the same approach and allowed, for example, the bars of the trusses shown in Figure 10.1(b) and (c) to deform, we would have to examine the geometry in detail at all the pins to determine the relations between those deformations. One way to proceed is to introduce deformation–displacement relations and use them to formulate and solve the problem directly in terms of the displacements. In this case, the compatibility equations are automatically satisfied but the forces can only be obtained from the displacements as secondary quantities. We will say more about that approach shortly. In this section, we will describe a force-based method that can instead be used to solve directly for the forces in a manner that is a natural extension of how we have learned to solve statically determinate problems. This method is best described by placing all our equations in matrix-vector form, which we will now do.

We had already placed the moment equilibrium equation of Eq. (10.1) in matrix-vector form:

$$[E_1]\{F_1\} = \{P_1\} \quad \rightarrow \quad [4 \ \ 6 \ \ 8]\begin{Bmatrix} T_B \\ T_C \\ T_D \end{Bmatrix} = \{250\} \qquad (10.7)$$

The compatibility equations we will write as:

$$[S_1]\{\Delta_1\} = \{0\} \quad \rightarrow \quad \begin{bmatrix} 1 & -2/3 & 0 \\ 0 & 1 & -3/4 \end{bmatrix}\begin{Bmatrix} \Delta_B \\ \Delta_C \\ \Delta_D \end{Bmatrix} = \begin{Bmatrix} 0 \\ 0 \end{Bmatrix} \qquad (10.8)$$

The deformation–force relations can be written in terms of a flexibility matrix $[G_1]$:

$$\{\Delta_1\} = [G_1]\{F_1\} \quad \rightarrow \quad \begin{Bmatrix} \Delta_B \\ \Delta_C \\ \Delta_D \end{Bmatrix} = \begin{bmatrix} 1/k_B & 0 & 0 \\ 0 & 1/k_C & 0 \\ 0 & 0 & 1/k_D \end{bmatrix}\begin{Bmatrix} T_B \\ T_C \\ T_D \end{Bmatrix} \qquad (10.9)$$

Combining equilibrium and the compatibility equations (written in terms of the forces), then we can form a solvable system:

$$\begin{bmatrix} [E_1] \\ - \\ [S_1][G_1] \end{bmatrix}\{F_1\} = \begin{Bmatrix} \{P_1\} \\ - \\ \{0\} \end{Bmatrix} \quad \rightarrow \quad \begin{bmatrix} 4 & 6 & 8 \\ 1/k_B & -2/3k_C & 0 \\ 0 & 1/k_C & -3/4k_D \end{bmatrix}\begin{Bmatrix} T_B \\ T_C \\ T_D \end{Bmatrix} = \begin{Bmatrix} 250 \\ 0 \\ 0 \end{Bmatrix}$$

$$(10.10)$$

Equation (10.10) shows that to solve statically indeterminate problems, we need three things: (1) the equilibrium equations for the deformable elements of the system, (2) the flexibility matrix that describes the relationships between the internal forces of the non-rigid members and their deformations, and (3) the compatibility matrix that describes the compatibility equations for the deformations.

A key part of obtaining this solution, therefore, was our ability to define the compatibility matrix $[S_1]$. Note that we can write the relationship between the deformations and the small rotational displacement $U = \theta$ of the hanging sign (see Figure 10.5) as

$$\begin{Bmatrix} \Delta_B \\ \Delta_C \\ \Delta_D \end{Bmatrix} = \begin{bmatrix} 4 \\ 6 \\ 8 \end{bmatrix}\{U\} \qquad (10.11)$$

But we recognize the matrix on the right-hand side of Eq. (10.10) as just the transpose of the equilibrium matrix $[E_1]$. Thus, we can write the this deformation–displacement relationship as

$$\{\Delta_1\} = [E_1]^T \{U\} \tag{10.12}$$

Since the three deformations are related to a single angular displacement, it is apparent that there must be two relations between those three deformations. Those two relations are the compatibility equations. When we are trying to find the compatibility equations, we are trying to find a compatibility matrix $[S_1]$ which satisfies

$$[S_1]\{\Delta_1\} = [S_1][E_1]^T \{U\} = \{0\} \tag{10.13}$$

and which can be satisfied for all possible rotational displacements if

$$[S_1][E_1]^T = [0] \tag{10.14}$$

Taking the transpose of Eq. (10.14), we also must have

$$[E_1][S_1]^T = [0]^T \tag{10.15}$$

For our sign problem, we specifically have

$$[4 \quad 6 \quad 8] \begin{bmatrix} 1 & 0 \\ -2/3 & 1 \\ 0 & -3/4 \end{bmatrix} = [0 \quad 0] \tag{10.16}$$

which is obviously satisfied. Thus, we see that the two column vectors in the transpose of the compatibility matrix, $[S_1]^T$, are

$$\{F_{h1}\} = \left\{ \begin{array}{c} 1 \\ -2/3 \\ 0 \end{array} \right\}, \quad \{F_{h2}\} = \left\{ \begin{array}{c} 0 \\ 1 \\ -3/4 \end{array} \right\} \tag{10.17}$$

These vectors are both solutions to the homogeneous equilibrium equation, i.e.,

$$[E_1]\{F_{h1}\} = \{0\}, \quad [E_1]\{F_{h2}\} = \{0\} \tag{10.18}$$

Note that we can multiply these homogeneous solutions by any constant or take any combinations of these solutions and still have a homogeneous solution. Similarly, we could multiply either of the two compatibility equations by constants or take different combinations and still have legitimate compatibility equations which will lead to the same solution.

Thus, to obtain the compatibility matrix $[S_1]$, we determine the homogeneous solutions of the equilibrium equations, Eq. (10.15), and then simply take the transpose of the matrix of those solutions. The compatibility equations, therefore, can be found directly from the equilibrium equations. Multiplying the compatibility matrix by the flexibility matrix gives us the additional equations needed to solve for the forces. Although this approach of obtaining the compatibility equation has only been shown for this problem, one can show it is true for all systems that have deformable elements that behave in a linear elastic manner.

We can obtain non-trivial solutions of the homogeneous equilibrium equations since for statically indeterminate problems the equilibrium equations form a underdetermined system of equations, i.e., there are fewer equations than there are unknown forces. In our present case, for example, the equilibrium matrix is a 1×3 matrix. For statically determinate problems, the equilibrium matrix is a square matrix and the only homogeneous solution of the system of equilibrium equations (where the known forces/moments are all zero) is the zero vector.

Obtaining homogeneous solutions to a system of underdetermined linear equations is no more difficult than finding solutions to a statically determinate system of equations. MATLAB has a built-in function named null that does all the work of obtaining homogeneous solutions for us. Appendix A gives more details of the solution process. Here is how we can obtain the homogeneous solutions of the equilibrium equations for the hanging sign in MATLAB:

```
E1 = [4 6 8];           % the equilibrium equation matrix (vector)
S1 = null(E1, 'r'). '   % determine homogeneous solutions
                        % and take the transpose
S1 = -1.5000  1.0000  0
     -2.0000  0       1.0000
```

The compatibility equations generated by this compatibility matrix look different from the ones we have been using. Explicitly, we have:

$$-\frac{3}{2}\Delta_B + \Delta_C = 0$$
$$-2\Delta_B + \Delta_D = 0$$

(10.19)

However, the first compatibility equation is just our previous first equation obtained from similar triangles ABE and ACF multiplied by –3/2. The second equation is obtained by adding 2/3 times our previous second compatibility equation to the first equation and then multiplying that result by −2. We see that the second equation in Eq. (10.19) also can be obtained from similar triangles ABE and ADG in Figure 10.5. Thus, the compatibility equations are not unique, as stated earlier. When the equilibrium matrix is given in terms of whole numbers or fractions, the 'r' option in the null function also gives a result in terms of whole numbers or fractions. If we use the null function without this option, a different solution procedure is used by the null function and the row vectors in the compatibility matrix are normalized:

```
S1 = null(E1). '
S1 = -0.5571    0.7737   -0.3017
     -0.7428   -0.3017    0.5977
```

but this is still a legitimate compatibility matrix. You can verify this by computing $[E_1][S_1]^T$ in MATLAB:

```
E1*S1. '
ans = 1.0e - 14 *
-0.1776    0
```

which gives a set of values approximately equal to zero. [Note that if we use a symbolic equilibrium matrix in the null function, then we cannot use the 'r' option.] Now, let us solve the hanging sign problem in MATLAB by using the null function and again assume all the flexibilites are the same:

```
syms k                           % symbolic spring constant
E1 = [4 6 8];                    % equilibrium matrix
P1 = 250;                        % known moment in the equilibrium equation
S1 = null(E1).';                 % compatibility matrix
G1 = [1/k 0 0; 0 1/k 0; 0 0 1/k]; % flexibility matrix
Esys = [E1; S1*G1];              % form system of equations
Psys = [P1; 0; 0];
T = Esys\Psys;                   % solve for the tensions
double(T)                        % numerical values for Tb, Tc, Td(in lb)
ans = 8.6207
     12.9310
     17.2414
```

which obviously gives the same solution as before.

Displacements
Once we have found the forces in a problem, if we want to find the displacements there are several ways to proceed. The approach we will show here is perhaps the simplest to describe. First, we write the deformation–displacement relations in terms of the forces rather than the deformations:

$$\{F_1\} = [G_1]^{-1}\{\Delta_1\} = [G_1]^{-1}[E_1]^T\{U\} \qquad (10.20)$$

Placing Eq. (10.20) into the equilibrium equation $[E_1]\{F_1\} = \{P_1\}$, we obtain the equilibrium equations written in terms of the displacements as

$$[E_1][G_1]^{-1}[E_1]^T\{U\} = \{P_1\} \qquad (10.21)$$

where $[G_1]^{-1}$ is the inverse of the flexibility matrix. These equations are called *Navier's Equations*. The matrix $[K] = [E_1][G_1]^{-1}[E_1]^T$ is called a *stiffness matrix* since it contains the stiffnesses of the wires, k, rather than the flexibilities $1/k$. We can then solve for the displacements (which in this case is the single angle of rotation). In MATLAB, we have

```
K = E1*inv(G1)*E1 .';   % stiffness matrix
U = K\P1                 % solve for the displacement (angular rotation)
U = 125/(58*k)           % angular displacement in radians
```

While the tensions in the wires did not depend on their common flexibility, $1/k$, we see the displacement U (which is the small rotation θ of the top of the sign) does depend on it. Once we have solved for the displacement, we can also find the tensions since they are related to the displacement by $T_B = 4k\theta$, $T_C = 6k\theta$, $T_D = 8k\theta$. [Note: here the spring constant k is measured in lb/ft since we specified the dimensions in feet in obtaining the moment equation, but the tensions are independent of k so they are independent of the units used to measure k. This is not true, however, for the displacement.] In MATLAB, we find

```
Tb = double(4*k*U)  % the tensions (in lb) obtained from the displacement
Tb = 8.6207
Tc = double(6*k*U)
Tc = 12.9310
Td = double(8*k*U)
Td = 17.2414
```

Note that we can also get the tensions more directly in matrix-vector terms from $\{T\} = [G_1]^{-1}[E_1]^T\{U\}$:

T = inv(G1)*E1 .'*U;

double(T)

ans = 8.6207

    12.9310

    17.2414

Using displacements is in fact an alternate method for completely solving our statically indeterminate problem. In this approach, we write our equilibrium equations in terms of a stiffness matrix and the displacements and solve that system of equations. We then determine the forces in the deformable elements from their relations to the displacements. Note that we never have to use compatibility in a displacement-based approach since we can define the deformations from the displacements so that they will inherently be compatible with those displacements. In a force-based approach, we augment the equilibrium equations with the compatibility equations so that we can solve that augmented system directly for the forces. Since statics is primarily interested in obtaining forces/moments, we will use the force-based method in this book and only use the stiffness matrix to find the displacements when they are needed.

Since the stiffness matrix is formed from the equilibrium matrices and the flexibility matrix, it is important that the units in those matrices be consistent. Thus, if the force equations are given, for example, in units of lb and the moment equations in units of ft-lb, the flexibilities must be given in units of ft/lb. However, if the flexibilities are given in units of in/lb (which is often the case), the force equations must be given in units of lb and moment equations given in units of in-lb. *Inconsistent units do not affect the solution for the forces since we can always multiply all the flexibilities by an arbitrary constant and this change still gives an appropriate compatibility equation but consistent units must be used in obtaining displacements.*

In Chapter 15 of the addendum to this book (see www.eng-statics. org), an alternative way to determine displacements in a force-based approach will be given that does not require us to determine a stiffness matrix. In that chapter, we will also discuss both displacement-based and force-based finite element methods. The finite element method (using displacements) has been developed for over 60 years and is now the method of choice for engineers to solve complex statics and dynamics problems ranging from buildings to bridges and airframes and even biological structures such as the heart. The force-based finite element method is more recent, but since it solves directly for forces/ moments, it is a generalization of the force-based approach considered in this chapter.

Before we leave this problem, we want to show how we can use both equations of equilibrium in Eq. (10.1). This will demonstrate how we can handle statically indeterminate problems with a mixture of unknown forces for deformable elements as well as reactions associated with non-deformable connections.

Equilibrium
In this case, we can write the equilibrium equations as

$$[E_2]\{F_2\} = \{P_2\} \quad \rightarrow \quad \begin{bmatrix} 1 & 1 & 1 & 1 \\ 0 & 4 & 6 & 8 \end{bmatrix} \begin{Bmatrix} A_y \\ T_B \\ T_C \\ T_D \end{Bmatrix} = \begin{Bmatrix} 50 \\ 250 \end{Bmatrix} \qquad (10.22)$$

Compatibility
If we use the null function to solve for the compatibility matrix, we have

E2 = [1 1 1 1; 0 4 6 8];
S2 = null(E2,'r').'
S2 =

    0.5000 −1.5000 1.0000   0
    1.0000 −2.0000   0   1.0000

so in this case the compatibility equations are:

$$[S_2]\{\Delta_2\} = \{0\} \quad \rightarrow \quad \begin{bmatrix} 1/2 & -3/2 & 1 & 0 \\ 1 & -2 & 0 & 1 \end{bmatrix} \begin{Bmatrix} \Delta_A \\ \Delta_B \\ \Delta_C \\ \Delta_D \end{Bmatrix} = \begin{Bmatrix} 0 \\ 0 \end{Bmatrix} \qquad (10.23)$$

where Δ_A is a deformation associated with the reaction force A_y. But there is no deformation (or displacement) at the fixed pin A, so we must set $\Delta_A = 0$. If we do that and examine the resulting compatibility equations in MATLAB, we find

```
syms Db Dc Dd        % the deformations of the wires at B, C, and D
S2*[0; Db; Dc; Dd]   % the compatibility equations
ans = Dc - (3*Db)/2
       Dd - 2*Db
```

where Db = Δ_B, Dc = Δ_C, Dd = Δ_D. Comparing with Eq. (10.19), we see that these are identical to the compatibility equations obtained previously.

Flexibility
To ensure that the deformation Δ_A is zero, we must set the flexibility associated with the reaction force A_y equal to zero, so we have

$$\{\Delta_2\} = [G_2]\{F_2\} \rightarrow \begin{Bmatrix} \Delta_A \\ \Delta_B \\ \Delta_C \\ \Delta_D \end{Bmatrix} = \begin{bmatrix} 0 & 0 & 0 & 0 \\ 0 & 1/k_B & 0 & 0 \\ 0 & 0 & 1/k_C & 0 \\ 0 & 0 & 0 & 1/k_D \end{bmatrix} \begin{Bmatrix} A_y \\ T_B \\ T_C \\ T_D \end{Bmatrix} \quad (10.24)$$

Equilibrium and Compatibility
With all these results, we then can augment the equilibrium equations with the compatibility equations as

$$\begin{bmatrix} [E_2] \\ - \\ [S_2][G_2] \end{bmatrix} \{F_2\} = \begin{Bmatrix} \{P_2\} \\ - \\ \{0\} \end{Bmatrix} \rightarrow \begin{bmatrix} 1 & 1 & 1 & 1 \\ 0 & 4 & 6 & 8 \\ 0 & -3/2k_B & 1/k_C & 0 \\ 0 & -2/k_B & 0 & 1/k_D \end{bmatrix} \begin{Bmatrix} A_y \\ T_B \\ T_C \\ T_D \end{Bmatrix} = \begin{Bmatrix} 50 \\ 250 \\ 0 \\ 0 \end{Bmatrix}$$

$$(10.25)$$

and solve. Here are the calculations in MATLAB, where again all the flexibilities are the same:

```
syms k                 % symbolic spring constant
E2 = [1 1 1 1;  0 4 6 8];   % both equilibrium equations
P2 = [50;  250];       % known force/moment
S2 = null(E2).';       % compatibility matrix  (not using 'r' option)
G2 = [0 0 0 0;  0 1/k 0 0;  0 0 1/k 0;  0 0 0 1/k];   % flexibility matrix (note zero)
Esys = [E2; S2*G2];    % form system matrix and augmented known load
Psys = [P2; 0; 0];
double(Esys\Psys)      % solve and convert to numerical values
ans = 11.2069          % solution for Ay, Tb, Tc, Td (in lb, independent of k)
       8.6207
      12.9310
      17.2414
```

In this case, we obtain the unknown reaction force as well as the tensions.

Displacements
If we want to compute the displacements, as before, by placing the compatibility equation into the equilibrium equation, we will formally find

$$[E_2][G_2]^{-1}[E_2]^T \{u\} = \{P_2\} \tag{10.26}$$

But we cannot solve this equation when we include zero flexibilities. You can verify this by computing the determinant of the flexibility matrix:

```
det(G2)
ans = 0
```

and if the determinant of a matrix is zero, we cannot compute the inverse of that matrix. However, there is a simple work-around to this problem. If we replace the zero flexibilities with very small but non-zero entries, then we can compute the inverse and obtain the displacements. To see this, let's make $k = 10^3$ lb/ft so the wire flexibilities are $1/k = 10^{-3}$ but let's make the flexibility associated with A_y be a much smaller value of 10^{-9}.

```
G2(1,1) = 10^-9;
G2(2,2) = 10^-3;
G2(3,3) = 10^-3;
G2(4,4) = 10^-3.
```

Now we can obtain the solution (approximately):

```
u2 = E2*inv(G2)*E2.'\P2;
double(u2)     % displacements (in ft)
ans =
      0.0000
      0.0022
```

The first displacement is the displacement associated with the reaction at A. It is not identically zero but is very small. If we change the format in MATLAB so we can see more decimal places, we have

```
format long
double(u2)
ans = 0.000000010206894
      0.002155170674792
```

The first displacement is zero to seven decimal places. The second displacement is the small rotation θ (in radians) which we found before was $\theta = 125/(58k)$. In this case, MATLAB gives

```
125/(58*10^3)
ans = 0.002155172413793
```

in agreement with our above result to eight decimal places.

10.3 Steps in the Force-Based Method

The methods we used to solve the example of the last section is a general approach for solving statically indeterminate problems. In more complex cases, the flexibilities will be described by multiple constants, but the general procedures remain the same. In this text, we will only examine problems where the flexibility matrix is a diagonal matrix composed of spring-like flexibilities. This means that we will not examine statically indeterminate problems like the beam problem of Figure 10.1(a) or the frame problem of Figure 10.1(d) since the flexibilities in those cases are more complex and would require us to deal with topics that are beyond a statics course. We will be able to analyze the trusses of Figure 10.1(b) and (c) and other problems that are extensions of the commonly solved statically determinate problems seen in statics. [Note: while we cannot use a diagonal flexibility matrix when a beam or frame is deformable, we will be able to examine such problems where the beam or frame is still assumed to be rigid but where the supports are allowed to be deformable in a spring-like manner. We will see such examples in the problems at the end of this chapter.]

We can outline the force-based method as a five-step procedure:

A Force-Based Method for Solving Statically Indeterminate Problems

1. Use free body diagrams to write the equilibrium equations in terms of the unknown forces of the deformable elements and reactions, $\{F\}$, the known applied forces, $\{P\}$, and an equilibrium matrix, $[E]$:

$$[E]\{F\} = \{P\}$$

2. Solve the homogeneous equilibrium equations $[E][S]^T = 0$ for the compatibility matrix $[S]^T$. The transpose of this solution, $[S]$, satisfies the compatibility equations $[S]\{\Delta\} = 0$ for a vector, $\{\Delta\}$, of deformations.

3. Write the flexibility matrix $[G]$ in terms of the flexibilities associated with the deformable elements and zero flexibilities for the reactions at rigid supports, where $\{\Delta\} = [G]\{F\}$.

4. Write the compatibility equations in terms of the forces $[S][G]\{F\} = \{0\}$

5. Append the compatibility equations to the equilibrium equations and solve, which gives us both the forces associated with the deformable elements and the reactions:

$$\begin{bmatrix} [E] \\ - \\ [S][G] \end{bmatrix} \{F\} = \begin{Bmatrix} \{P\} \\ - \\ \{0\} \end{Bmatrix} \rightarrow \{F\}$$

The force-based method solves for the forces without dealing directly with the displacements but, as shown previously, if we can form a non-singular stiffness matrix $[K] = [E][G']^{-1}[E]^T$ that uses a modified flexibility matrix $[G']$, we can also solve for the displacements of the deformable members (as well as zero displacements for the supports at which the reaction forces act) that are contained in the vector $\{U\}$:

$$[K]\{U\} = \{P\} \rightarrow \{U\}$$

In the force-based method described here, the forces/moments in the non-rigid parts of the structure are assumed to be proportional to the deformations (this is called linear elastic behavior) and the deformations in the problem are assumed to be very small so that we can treat the geometries in the free body diagrams when the structure is loaded and unloaded as essentially the same. The deformations of the structures considered in many applications are indeed small and many structures do deform in a linear elastic manner when the deformations are small so that these requirements can often be satisfied in practice.

10.4 Examples of the Force-Based Method

The steps outlined in the last section are best illustrated by solving some statics problem for common structures that are statically indeterminate. One classic example is called *Navier's table problem*.

Claude-Louis Navier (1785–1836) was a French mathematician/ engineer who was noted in his day for his expertise in bridge-building but today his name is usually associated with the Navier–Stokes equations for the motion of fluids and Navier's equations for the displacements of an elastically deformable solid.

As seen in Figure 10.6, the problem is for a four-legged table where a downward force is applied to the tabletop, whose own weight will be neglected. We want to determine the forces in the legs. We previously solved a similar problem in Chapter 4 for a three-legged table, which was statically determinate. However, the table with four legs is statically indeterminate since it represents a parallel force system where there are three equations of equilibrium (one force equation and two moment equations) but four unknown forces in the table legs. If we assume the legs are deformable and tabletop is rigid, it is a problem that is not difficult to solve with our force-based method, as shown in the next example.

Example 10.1

In Navier's table problem, assume the legs deform while the tabletop is rigid. Each leg carries a uniaxial compressive force, F, and experiences a

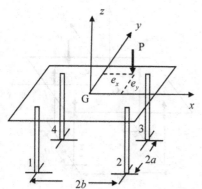

FIGURE 10.6
A table with four legs supporting an eccentric load.

shortening deformation, Δ, just like a spring. If the legs behave elastically, like the hanging sign problem, the force and deformation are related through a spring constant, $k = EA/L$, where E is Young's modulus, A is the cross-sectional area of the leg, and L is the leg length, i.e., $F = k\Delta$. Thus, the legs act as extremely stiff springs where the deformation of the legs under any normal loads will be very small. In this problem, we will take all the leg spring constants to be identical and determine the forces in the legs.

Free Body Diagram
Figure 10.7 shows the free body diagram of the table. We will neglect the weight of the table so the only known force is the force P which is applied eccentrically to the tabletop.

Equations of Equilibrium
The equations of equilibrium from Figure 10.7 are

$$\sum F_z = 0 \quad F_1 + F_2 + F_3 + F_4 - P = 0$$
$$\sum M_{Gx} = 0 \quad F_3a + F_4a - F_2a - F_1a - e_yP = 0 \qquad (10.27)$$
$$\sum M_{Gy} = 0 \quad F_1b + F_4b - F_2b - F_3b + e_xP = 0$$

which can be put in matrix-vector form as

$$\begin{bmatrix} 1 & 1 & 1 & 1 \\ -a & -a & a & a \\ b & -b & -b & b \end{bmatrix} \begin{Bmatrix} F_1 \\ F_2 \\ F_3 \\ F_4 \end{Bmatrix} = \begin{Bmatrix} P \\ Pe_y \\ -Pe_x \end{Bmatrix} \qquad (10.28)$$

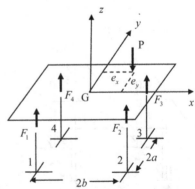

FIGURE 10.7
A free body diagram of the table in Figure 10.6.

Compatibility Equations
We have three equations and four unknowns, so there will only be a single compatibility equation and the compatibility matrix will be a vector. If we obtain the compatibility matrix in MATLAB, we find a simple result:

```
syms a b                              % declare symbolic values
E = [ 1 1 1 1;  −a −a a a;  b −b −b b];  % equilibrium matrix
S = null(E).'                         % determine compatibility matrix
S = [−1, 1, −1, 1]
```

We can explain this compatibility matrix and the compatibility equation physically. In Figure 10.7, we showed the forces as pushing on the tabletop so those are compressive forces. The corresponding deformations would, therefore, represent shortening of the legs, not elongations. Since the table is assumed to be rigid, it can only translate and rotate. If we let w be the translation of the table in the negative z-direction and let (θ_x, θ_y) represent small rotations about the x- and y-axes, respectively, then these rotations will produce additional displacements of the legs (Figure 10.8). The total shortening of the legs will be the displacements in the negative z-direction at the tabletop since the displacements at the floor are assumed to be zero. The displacements of the tabletop are shown in Figure 10.8 for all the legs (1,2,3,4). From this figure, we find

$$\Delta_1 = w - b\theta_y + a\theta_x$$
$$\Delta_2 = w + b\theta_y + a\theta_x$$
$$\Delta_3 = w + b\theta_y - a\theta_x \qquad (10.29)$$
$$\Delta_4 = w - b\theta_y - a\theta_x$$

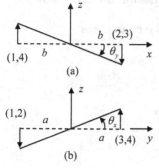

FIGURE 10.8
Small displacements at the legs due to the rotation of the tabletop, which is assumed to be rigid (a) looking down the negative y-axis, and (b) looking down the x-axis. A shortening of a leg will be a displacement arrow acting down, while an elongation (a negative shortening) will be a displacement arrow acting up.

In Eq. (10.29), the four deformations are related to three displacements so there must be one relationship between the deformations. This is the compatibility equation. From Eq. (10.29), it easy to see that the compatibility equation is

$$-\Delta_1 + \Delta_2 - \Delta_3 + \Delta_4 = 0 \tag{10.30}$$

which when written in matrix-vector form as $[S]\{\Delta\} = 0$ gives the same compatibility matrix found with MATLAB. However, we did not have to go through the deformation–displacement relations to get this result, since we found it directly from the equilibrium equations in the force-based method.

Since we have assumed all the legs have the same flexibility, the flexibility matrix is given by

$$[G] = \begin{bmatrix} 1/k & 0 & 0 & 0 \\ 0 & 1/k & 0 & 0 \\ 0 & 0 & 1/k & 0 \\ 0 & 0 & 0 & 1/k \end{bmatrix} \tag{10.31}$$

with no zero flexibilities since all the forces here are forces associated with deformations.

Equilibrium and Compatibility
We now can combine the equilibrium and compatibility equations and solve this problem symbolically in MATLAB:

```
syms a b P ex ey k            % define symbolically the variables of the problem
E = [1 1 1 1; -a -a a a;  b -b -b b];   % give the equilibrium matrix
Pv = [P; P*ey; -P*ex];        % give the external "force" vector
S = null(E).';                % find the compatibility matrix (here a vector)
flex = [ 1/k 1/k 1/k 1/k];    % specify the flexibility matrix
G = diag(flex);
Esys = [E; S*G];              % form up the system equations and solve
Psys = [Pv; 0];
F = Esys\Psys                 % the four forces in the table legs

F = -(P*a*ex - P*a*b + P*b*ey)/(4*a*b)
    (P*a*b + P*a*ex - P*b*ey)/(4*a*b)
    (P*a*b + P*a*ex + P*b*ey)/(4*a*b)
    (P*a*b - P*a*ex + P*b*ey)/(4*a*b)
```

We can write this solution as

$$F_1 = \frac{P}{4}\left(1 - \frac{e_x}{b} - \frac{e_y}{a}\right)$$

$$F_2 = \frac{P}{4}\left(1 + \frac{e_x}{b} - \frac{e_y}{a}\right)$$

$$F_3 = \frac{P}{4}\left(1 + \frac{e_x}{b} + \frac{e_y}{a}\right) \tag{10.32}$$

$$F_4 = \frac{P}{4}\left(1 - \frac{e_x}{b} + \frac{e_y}{a}\right)$$

This result is independent of k since all the flexibilities were the same. If the force P is applied at the center of the plate ($e_x = e_y = 0$), then we find $F_1 = F_2 = F_3 = F_4 = P/4$, as expected.

Displacements

Now let us find the displacements. First, note that the deformation–displacement relations from Eq. (10.29) give

$$\begin{Bmatrix} \Delta_1 \\ \Delta_2 \\ \Delta_3 \\ \Delta_4 \end{Bmatrix} = \begin{bmatrix} 1 & -a & b \\ 1 & -a & -b \\ 1 & a & -b \\ 1 & a & b \end{bmatrix} \begin{Bmatrix} -u_z \\ -\theta_x \\ -\theta_y \end{Bmatrix} \tag{10.33}$$

which is of the form $\{\Delta\} = [E]^T\{U\}$, where $u_z = -w$, and $\{U\}$ is given by

$$\{U\} = \begin{Bmatrix} -u_z \\ -\theta_x \\ -\theta_y \end{Bmatrix} \tag{10.34}$$

so these are generalized displacements where there is one displacement and two rotations. The stiffness matrix can be formed directly from the inverse of the flexibility matrix since there are no zero flexibilities here. We can then solve for the displacements. The steps are:

```
K = E*inv(G)*E.';      % form up stiffness matrix
U = K\Pv                % solve for displacements. Note: using Pv, not
                        % the extended vector Psys

U = P/(4*k)
    (P*ey)/(4*a^2*k)
   -(P*ex)/(4*b^2*k)
```

which, changing the signs for the rotations (the sign on the translation does not need to be changed since w is already defined to be the translation in the negative z-direction), gives the generalized displacements as

$$w = \frac{P}{4k}$$

$$\theta_x = -\frac{Pe_y}{4a^2k} \tag{10.35}$$

$$\theta_y = \frac{Pe_x}{4b^2k}$$

Unlike the forces, the displacements do depend on the constant k. In the force-based method, the forces can all be found without reference to the displacements. Forces are often the quantities of most interest, so that feature of the force-based method is important. If displacements are needed, they can be found through the stiffness matrix formed from the equilibrium equations and the flexibilities for deformable elements and very small (but not zero) flexibilities for reactions, if they are part of the problem, as illustrated previously.

Before we leave this problem, we need to say a few words about why we found the negative of the generalized displacements here. In setting up equilibrium equations, we apply those equations to all the forces, including forces of deformable elements, $\{F\}$, other forces, such as reactions, $\{R\}$ (if they are present), and the known external applied loads, $\{P_e\}$. If we obtain all those equilibrium equations in the positive (x, y, z) directions, then they are of the form:

$$\begin{matrix} \sum\limits_{x,y,z} \mathbf{F} = 0 \\ \sum\limits_{x,y,z} \mathbf{M} = 0 \end{matrix} \quad \rightarrow \quad [E]\left\{\begin{matrix} F \\ R \end{matrix}\right\}_{x,y,z} + \left\{\begin{matrix} P_e \\ M_e \end{matrix}\right\}_{x,y,z} = 0$$

[Note: $\{M_e\}$ are often the moments generated by the applied loads P_e. In some problems, the moments could also include known couples]. If we place all the applied load terms on the right-hand side of these equations, then we have

$$[E]\left\{\begin{matrix} F \\ R \end{matrix}\right\}_{x,y,z} = -\left\{\begin{matrix} P_e \\ M_e \end{matrix}\right\}_{x,y,z}$$

These equations in the matrix-vector form are

$$[E]\{F\} = \{P\} \quad \text{where} \quad \{P\} = -\begin{Bmatrix} P_e \\ M_e \end{Bmatrix}_{x,y,z}$$

In solving for the displacements, we wrote the equilibrium equations in terms of the displacements and the stiffness matrix:

$$[K]\{U\} = \{P\}$$

In this equation, if the displacements and applied forces both act in the positive (x, y, z) directions, then this equation is

$$[K]\{U\}_{x,y,z} = \begin{Bmatrix} P_e \\ M_e \end{Bmatrix}_{x,y,z}$$

Thus, if we place the negative of the applied external forces/moments acting along the positive (x, y, z) directions in the vector $\{P\}$ (which is the same applied force/moment vector obtained when those applied forces/moments are placed on the right-hand side of the equilibrium equations), we will find instead:

$$[K]\{-U\}_{x,y,z} = -\begin{Bmatrix} P_e \\ M_e \end{Bmatrix}_{x,y,z}$$

These are precisely the equations we solved for the displacements, so the solution we obtained was for the negative of the displacements acting in (x, y, z) directions. In the hanging sign problem solved previously, this behavior was also present since the angular displacement θ that was solved for was the negative of the rotation in the z-direction, although we did not note that fact explicitly in that solution.

All concurrent cable-supported systems are statically indeterminate if there are more than two cables in two-dimensional problems and more than three cables in three-dimensional cases. Let's consider a two-dimensional example.

Example 10.2

A 150 lb weight W is suspended by three cables as shown in the two-dimensional problem of Figure 10.9. Assume all cables behave in a linear elastic manner and their angles are $\theta_1 = 30°$, $\theta_2 = 10°$, and $\theta_3 = 45°$. Let AE be the same for all the cables, where the spring constant $k = AE/L$, but their lengths L are different, as can be found from Figure 10.9. Determine the tensions in the cables.

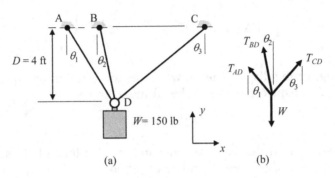

(a)   (b)

FIGURE 10.9
(a) A weight supported by three cables. (b) The free body diagram.

Free Body Diagram
The free body diagram is shown in Figure 10.9. There are three unknown tensions and only two equations of equilibrium, so the problem is statically indeterminate.

Equations of Equilibrium
Summing forces in the x- and y-directions, we find

$$\begin{aligned} \sum F_x = 0 & \quad -T_{AD}\sin(30°) - T_{BD}\sin(10°) + T_{CD}\sin(45°) = 0 \\ \sum F_y = 0 & \quad T_{AD}\cos(30°) + T_{BD}\cos(10°) + T_{CD}\cos(45°) - 150 = 0 \end{aligned} \tag{10.36}$$

Placing this in matrix-vector form, we have

$$[E]\{F\} = \{P\} \;\; \rightarrow \;\; \begin{bmatrix} -\sin(30°) & -\sin(10°) & \sin(45°) \\ \cos(30°) & \cos(10°) & \cos(45°) \end{bmatrix} \begin{Bmatrix} T_{AD} \\ T_{BD} \\ T_{CD} \end{Bmatrix} = \begin{Bmatrix} 0 \\ 150 \end{Bmatrix}$$

$$\tag{10.37}$$

Setting the problem up symbolically in MATLAB, we have

```
syms k D AE
E = [-sind(30) - sind(10) sind(45); cosd(30) cosd(10) cosd(45)];
P = [0; 150];
```

Compatibility Equations
We now need to find the compatibility matrix $[S]$ (a vector in this case) that satisfies the homogeneous equilibrium equation $[E][S]^T = 0$. Using the null function in MATLAB and taking the transpose, we have

```
S = null(E).';
```

Since AE is the same for all of the cables and the vertical distance D is also the same, we can obtain the lengths in terms of D and write the flexibilities in terms of a common factor D/AE which can be factored out of the flexibility matrix, leaving terms that only depend on the L/D ratios for the cables:

$$[G] = \frac{D}{AE} \begin{bmatrix} L_{AD}/D & 0 & 0 \\ 0 & L_{BD}/D & 0 \\ 0 & 0 & L_{CD}/D \end{bmatrix} \rightarrow \frac{D}{AE} \begin{bmatrix} 1/\cos(30°) & 0 & 0 \\ 0 & 1/\cos(10°) & 0 \\ 0 & 0 & 1/\cos(45°) \end{bmatrix}$$

which is generated in MATLAB as

G = (D/AE)*[1/cosd(30) 0 0; 0 1/cosd(10) 0; 0 0 1/cosd(45)];

Equilibrium and Compatibility
We can combine the equilibrium equations with the compatibility equations and solve the resulting system of equations:

Esys = [E; S*G];   % set up the system of equations to solve
Psys = [P; 0];
F = Esys\Psys;   % solve for the tensions
Fn = double(F)   % the numerical values (in lb) for Tad, Tbd, and Tcd
Fn = 48.4377
     72.3660
     52.0220

These results are independent of D/AE since it is a common factor in the compatibility equation. We will not determine the displacements (which are the displacements (u_x, u_y) at point D for this problem) but they are easily generated from a stiffness matrix, as discussed earlier, if D/AE is given.

Now, consider a more challenging truss problem where we will have a mixture of deformable truss elements and fixed supports that produce unknown internal forces of deformation and reactions.

Example 10.3

Consider the truss shown in Figure 10.10(a). There are six members in the truss, so there are six internal forces in those members as well as three external reaction forces acting on the truss at pins A and D. There are a total of nine unknown forces but only eight equations of equilibrium at

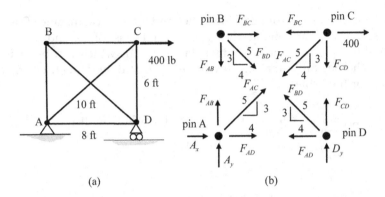

(a)   (b)

FIGURE 10.10
(a) A statically indeterminate truss. Members BD and AC are not connected where they cross. (b) Free body diagrams for the forces at the pins.

the four pins. Thus, the problem is statically indeterminate. In this case, we can solve for all the external reactions by using three equilibrium equations for the entire structure but that still leaves only five independent equilibrium equations to solve for the six internal forces, so the problem cannot be solved with equilibrium alone. Determine the forces in all the members as well as the reaction forces at the supports and determine the displacements at the pins.

Free Body Diagrams
Let us use the method of pins and solve for all nine forces with our force-based method. The forces at the four pins are shown in the free body diagrams in Figure 10.10(b).

Equations of Equilibrium
The eight equilibrium equations for the nine unknown forces at pins A, B, C, and D are

$$
\begin{aligned}
{}_A\textstyle\sum F_x &= 0 & A_x + 4F_{AC}/5 + F_{AD} &= 0 \\
{}_A\textstyle\sum F_y &= 0 & A_y + 3F_{AC}/5 + F_{AB} &= 0 \\
{}_B\textstyle\sum F_x &= 0 & F_{BC} + 4F_{BD}/5 &= 0 \\
{}_B\textstyle\sum F_y &= 0 & -F_{AB} - 3F_{BD}/5 &= 0 \\
{}_C\textstyle\sum F_x &= 0 & -F_{BC} - 4F_{AC}/5 + 400 &= 0 \\
{}_C\textstyle\sum F_y &= 0 & -F_{CD} - 3F_{AC}/5 &= 0 \\
{}_D\textstyle\sum F_x &= 0 & -F_{AD} - 4F_{BD}/5 &= 0 \\
{}_D\textstyle\sum F_y &= 0 & D_y + F_{CD} + 3F_{BD}/5 &= 0
\end{aligned}
\tag{10.38}
$$

Thus, the equations of equilibrium in matrix-vector form are:

$$\begin{bmatrix} 1 & 0 & 1 & 4/5 & 0 & 0 & 0 & 0 & 0 \\ 0 & 1 & 0 & 3/5 & 1 & 0 & 0 & 0 & 0 \\ 0 & 0 & 0 & 0 & 0 & 4/5 & 1 & 0 & 0 \\ 0 & 0 & 0 & 0 & -1 & -3/5 & 0 & 0 & 0 \\ 0 & 0 & 0 & -4/5 & 0 & 0 & -1 & 0 & 0 \\ 0 & 0 & 0 & -3/5 & 0 & 0 & 0 & -1 & 0 \\ 0 & 0 & -1 & 0 & 0 & -4/5 & 0 & 0 & 0 \\ 0 & 0 & 0 & 0 & 0 & 3/5 & 0 & 1 & 1 \end{bmatrix} \begin{Bmatrix} A_x \\ A_y \\ F_{AD} \\ F_{AC} \\ F_{AB} \\ F_{BD} \\ F_{BC} \\ F_{CD} \\ D_y \end{Bmatrix} = \begin{Bmatrix} 0 \\ 0 \\ 0 \\ 0 \\ -400 \\ 0 \\ 0 \\ 0 \end{Bmatrix} \qquad (10.39)$$

The flexibility matrix is:

$$[G] = \frac{L}{AE} \begin{bmatrix} 0 & 0 & 0 & 0 & 0 & 0 & 0 & 0 & 0 \\ 0 & 0 & 0 & 0 & 0 & 0 & 0 & 0 & 0 \\ 0 & 0 & 4/5 & 0 & 0 & 0 & 0 & 0 & 0 \\ 0 & 0 & 0 & 1 & 0 & 0 & 0 & 0 & 0 \\ 0 & 0 & 0 & 0 & 3/5 & 0 & 0 & 0 & 0 \\ 0 & 0 & 0 & 0 & 0 & 1 & 0 & 0 & 0 \\ 0 & 0 & 0 & 0 & 0 & 0 & 4/5 & 0 & 0 \\ 0 & 0 & 0 & 0 & 0 & 0 & 0 & 3/5 & 0 \\ 0 & 0 & 0 & 0 & 0 & 0 & 0 & 0 & 0 \end{bmatrix} \qquad (10.40)$$

since the truss is composed of straight two-force members that can only be in tension or compression, just like a spring. Like the legs of Navier's table problem, the flexibility of a straight truss member is given by $1/k = L/AE$ where A is the cross-sectional area of a bar, E is Young's modulus, a material property, and L is the bar length. Here, the bars are assumed to have the same cross-sectional area, A, and Young's modulus, E, but their lengths are obviously different, so these differences must be accounted for. The length of the diagonal members is taken to be L (which in this case is 10 ft), so the other member lengths are either $3L/5$ or $4L/5$, as seen in Eq. (10.40). There is a common flexibility factor of L/AE in the flexibility matrix in Eq. (10.40) that can be canceled out in the compatibility equation and so it can be omitted. The flexibilities associated with the reaction forces (A_x, A_y, D_y) are set to be zero in $[G]$ because they act at the rigid pin and at a rigid roller support. We now can implement the steps in the force-based method in MATLAB. We will illustrate the solution steps using symbolic expressions as well as a solution that is purely numeric. First, consider the symbolic solution

Begin by defining the symbolic variables and placing them in a symbolic vector, T:

```
syms Ax Ay Fad Fac Fab Fbd Fbc Fcd Dy
T = [Ax Ay Fad Fac Fab Fbd Fbc Fcd Dy];
```

Then define all the flexibilities for the symbolic forces contained in the vector T and generate the diagonal flexibility matrix [G]. The common *L/AE* flexibility will be omitted so that the flexibility matrix will be a purely numeric matrix:

```
flex = [0 0 4/5 1 3/5 1 4/5 3/5 0];
G = diag(flex);
```

Here, we placed all the flexibilities in a MATLAB vector flex and then used the MATLAB function diag to construct a matrix with those flexibilities along the main diagonal and with zeros for all other elements. We used this same procedure previously for the table problem. For problems with many forces, this is much easier than constructing the flexibility matrix element by element.

Next, define the equilibrium equations symbolically, as done previously in earlier chapters, but omit the symbolic == 0 in these equations, so we only entering the left-hand side of the equilibrium equations (you can leave the == 0 in the equations but it simply adds more typing):

```
Eq(1) = Ax + 4*Fac/5 + Fad;
Eq(2) = Ay + 3*Fac/5 + Fab;
Eq(3) = Fbc + 4*Fbd/5;
Eq(4) = -Fab - 3*Fbd/5;
Eq(5) = -Fbc - 4*Fac/5 + 400;
Eq(6) = -Fcd - 3*Fac/5;
Eq(7) = -Fad - 4*Fbd/5;
Eq(8) = Dy + Fcd + 3*Fbd/5;
```

To get the equilibrium matrix, we could extract it by hand from these equations, as we did to get Eq. (10.39), and then enter that result in MATLAB but that approach becomes progressively more burdensome as the number of equilibrium equations increases. There is a MATLAB function equationsToMatrix that does the work for us. This function is called in MATLAB by:

```
[E, P] = equationsToMatrix( Eq, T);
```

where Eq is a row vector containing the symbolic equations, and T is a row vector of unknowns, just as present in our current setup. The output of the function is E, the equilibrium matrix of coefficients, and the known right-hand side column vector of the equilibrium equations, P, where we have [E]{T} = {P}.

Compatibility Equations
We need the coefficient matrix for the equilibrium equations, so we can use it to determine the compatibility matrix. The result of using the function equationsToMatrix is shown here (compare with Eq. (10.39)):

```
[E, P] = equationsToMatrix(Eq, T);
E
E = [1, 0,  1,  4/5,  0,   0, 0, 0, 0]
    [0, 1,  0,  3/5,  1,   0, 0, 0, 0]
    [0, 0,  0,   0,   0, 4/5, 1, 0, 0]
    [0, 0,  0,   0,  -1, -3/5, 0, 0, 0]
    [0, 0,  0, -4/5,  0,   0, -1, 0, 0]
    [0, 0,  0, -3/5,  0,   0, 0, -1, 0]
    [0, 0, -1,   0,   0, -4/5, 0, 0, 0]
    [0, 0,  0,   0,   0, 3/5, 0, 1, 1]
```

P.'

```
ans =
   0 0 0 0 -400 0 0 0
```

Then, we compute the compatibility matrix:

```
S = null(E).';
```

Equilibrium and Compatibility
Next, append the compatibility equations to the equilibrium equation matrix:

```
Esys = [E; S*G];
```

and append the compatibility equation right-hand side zero to the row P vector, and solve the system of equations:

```
Psys = [P; 0];
F = Esys\Psys;
```

324   *Engineering Statics with MATLAB®*

This is a symbolic solution, which when converted into numerical values gives:

```
double(F)              % unknown forces (in lb)
ans = -400.0000        % Ax = -400 lb
      -300.0000        % Ay = -300 lb
       140.7407        % Fad = 140.7 lb T
       324.0741        % Fac = 324.1 lb T
       105.5556        % Fab = 105.6 lb T
      -175.9259        % Fbd = 175.9 lb C
       140.7407        % Fbc = 140.7 lb T
      -194.4444        % Fcd = 194.4 lb C
       300.0000        % Dy = 300 lb
```

To obtain the same solution in purely numerical terms (i.e., without the use of symbolic algebra), we need to enter by hand all the components of the equilibrium matrix from the equations in Eq. (10.39):

```
flex = [0 0 4/5 1 3/5 1 4/5 3/5 0];   % the flexibilities, omitting L/EA factor
G = diag(flex);                        % form up the flexibility matrix
E = [1 0 1 4/5 0 0  0 0 0;             % enter the equilibrium matrix
0 1 0 3/5 1 0 0 0 0;
0 0 0 0 0 4/5 1 0 0;
0 0 0 0 -1 -3/5 0 0 0;
0 0 0 - 4/5 0 0 -1 0 0;
0 0 0 -3/5 0 0 0 -1 0;
0 0 -1 0 0 -4/5 0 0 0;
0 0 0 0 0 3/5 0 1 1];
```

In computing the compatibility matrix, the "rational" option is used by entering an additional argument 'r' to the null function. As indicated earlier, this gives a compatibility matrix similar to what we obtain with simple physical arguments when the elements of the equilibrium matrix are integers or simple fractions, as they are here (which you can verify by looking at S being calculated with and without the 'r' option to the null function). Omitting this option still gives a legitimate compatibility matrix. [Note: when using the null function with a symbolic equilibrium matrix, we cannot use the "r" option, as we have mentioned previously.]

```
S = null(E, 'r').';  % determine the compatibility matrix
```

Forming up the set of equations and solving them follows the same steps taken in the symbolic solution, except now everything is calculated numerically:

```
Esys = [E; S*G];              % combine, equilibrium, compatibility
P = [0; 0; 0; 0; −400; 0; 0; 0];   % original force vector
Psys = [P; 0];                % known force vector and a zero from compatibility
F = Esys\Psys                 % solve [Es]{F} = {Ps} for the forces (in lb)
F = −400.0000
    −300.0000
     140.7407
     324.0741
     105.5556
    −175.9259
     140.7407
    −194.4444
     300.0000
```

The solution is shown in Figure 10.11. The external reactions are seen to be those that can be found by solving the equilibrium equations for the entire truss, as they are in equilibrium themselves with the applied force.

Displacements

To find the displacements at the pins, we must place the L/EA factor back into the flexibility matrix and modify the flexibility matrix so that it contains small but not zero flexibilities. Then, we can compute the stiffness matrix and solve for the displacements. In MATLAB we have, letting $L = 120$ in., $A = 3$ in^2, and $E = 10^7$ lb/in^2:

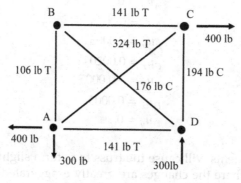

FIGURE 10.11
The forces acting in the truss problem of Figure 10.10(a).

```
A = 3;                    % area (in^2)
Em = 10^7;                % Young's modulus (lb/in^2)
L = 120;                  % length (in.)
G = (L/(A*Em))*G;         % flexibility matrix now includes L/AE factor
G(1,1) = 10^-9;           % replace zero fexibilities with small values
G(2,2) = 10^-9;
G(9,9) = 10^-9;
K = E*inv(G)*E.';         % compute the stiffness matrix and solve
U = K\P;
double(U)                 % displacements (in inches)
ans = -0.0000             % - ux at A
      -0.0000             % - uy at A
      -0.0015             % - ux at B
      -0.0003             % - uy at B
      -0.0020             % - ux at C
       0.0005             % - uy at C
      -0.0005             % - ux at D
       0.0000             % - uy at D
```

Here, the $\{P\}$ vector was [0; 0; 0; 0; −400; 0; 0; 0] but the 400 lb force was applied in the plus x-direction. Thus, the displacements that are obtained from the stiffness matrix equations $[K]\{U\} = \{P\}$ will be displacements in the negative x- and y-directions. We discussed the reason for this behavior in the table problem. Thus, the displacements at pins A, B, C, D are the negatives of the ones found in MATLAB and we have

$$_Au_x = 0$$
$$_Au_y = 0$$
$$_Bu_x = 0.0015$$
$$_Bu_y = 0.0003$$
$$_Cu_x = 0.0020$$
$$_Cu_y = -0.0005$$
$$_Du_x = 0.0005$$
$$_Du_y = 0$$

These displacements will cause the truss to deform slightly, as shown in Figure 10.12 (where the changes are greatly exaggerated).

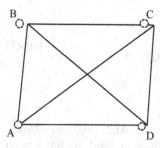

FIGURE 10.12
The deformed shape of the truss, with the displacements greatly exaggerated. The original pin positions are shown as dashed circles. Point A remains fixed and point D only displaces in the horizontal direction.

Now, consider a problem where a set of deformable elements (springs) are part of a system of otherwise rigid elements.

Example 10.4

Consider the X-frame shown in Figure 10.13(a), where the rigid pinned structure is supported by a pin and a roller and is held in the configuration shown by two springs that are assumed to be very stiff. Let the springs constants be identical, i.e., $k_1 = k_2 = k$. The lengths of CE and CD are both 5 ft. We want to (a) find the tensions in the springs and (b) determine all the forces present in the structure. We will also discuss the displacements in this problem.

Free Body Diagram
If we examine a free body diagram of the entire structure (Figure 10.13(b)), we see that we can solve for all the reaction forces.

Equations of Equilibrium
For the entire structure, summing forces and taking moments about point B, we have

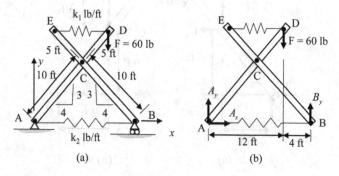

(a)   (b)

FIGURE 10.13
(a) An X-frame. (b) The free body diagram of the entire frame.

$$\sum F_x = 0 \quad A_x = 0$$
$$\sum F_y = 0 \quad A_y + B_y - 60 = 0$$
$$\sum M_{zB} = 0 \quad -16A_y + (4)(60) = 0 \qquad (10.41)$$
$$\rightarrow A_y = 15 \text{ lb}, B_y = 45 \text{ lb}$$

Free Body Diagram

If we next examine the free body diagram of member ACD (Figure 10.14(a)), we see that there are now four unknowns, since A_x and A_y have been obtained, but only three equations of equilibrium so the problem is statically indeterminate of order one. We can also see this from both free body diagrams of Figure 10.14 since there are seven unknowns and six equations of equilibrium.

Equations of Equilibrium

We could write all three equations of equilibrium for member ACD but if we take moments about C, we obtain an equation involving the unknown tensions (T_1, T_2) only:

$$\sum M_{zC} = 0 \quad 6T_2 + 3T_1 - (4)(60) - (8)(15) = 0$$
$$\rightarrow 6T_2 + 3T_1 = 360 \qquad (10.42)$$

which in matrix-vector form is

$$[E]\{F\} = \{P\} \rightarrow [3 \quad 6]\begin{Bmatrix} T_1 \\ T_2 \end{Bmatrix} = \{360\} \qquad (10.43)$$

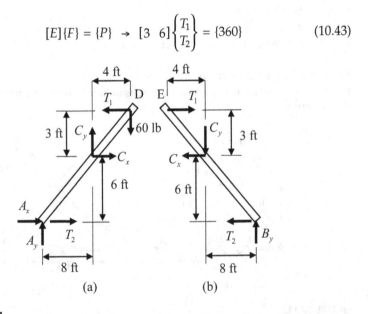

FIGURE 10.14
(a) Free body diagram of member ACD. (b) Free body diagram of member BCD.

Compatibility Equations

We can obtain the compatibility equation from the equilibrium equation of Eq. (10.42) but the problem is very simple, so we can also obtain it directly from the geometry. We see from Figure 10.15 that regardless of the angle of the X-frame, we have the lengths of the springs given by $L_1 = 10\sin\theta$ and $L_2 = 20\sin\theta$ so that for small changes of those lengths, we also have the deformations (elongations) of the springs given by $\Delta_1 = \Delta L_1 = 10\cos\theta\Delta\theta$ and $\Delta_2 = \Delta L_2 = 20\cos\theta\Delta\theta$, which imply the compatibility equation:

$$\Delta_2 - 2\Delta_1 = 0 \rightarrow [-2 \ \ 1]\begin{Bmatrix} \Delta_1 \\ \Delta_2 \end{Bmatrix} = 0 \tag{10.44}$$

so the compatibility matrix is $[S] = [-2 \ 1]$. Since:

$$\{\Delta\} = [G]\{F\} \rightarrow \begin{Bmatrix} \Delta_1 \\ \Delta_2 \end{Bmatrix} = \begin{bmatrix} 1/k & 0 \\ 0 & 1/k \end{bmatrix}\begin{Bmatrix} T_1 \\ T_2 \end{Bmatrix} \tag{10.45}$$

the compatibility equation is (canceling the common flexibility):

$$[S][G]\{F\} = \{0\} \rightarrow [-2 \ \ 1]\begin{Bmatrix} T_1 \\ T_2 \end{Bmatrix} = 0 \tag{10.46}$$

Equilibrium and Compatibility

It is easy to solve Eq. (10.43) and Eq. (10.46) for the tensions:

$$\begin{bmatrix} [E] \\ - \\ [S][G] \end{bmatrix}\{F\} = \begin{Bmatrix} \{P\} \\ - \\ \{0\} \end{Bmatrix} \rightarrow \begin{bmatrix} 3 & 6 \\ -2 & 1 \end{bmatrix}\begin{Bmatrix} T_1 \\ T_2 \end{Bmatrix} = \begin{Bmatrix} 360 \\ 0 \end{Bmatrix} \tag{10.47}$$

we find $T_1 = 24$ lb and $T_2 = 48$ lb.

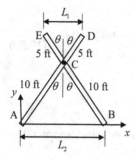

FIGURE 10.15
The lengths of the two springs in the X-frame in a general position.

Displacements

If we also want to find the displacement, we need to solve

$$[E][G]^{-1}[E]^T\{U\} = \{P\} \rightarrow [3 \;\; 6]\begin{bmatrix} k & 0 \\ 0 & k \end{bmatrix}\begin{bmatrix} 3 \\ 6 \end{bmatrix}\{U\} = \{360\} \qquad (10.48)$$

which gives $\{U\} = 8/k$. What, physically, is the displacement here? The deformations are $\Delta_1 = 10\cos\theta\Delta\theta = 6\Delta\theta$ and $\Delta_2 = 20\cos\theta\Delta\theta = 12\Delta\theta$ for the configuration of Figure 10.13(a) where $\cos\theta = 3/5$. Comparing this with the strain-displacement relation, we find

$$\{\Delta\} = [E]^T\{U\} \rightarrow \begin{Bmatrix} \Delta_1 \\ \Delta_2 \end{Bmatrix} = \begin{bmatrix} 3 \\ 6 \end{bmatrix}\{U\} = \begin{bmatrix} 3 \\ 6 \end{bmatrix}(2\Delta\theta) \qquad (10.49)$$

so $\{U\} = 2\Delta\theta$, and it is the change of the total angle $\Delta\alpha = 2\Delta\theta$ between the two arms of the X-frame (measured in radians). This is a rotation about the negative z-axis, i.e., $\Delta\alpha = 2\Delta\theta = -2\Delta\theta_z$, which shows a negative sign again, as explained earlier.

This solution was all very easy and understandable. Now, let's solve for all the forces in the frame at once.

Free Body Diagrams

We will use the two free body diagrams in Figure 10.14. There are seven unknowns in these diagrams and only six equations of equilibrium so again we must have one compatibility equation.

Equations of Equilibrium

Summing forces in the two diagrams and summing moments about C, we have

$$
\begin{aligned}
_{ACD}\Sigma F_x = 0 &\quad A_x + C_x + T_2 - T_1 = 0 \\
_{ACD}\Sigma F_y = 0 &\quad A_y + C_y - 60 = 0 \\
_{ACD}\Sigma M_{zC} = 0 &\quad 3T_1 - (4)(60) + 6A_x + 6T_2 - 8A_y = 0 \\
_{BCE}\Sigma F_x = 0 &\quad T_1 - C_x - T_2 = 0 \\
_{BCE}\Sigma F_y = 0 &\quad B_y - C_y = 0 \\
_{BCE}\Sigma M_{zC} = 0 &\quad 8B_y - 6T_2 - 3T_1 = 0
\end{aligned} \qquad (10.50)
$$

Let's place these equilibrium equations into MATLAB and extract the equilibrium matrix and the vector of known loads:

```
syms Ax Ay By Cx Cy T1 T2 k        % symbolic variables
Eq(1) = Ax + Cx + T2 - T1;          % equations of equilibrium
Eq(2) = Ay + Cy -60;
Eq(3) = 3*T1 -240 + 6*Ax + 6*T2 - 8*Ay ;
Eq(4) = T1 - Cx - T2;
Eq(5) = By - Cy;
Eq(6) = 8*By -6*T2 - 3*T1;
T = [T1 T2 Ax Ay Cx Cy By];         % vector of unknown forces
[E, P] = equationsToMatrix(Eq, T)   % extract the equilibrium matrix
                                    % and load vector


E = [-1,   1,   1,   0,   1,   0,   0]
    [0,    0,   0,   1,   0,   1,   0]
    [3,    6,   6,  -8,   0,   0,   0]
    [1,   -1,   0,   0,  -1,   0,   0]
    [0,    0,   0,   0,   0,  -1,   1]
    [-3,  -6,   0,   0,   0,   0,   8]
P.'
ans = 0 60 240 0 0 0
```

Compatibility Equation

Now, we can generate the compatibility matrix, the flexibility matrix, and determine the compatibility equation:

```
S = null(E).';                      % determine the compatibility matrix
flex = [ 1/k 1/k 0 0 0 0 0];         % form up the flexibility matrix with zero flexibilities

G = diag(flex)
G = [1/k, 0, 0, 0, 0, 0, 0]
    [0, 1/k, 0, 0, 0, 0, 0]
    [0, 0, 0, 0, 0, 0, 0]
    [0, 0, 0, 0, 0, 0, 0]
    [0, 0, 0, 0, 0, 0, 0]
    [0, 0, 0, 0, 0, 0, 0]
    [0, 0, 0, 0, 0, 0, 0]

S*G                                 % display the compatibility equation
                                    % in terms of the forces
ans = [2/(3*k), -1/(3*k), 0, 0, 0, 0, 0]
```

If we multiply this compatibility equation by $-3k$, we recover the same equation in terms of (T_1, T_2) as before.

Equilibrium and Compatibility

Now, we can combine the equilibrium and compatibility equations into a system of equations we can solve

```
Esys = [E; S*G];        % form up the system of equations and solve
Psys = [P; 0];
F = Esys\Psys;
F = 24      % T1 = 24 lb
    48      % T2 = 48 lb
     0      % Ax = 0
    15      % Ay = 15 lb
   -24      % Cx = -24 lb
    45      % Cy = 45 lb
    45      % By = 45 lb
```

The overall reactions on the X-frame are the same as before as well as the tensions in the springs. The reactions at C are also compatible with our free body diagrams. Unlike the reactions at A and B, while it is true that there are no deformations of the X-frame associated with the forces at C, there are non-zero displacements at that point. In fact, in a general configuration such as seen in Figure 10.15, the coordinates of C are $x_C = 10\sin\theta$ and $y_C = 10\cos\theta$ so as θ changes, so do these coordinates. These changes of coordinates represent the motion of point C due to the rigid body rotation of the arm ACD about the fixed point A of the X-frame. However, as we have seen, we can still generate a solution for all the forces with such rigid body displacements present since the force-based method does not require us to consider the displacements, only the deformations.

If we do solve for the displacements, they will satisfy zero displacements at A and have no y-component at B but they will not constrain the rigid body displacements at C. We will not show those results here as the discussion of displacements is rather involved and the angular displacement found previously using only the moment equation is all that we need to determine the displacements at all points of the X-frame.

10.5 Problems

P10.1 A rigid beam is supported by three linear springs at A, B, and C, as shown in Figure P10.1. The length $L = 8$ ft and the magnitude of the distributed load is $w = 50$ lb/ft. The spring constants are $k_A = 1000$ lb/in, $k_C = 2000$ lb/in, $k_B = 1500$ lb/in, and the springs are undeformed when the beam is horizontal. Determine the forces in the springs and the vertical displacement of the beam at A and its angular rotation.

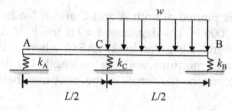

Fig. P10.1

P10.2 The three-bar truss in Figure P10.2 supports forces $P_1 = 350$ lb and $P_2 = 1350$ lb. The distances $a = 5$ ft and $b = 12$ ft. The spring constant of a truss member is given by $k = AE/L$ where A is the cross-sectional area, E is Young's modulus, and L is the length. Let AE be the same for all members. Determine the forces in the three members.

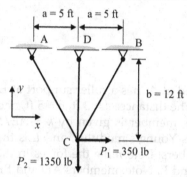

Fig. P10.2

P10.3 The truss of Figure P10.3 is pinned at both A and C and is loaded by the forces $F = 2000$ lb and $P = 5000$ lb. The distances $a = 8$ ft. and $b = 6$ ft. The spring constant of a truss member is given by $k = AE/L$ where A is the cross-sectional area, E is Young's modulus, and L is the length. Let AE be the same for all members. Determine the forces in the truss members and the reactions at A and C.

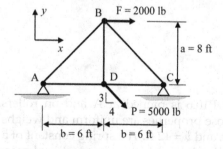

Fig. P10.3

P10.4 The truss of Figure P10.4 is pinned at both A and C and is loaded by the forces $F = 1000$ lb and $P = 3000$ lb. The distances $a = 9$ ft and $b = 6$ ft. The spring constant of a truss member is given by $k = AE/L$ where A is the cross-sectional area, E is Young's modulus, and L is the length. Let AE be the same for all members. Determine the forces in the truss members and the reactions at A and C.

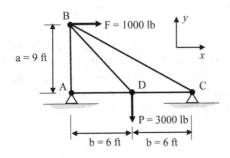

Fig. P10.4

P10.5 The truss of Figure P10.5 is pinned at A, has a roller support at E, and is loaded by the force $P = 5000$ lb. The distances $a = 3$ ft, $b = 5$ ft, and $c = 5$ ft. The spring constant of a truss member is given by $k = AE/L$ where A is the cross-sectional area, E is Young's modulus, and L is the length. Let AE be the same for all members. Determine the forces in the truss members and the reactions at A and E. Note: members AD and BE are not connected where they cross.

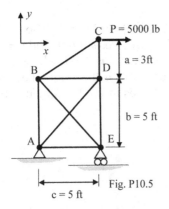

Fig. P10.5

P10.6 The truss shown in Figure P10.6 is pinned at A and on rollers at E. The truss supports a sign whose properties are uniform and weighs $W = 1600$ lb. The distances $a = 10$ ft and $b = 12$ ft. The spring constant of a truss member is given by $k = AE/L$ where A is the cross-sectional area,

L is the length, and E is Young's modulus. $A = 1\,\text{in}^2$ and $E = 10 \times 10^6\,\text{lb/in}^2$ for each member. Determine the forces in the truss members, the reactions at A and E, and the displacements at the pins. Note: members AD and BE are not connected where they cross.

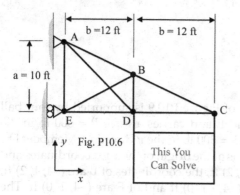

Fig. P10.6

This You
Can Solve

P10.7 The truss shown in Figure P10.7 is pinned at A and C and carries a load $P = 400$ N. The distances $a = 3$ m and $b = 4$ m. The spring constant of a truss member is given by $k = AE/L$ where A is the cross-sectional area, E is Young's modulus, and L is the length. Let AE be the same for all members. Determine the forces in the truss members and the reactions at A and C.

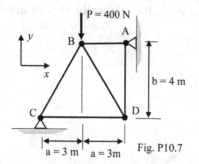

Fig. P10.7

P10.8 The truss shown in Figure P10.8 is pinned at A and on rollers at C and carries a load $P = 400$ N. The distances $a = 3$ m and $b = 4$ m. The spring constant of a truss member is given by $k = AE/L$ where A is the cross-sectional area, E is Young's modulus, and L is the length. Let AE be the same for all members. Determine the forces in the truss members and the reactions at A and C. Note: members AC and BD are not connected where they cross.

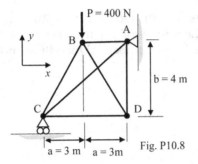

Fig. P10.8

P10.9 The four-bar spatial truss of Figure P10.9 is supported by four ball and socket joints at A, B, D, and E and carries a force $P = 600$ lb in the negative z-direction and a force $F = 500$ lb in the y-direction. Members DC and EC lie in the x-y plane. Pin C is at the origin of the x-y-z coordinates and the coordinates of A are $(-4, -1, 2)$ ft, the coordinates of B are $(-4, 1, 2)$ ft, while the coordinates of D are $(-4, -1, 0)$ ft and of E are $(-4, 1, 0)$ ft. The spring constant of a truss member is given by $k = AE/L$ where A is the cross-sectional area, E is Young's modulus, and L is the length. Let AE be the same for all members. (a) Determine the forces in the truss members and the reactions at A, B, D, and E. Note that if $F = 0$, we can use the symmetry of the problem to directly obtain the forces in the members from equilibrium at C, i.e., the problem is statically determinate. (b) Show that your solution to the statically indeterminate problem where $F = 0$ but where symmetry is not used agrees with the statically determinate solution.

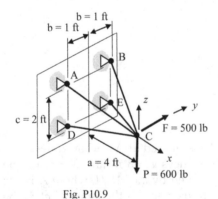

Fig. P10.9

10.5.1 Review Problems

These statically indeterminate problems are typically the types of problems that might be put in exams so they should be done by hand (i.e., with a calculator). The compatibility equations can be obtained by using geometry changes to directly relate the deformations to each other. You can check your answers by using MATLAB and the force-based method described in this chapter.

R10.1 A rigid beam is supported by a roller at C and two linear springs at A and B, as shown in Figure R10.1. The length L = 6 ft and the magnitude of the distributed load w = 25 lb/ft. The spring constants are k = 2000 lb/in and the springs are undeformed when the beam is horizontal. Determine the forces in the springs and at support C. Neglect the weight of the beam.

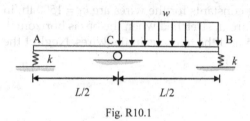

Fig. R10.1

R10.2 An L-shaped rigid bar is supported by a pin at B and linear springs at A and C. The length L = 4 ft and the force P = 120 lb. The spring constants are k = 2400 lb/in and the springs are undeformed when the assembly is in the position shown in Figure R10.2. Determine the forces in the springs and the reaction forces at B. Neglect the weight of the bar.

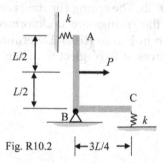

Fig. R10.2 |←3L/4→|

R10.3 Two linear springs are attached to each other through a movable rigid support at C which has a total force of P lb applied to it. The ends of the springs at A and B are fixed. The distance L = 12 in, the spring constants are k_1 = 2000 lb/in and k_2 = 4000 lb/in, and the force P = 75 lb. The springs are undeformed in the position shown in Figure R10.3. Determine the forces in the springs and the reaction forces at A and B.

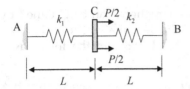

Fig. R10.3

R10.4 A rigid bar is supported by a pin at A and two flexible wires that are attached to the bar at B and C (see Figure R10.4). The distance $L = 10$ ft. The force $P = 150$ lb and the spring constants for the wires are $k_1 = 1500$ lb/in and $k_2 = 1000$ lb/in. The wires are unstretched when the bar is horizontal. Determine the forces at the pin and the tensions in the wires. Neglect the weight of the bar.

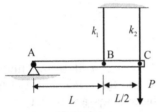

Fig. R10.4

R10.5 An L-shaped rigid bar is supported by a pin at B and linear springs at A. The length $L = 3$ ft and the force $P = 200$ lb. The spring constants are $k_1 = 2000$ lb/in and $k_2 = 1000$ lb/in, and the springs are undeformed when the assembly is in the position shown in Figure R10.5. Determine the forces in the springs and the reaction forces at B. Neglect the weight of the bar.

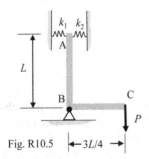

Fig. R10.5

11

Area Moments and Mass Moments of Inertia

OBJECTIVES

- To define area moments and mixed area moments.
- To compute area and mixed area moments by analytical integration and with MATLAB®.
- To obtain the parallel axis theorem and show its use in area moment problems.
- To define the radius of gyration.
- To define and use the composite body relationship for more complex geometries.
- To show how area moments and mixed area moments change under coordinate rotation and obtain the principal area moments and the principal directions.
- To demonstrate the use of matrices to solve area moment problems.
- To extend the relationships for area moments to mass moments of inertia.

Centroids are geometrical properties that appear frequently in engineering. Area moments also play an important role in problems such as bending of beams, where they represent "higher order" moments of cross-sectional areas than the expressions found for the centroids of areas. In this chapter, we will define area moments and mixed area moments and examine a range of topics including their calculation by integration and the use of the concept of composite bodies. We will see how area moments and mixed area moments change with translation and rotation of the coordinate axes and compute the axes about which the area moments have their largest and smallest values called the principal moments. We will use matrices and MATLAB to demonstrate effective

DOI: 10.1201/9781003372592-11

ways to determine principal moments and their directions. Finally, we will examine related concepts such as the mass moments of inertia.

11.1 Area Moments

In dealing with the concept of the centroid of an area, we introduced the first moments of the area, defined by quantities such as

$$\int_A x\, dA, \quad \int_A y\, dA$$

The location of the centroid of the area, (x_c, y_c) then was defined as these first moments divided by the total area, A:

$$x_c = \frac{\int_A x\, dA}{A}, \quad y_c = \frac{\int_A y\, dA}{A} \tag{11.1}$$

We can also define the second moments of an area in an x-y plane by the quantities

$$I_{xx} = \int_A y^2\, dA, \quad I_{yy} = \int_A x^2\, dA, \quad I_{xy} = \int_A xy\, dA \tag{11.2}$$

The second moments (I_{xx}, I_{yy}) are also sometimes written as (I_x, I_y). The area moment I_{xy} is called a mixed area moment. Area moments play an important role in defining the deformation and strength of beams, where the internal bending moment, $M(x)$, acting in a beam is related to the deflection of the beam in the y-direction, $v(x)$, through what is called the moment–curvature relationship:

$$M(x) = EI_{zz} \frac{d^2 v}{dx^2} \tag{11.3}$$

E is a material property called Young's modulus, and $I_{zz} = \int_A y^2\, dA$ is an area moment of the cross-sectional area of the beam in the y-z plane, where the z-axis is a centroidal axis of the cross section (see Figure 11.1).

For simple shapes such as the rectangular area shown in Figure 11.2(a), it is easy to calculate the area moments by integration. (Note the difference in the (x, y, z) coordinates from Figure 11.1. We will use the coordinates of Figure 11.2 from now on where the area lies in the x-y

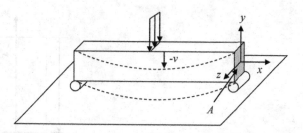

FIGURE 11.1
Bending of a beam (shown by the dashed lines, where the deflection is greatly exaggerated).

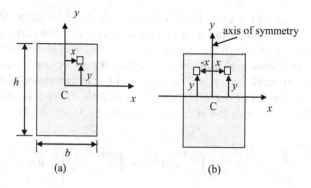

FIGURE 11.2
Geometry for determining an area moment about a set of centroidal axes. (b) The mixed area moment is zero if an area has an axis of symmetry about one of the coordinates because of canceling area terms.

plane.) For example, the area moment I_{xx} calculated for a set of centroidal x-y axes is

$$I_{xx} = \int_{x=-b/2}^{x=+b/2} \int_{y=-h/2}^{y=+h/2} y^2 \, dx \, dy = \int_{x=-b/2}^{x=+b/2} \left. \frac{y^3}{3} \right|_{y=-h/2}^{y=+h/2} dx \tag{11.4}$$

$$= \frac{bh^3}{12}$$

Similarly, just interchanging the roles of b and h, we have

$$I_{yy} = \frac{hb^3}{12} \tag{11.5}$$

It is not necessary to calculate I_{xy} because $I_{xy} = 0$. This is true because if either the x- or y-axes is an axis of symmetry for an area, the mixed area

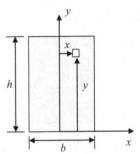

FIGURE 11.3
Geometry for determining the area moment about an x-axis
that is along the base of a rectangle.

moment will vanish. Figure 11.2(b) shows that there are area elements on
either side of the y-axis (an axis of symmetry) that have canceling terms
of $xydA$ and $- xydA$ so that the total area integral will also vanish. For the
rectangle, the x-axis is also an axis of symmetry so we could see the same
cancellation around that axis. The values of the area moments depend on
our choice of axes. For the rectangle, for example, if we take the x-axis to
be along the bottom of the rectangle (Figure 11.3), we find

$$I_{xx} = \int_{x=-b/2}^{x=+b/2} \int_{y=0}^{y=h} y^2 dx dy = \int_{x=-b/2}^{x=+b/2} \frac{y^3}{3}\Big|_{y=0}^{y=h} dx \tag{11.6}$$

$$= \frac{h^3}{3} \int_{x=-b/2}^{x=+b/2} dx = \frac{bh^3}{3}$$

Rectangles are important building blocks for obtaining the area moments
for other shapes so that it is good to memorize the centroidal and base
values given by Eq. (11.4) and Eq. (11.6). The values for other simple
shapes can often be found in tables. Usually, tables use centroidal axes
for the tabulated area moments. If we want to find area moments that are
with respect to a set of non-centroidal axes but where the axes are
parallel to a set of centroidal axes where the area moments are known,
then we do not have to obtain the area moments by integration. Instead,
we can use the *parallel axis theorem*. Consider, for example, the area
shown in Figure 11.4 where there are a set of centroidal (x', y') axes and a
parallel set of (x, y) axes. From the geometry, we have

$$I_{xx} = \int_A y^2 dA = \int_A (y_c + y')^2 dA$$

$$= y_c^2 \int_A dA + 2y_c \int_A y' dA + \int_A (y')^2 dA \tag{11.7}$$

$$= y_c^2 A + I_{x'x'}$$

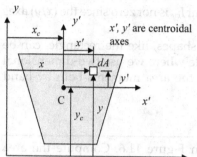

FIGURE 11.4
Geometry for deriving the parallel axis theorem.

In a similar fashion, we can obtain the parallel axis theorem relations for all the area moments:

$$I_{xx} = I_{x'x'} + Ay_c^2$$
$$I_{yy} = I_{y'y'} + Ax_c^2 \qquad (11.8)$$
$$I_{xy} = I_{x'y'} + Ax_c y_c$$

where it is important to remember that the primed coordinates in Eq. (11.8) must be centroidal. To illustrate the use of the parallel axis theorem, consider a set of (x, y) axes for a rectangle, as shown in Figure 11.5. Since we know all the area moments about the centroid, we have

$$I_{xx} = \frac{bh^3}{12} + (bh)\left(\frac{h}{2}\right)^2 = \frac{bh^3}{3}$$
$$I_{yy} = \frac{hb^3}{12} + (bh)\left(\frac{b}{2}\right)^2 = \frac{hb^3}{3} \qquad (11.9)$$
$$I_{xy} = 0 + (bh)\left(\frac{b}{2}\right)\left(\frac{h}{2}\right) = \frac{b^2h^2}{4}$$

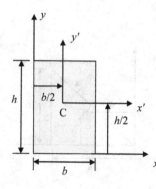

FIGURE 11.5
Geometry for determining the area moments of a rectangle about a set of non-centroidal axes using the parallel axis theorem.

where we see that the mixed area moment I_{xy} is not zero since the (x, y) axes here are not axes of symmetry.

The area moments for many simple shapes, like the rectangle, can be done by integration. Here is an example where we use two-dimensional integrals, strip integrals, and compute the area moments both by hand and with MATLAB.

Example 11.1

Consider the triangular area shown in Figure 11.6. Compute the area moments with respect to the x- and y-axes shown by (a) two-dimensional integrals and (b) one-dimensional strip integrals. Perform the integrations both by hand and with the use of MATLAB.

(a) Figure 11.6 shows the geometry as set up for performing two-dimensional integrations. Using that figure, we have for I_{xx}:

$$I_{xx} = \int_{x=0}^{x=b} \int_{y=0}^{y=hx/b} y^2 dx dy = \int_{x=0}^{x=b} \frac{y^3}{3}\Big|_{y=0}^{y=hx/b} dx$$

$$= \frac{h^3}{3b^3} \int_{x=0}^{x=b} x^3 dx = \frac{bh^3}{12}$$

(11.10a)

and for I_{yy}:

$$I_{yy} = \int_{x=0}^{x=b} \int_{y=0}^{y=hx/b} x^2 dx dy = \int_{x=0}^{x=b} x^2 y \Big|_{y=0}^{y=hx/b} dx$$

$$= \frac{h}{b} \int_{x=0}^{x=b} x^3 dx = \frac{b^3 h}{4}$$

(11.10b)

and finally, for I_{xy}:

$$I_{xy} = \int_{x=0}^{x=b} \int_{y=0}^{y=hx/b} xy dx dy = \int_{x=0}^{x=b} x\frac{y^2}{2}\Big|_{y=0}^{y=hx/b} dx$$

$$= \frac{h^2}{2b^2} \int_{x=0}^{x=b} x^3 dx = \frac{b^2 h^2}{8}$$

(11.10c)

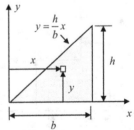

FIGURE 11.6
Geometry for obtaining the area moments of a triangular area by two-dimensional integrals.

These same calculations done symbolically in MATLAB are

```
syms b h x y
Ixx = int(int(y^2, y, [0, h*x/b]), x, [0, b])
Ixx = (b*h^3)/12
Iyy = int(int(x^2, x, [b*y/h, b]), y, [0, h])
Iyy = (b^3*h)/4
Ixy = int(int(y, y, [0, h*x/b])*x, x, [0, b])
Ixy = (b^2*h^2)/8
```

Examine the integral terms in the MATLAB expressions carefully to understand these integrations. You can do either the x- or y-integrations first, so both choices are shown. For the triangle, it makes little difference but in other cases, you should do the integration in the order that makes the integrals easiest to perform.

As in the case of calculating centroids if one uses strip area elements instead of two-dimensional elements, then one only needs to do single integrals. Consider first the use of a vertical strip as shown in Figure 11.7(a). The area of the strip is $dA_s = l_s dx$. We cannot use this area in the definition of I_{xx}, since by that definition:

$$I_{xx} = \int y^2 dA \tag{11.11}$$

and the strip area dA_s, unlike the small area dA in two-dimensions, has many y-values. Instead, we can use the parallel axis theorem for this strip element and write:

$$dI_{xx} = dI_{x'x'} + dA_s \left(\frac{l_s}{2}\right)^2$$

$$= \frac{1}{12} dx\, l_s^3 + dx \left(\frac{l_s^3}{4}\right) = \frac{1}{3} l_s^3 dx \tag{11.12}$$

(a) (b)

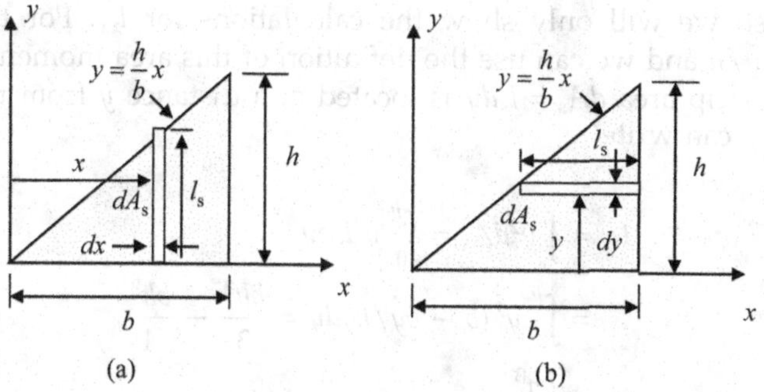

FIGURE 11.7
(a) The use of a vertical strip element, and (b) the use of a horizontal strip element.

and the integral can then be computed as

$$I_{xx} = \int dI_{xx} = \int_{x=0}^{x=b} \frac{1}{3}\left(\frac{hx}{b}\right)^3 dx$$

$$= \frac{h^3}{3b^3}\int_0^b x^3 dx = \frac{bh^3}{12}$$

(11.13)

For determining I_{yy}, we can use the definition of this area moment in conjunction with the strip element since the area of the strip, dA_s, is located at a definite distance x, as shown in Figure 11.7(a), so we can write

$$I_{yy} = \int x^2 dA_s = \int_0^b x^2 l_s dx$$

$$= \int_0^b x^2\left(\frac{hx}{b}\right)dx = \frac{hb^3}{4}$$

(11.14)

For the mixed area moment, we must again use the parallel axis theorem and write

$$dI_{xy} = dI_{x'y'} + dA_s x\frac{l_s}{2} = 0 + \frac{l_s^2 x dx}{2}$$

(11.15)

giving

$$I_{xy} = \int dI_{xy} = \frac{1}{2}\int_0^b l_s^2 x dx$$

$$= \frac{h^2}{2b^2}\int_0^b x^3 dx = \frac{h^2 b^2}{8}$$

(11.16)

The use of a horizontal strip (Figure 11.7(b)) follows in a very similar manner so we will only show the calculations for I_{xx}. For this strip, $l_s = b - by/h$ and we can use the definition of this area moment directly since the strip area $dA_s = l_s dy$ is located at a distance y from the x-axis and so we can write

$$I_{xx} = \int y^2 dA_s = \int_0^h y^2 l_s dy$$

$$= \int_0^h y^2 (b - by/h)dy = \frac{bh^3}{3} - \frac{bh^3}{4}$$

(11.17)

$$= \frac{bh^3}{12}$$

All these integrals can also be done in MATLAB. Here, we will only show the calculations for I_{xx} using a vertical and a horizontal strip and the calculations for I_{xy} using a vertical strip:

```
syms b h x y
Ixx = int(h^3*x^3/(3*b^3), x, [0 b])     % use of a vertical strip
Ixx = (b*h^3)/12
Ixx = int(y^2*(b − b*y/h), y, [0 h])     % use of a horizontal strip
Ixx = (b*h^3)/12
Ixy = int(h^2*x^3/(2*b^2), x, [0 b])     % use of a vertical strip
Ixy = (b^2*h^2)/8
```

11.2 Polar Area Moment and the Radius of Gyration

11.2.1 Polar Area Moment

An area moment that uses the distance $r^2 = x^2 + y^2$ in its definition is called the *polar area moment*, J_0, where

$$J_0 = \int_A r^2 dA = \int_A x^2 dA + \int_A y^2 dA = I_{yy} + I_{xx} \qquad (11.18)$$

Just as the area moments appear in problems involving the bending of beams, the polar area moment appears in the torsion of rods, as you will see in a later course on strength of materials.

11.2.2 Radius of Gyration

Area moments have the dimensions of an area times the square of a length so that one can define a length parameter associated with an area moment. This length is called a *radius of gyration*. If, for example, one calculates an area moment, I_b, about a b-b axis for an area A, as shown in Figure 11.8, then the radius of gyration, k_b, from that axis is defined as

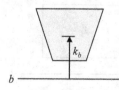

FIGURE 11.8
The geometry for relating an area moment about an axis b-b for an area A to a distance, k_b, called the radius of gyration.

$$k_b = \sqrt{\frac{I_b}{A}} \qquad (11.19)$$

Thus, instead of tabulating the area and area moment for a given cross-sectional area, one can tabulate the area and the radius of gyration. The properties of cross-sectional areas for beams, for example, are often tabulated in this manner. If one uses the polar area moment J_0 in Eq. (11.19) instead of I_b, we can define a radius of gyration associated with that area moment in the same fashion.

11.3 Composite Areas

We saw in the case of centroids in Chapter 7, we could decompose an area into simpler parts and calculate the centroid for such composite areas directly. In the case of area moments, we can often use the parallel axis theorem to deal with composite areas. Consider, for example, the composite area of Figure 11.9. If we want to calculate the area amount about the *b-b* axis, then we have

$$I_b = I_{b1} + I_{b2} + I_{b3} \qquad (11.20)$$

and the area moments for the individual pieces (I_{b1}, I_{b2}, I_{b3}) can be determined with the parallel axis theorem if area moments are known with respect to a parallel set of centroidal axes for each piece. This process is best illustrated with a specific example.

Example 11.2

Consider the I-beam cross-section shown in Figure 11.10. I-beams are often used in structures because the area in the upper and lower sections, called flanges, is concentrated at large distances from the centroid in the section, resulting in a very efficient use of material to produce a large

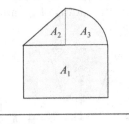

FIGURE 11.9
Geometry for determining the area moment for a composite area.

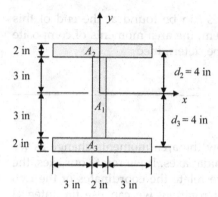

FIGURE 11.10
An I-beam cross section modeled as a composite area.

area moment that produces a beam of large strength. Determine the area moments with respect to the centroidal axes of this cross-section.

Break the I-beam cross section into the three areas shown in Figure 11.10, where areas (A_2, A_3) are the equal areas of the two flanges and A_1 is the area of the central section called the beam web. The centroidal axes are the x- and y-axes whose origin is at the beam center. From the parallel axis theorem, for $I_x = I_{xx}$:

$$
\begin{aligned}
I_x &= I_{x1} + I_{x2} + I_{x3} \\
&= [I_{c1} + A_1 d_1^2] + [I_{c2} + A_2 d_2^2] + [I_{c3} + A_3 d_3^2] \\
&= \frac{1}{12}(2)(6)^3 + 2\left\{\frac{1}{12}(8)(2)^3 + (8)(2)(4)^2\right\} \\
&= 558.7 \ in^4
\end{aligned}
$$
(11.21)

where (I_{c1}, I_{c2}, I_{c3}) are the area moments of each of the three sections about their own centroids. These areas are all rectangles, so from our previous results these moments are known. The distance $d_1 = 0$ since the x-centroidal axis of area A_1 is the x-axis, while the distances (d_2, d_3) are shown in Figure 11.10. Similarly, for $I_y = I_{yy}$, the parallel axis theorem gives

$$
\begin{aligned}
I_y &= I_{c1} + I_{c2} + I_{c3} \\
&= \frac{1}{12}(6)(2)^3 + 2\left\{\frac{1}{12}(2)(8)^3\right\} \\
&= 174.7 \ in^4
\end{aligned}
$$
(11.22)

since the y-axis is a centroidal axis for all three sections. Because of the symmetry of the cross section, we also have $I_{xy} = 0$.

Some area moments of simple shapes can be found at the end of this chapter. Using the parallel axis theorem, the area moments of composite bodies of more complex shapes can be determined.

11.4 Rotation of Coordinates

The parallel axis theorem tells us how the area moments change as we translate the axes from a set of centroidal axes. If we rotate our axes, the area moments will also change. If we relate the coordinates of the two sets of axes in terms of the angle of rotation, we can use the integral definitions to evaluate the area moments in the rotated axes in terms of those in the original axes. Consider, for example, a set of (x, y) axes and a set of (x', y') axes that have the same origin as the (x, y) axes but are rotated by an angle θ, as shown in Figure 11.11. The original and rotated axes are related through

$$x' = x \cos \theta + y \sin \theta$$
$$y' = -x \sin \theta + y \cos \theta \tag{11.23}$$

Now, let's compute the area moment $I_{x'x'}$ from its definition, using Eq. (11.23):

$$
\begin{aligned}
I_{x'x'} &= \int_A (y')^2 dA \\
&= \int_A (-x \sin \theta + y \cos \theta)^2 \, dA \\
&= \cos^2 \theta \int_A y^2 dA - 2 \sin \theta \cos \theta \int_A xy dA + \sin^2 \theta \int_A x^2 dA \\
&= I_{xx} \cos^2 \theta + I_{yy} \sin^2 \theta - 2 \sin \theta \cos \theta I_{xy}
\end{aligned}
\tag{11.24}
$$

Similar calculations can be done for $I_{y'y'}$ and $I_{x'y'}$, giving the transformation relations:

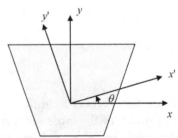

FIGURE 11.11
A set of (x, y) axes for an area and a set of rotated (x', y') axes.

$$I_{x'x'} = I_{xx} \cos^2 \theta + I_{yy} \sin^2 \theta - 2I_{xy} \sin \theta \cos \theta$$
$$I_{y'y'} = I_{xx} \sin^2 \theta + I_{yy} \cos^2 \theta + 2I_{xy} \sin \theta \cos \theta \qquad (11.25)$$
$$I_{x'y'} = I_{xx} \sin \theta \cos \theta - I_{yy} \sin \theta \cos \theta + I_{xy}(\cos^2 \theta - \sin^2 \theta)$$

These relations are often rewritten in terms of the angle 2θ, using the double angle formulae:

$$\cos 2\theta = \cos^2 \theta - \sin^2 \theta$$
$$\sin 2\theta = 2 \sin \theta \cos \theta \qquad (11.26)$$

The result is

$$I_{x'x'} = \frac{I_{xx} + I_{yy}}{2} + \frac{I_{xx} - I_{yy}}{2} \cos 2\theta - I_{xy} \sin 2\theta$$

$$I_{y'y'} = \frac{I_{xx} + I_{yy}}{2} + \frac{I_{yy} - I_{xx}}{2} \cos 2\theta + I_{xy} \sin 2\theta \qquad (11.27)$$

$$I_{x'y'} = I_{xy} \cos 2\theta + \frac{I_{xx} - I_{yy}}{2} \sin 2\theta$$

11.5 Principal Area Moments and Principal Axes

The relations of Eq. (11.27) tell us how the area moments change with a rotation of axes. There are, however, certain axes where these area moments have their largest and smallest values that are of particular interest. These axes are called the *principal axes* and the area moments with respect to the principal axes are called the *principal area moments*. Consider, for example, $I_{x'x'}$, where

$$I_{x'x'} = \frac{I_{xx} + I_{yy}}{2} + \frac{I_{xx} - I_{yy}}{2} \cos 2\theta - I_{xy} \sin 2\theta \qquad (11.28)$$

If $I_{x'x'}$ is a maximum or minimum, then

$$\frac{dI_{x'x'}}{d\theta} = (I_{yy} - I_{xx}) \sin 2\theta - 2I_{xy} \cos 2\theta = 0 \qquad (11.29)$$

so that

$$\tan 2\theta = \frac{-I_{xy}}{(I_{xx} - I_{yy})/2} \qquad (11.30)$$

We would get the same result, Eq. (11.30), if we had considered $I_{y'y'}$ instead. If $\theta = \beta$ is an angle that satisfies Eq. (11.30), then $\theta = \beta \pm \pi/2$ also satisfies Eq. (11.30). Thus, there are two principal directions, and they are orthogonal (at right angles) to each other. The principal area moments along the principal axes can be obtained by substituting Eq. (11.30) into the following equation:

$$I_{x'x'} = \frac{I_{xx} + I_{yy}}{2} + \frac{I_{xx} - I_{yy}}{2} \cos 2\theta - I_{xy} \sin 2\theta \qquad (11.31)$$

To help with that substitution, it is helpful to use the triangle shown in Figure 11.12. Note that Eq. (11.30) only gives us the ratio of the two straight sides of the triangle shown in Figure 11.12 so that the sides could have different signs but their ratio must always give the value shown in Eq. (11.30). This is the reason for including ± signs in Figure 11.12. Placing the sides seen in Figure 11.12 into Eq. (11.31), we find

$$I_{x'x'} = \frac{I_{xx} + I_{yy}}{2} + \frac{I_{xx} - I_{yy}}{2} \frac{\pm(I_{xx} - I_{yy})/2}{\sqrt{\left(\frac{I_{xx} - I_{yy}}{2}\right)^2 + I_{xy}^2}} - I_{xy} \frac{\mp I_{xy}}{\sqrt{\left(\frac{I_{xx} - I_{yy}}{2}\right)^2 + I_{xy}^2}} \qquad (11.32)$$

$$= \frac{I_{xx} + I_{yy}}{2} \pm \sqrt{\left(\frac{I_{xx} - I_{yy}}{2}\right)^2 + I_{xy}^2}$$

These are the two principal area moments which we will label as (I_{p1}, I_{p2}), where

$$I_{p1,p2} = \frac{I_{xx} + I_{yy}}{2} \pm \sqrt{\left(\frac{I_{xx} - I_{yy}}{2}\right)^2 + I_{xy}^2} \qquad (11.33)$$

If we examine the mixed area moment along the principal axes, we find

$$I_{x'y'} = I_{xy} \cos 2\theta + \frac{I_{xx} - I_{yy}}{2} \sin 2\theta$$

$$= \pm \frac{I_{xy}(I_{xx} - I_{yy})}{2\sqrt{\left(\frac{I_{xx} - I_{yy}}{2}\right)^2 + I_{xy}^2}} \mp \frac{(I_{xx} - I_{yy})I_{xy}}{2\sqrt{\left(\frac{I_{xx} - I_{yy}}{2}\right)^2 + I_{xy}^2}} = 0 \qquad (11.34)$$

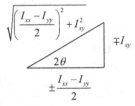

FIGURE 11.12
A triangle that helps determine the principal area moments.

so the mixed area moment is zero with respect to the principal axes. Since I_{xy} is zero if either the x- or y-axis is an axis of symmetry, it follows that if an area A has an axis of symmetry, it must also be a principal axis.

We now know the values of the two principal area moments and the angles of the two orthogonal principal directions, but we have not determined which principal value goes with which angle. One way to determine this is to place one of the principal directions into Eq. (11.31) to see which principal area moment is produced. However, it is not necessary to do this. As we will see shortly, the principal directions are also given by

$$\tan \theta_{p1} = \frac{(I_{xx} - I_{p1})}{I_{xy}}$$

$$\tan \theta_{p2} = \frac{(I_{xx} - I_{p2})}{I_{xy}} \tag{11.35}$$

and these expressions do identify the principal area moments with their respective angles. Of course, we only really need to use one of the relations in Eq. (11.35) since once, say, the principal direction θ_{p1} associated with I_{p1} is known, we know that I_{p2} lies along a direction that is at right angles to θ_{p1}.

Example 11.3

Determine the principal area moments and the principal directions with respect to the centroid of the triangle shown in Figure 11.13(a).

The area moments and mixed area moment for centroidal axes of the triangle (Figure 11.13(a)) are $I_{xx} = ab^3/36$, $I_{yy} = ba^3/36$, $I_{xy} = -a^2b^2/72$,

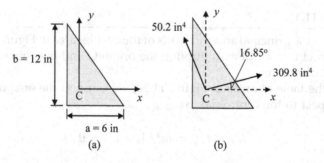

FIGURE 11.13
(a) The geometry of a triangular area and (b) the principal area moments and their directions at the centroid of the triangular area.

which can be obtained from tables (see Figure 11.20) or by integration
and the use of the parallel axis theorem. In this case, we have

$$I_{xx} = \frac{1}{36}(6)(12)^3 = 288 \ \text{in}^4$$

$$I_{yy} = \frac{1}{36}(12)(6)^3 = 72 \ \text{in}^4 \tag{11.36}$$

$$I_{xy} = \frac{-1}{72}(6)^2(12)^2 = -72 \ \text{in}^4$$

From Eq. (11.33):

$$\begin{aligned}
I_{p1,p2} &= \frac{I_{xx} + I_{yy}}{2} \pm \sqrt{\left(\frac{I_{xx} - I_{yy}}{2}\right)^2 + I_{xy}^2} \\
&= \frac{288 + 72}{2} \pm \sqrt{\left(\frac{288 - 72}{2}\right)^2 + (-72)^2} \\
&= 309.8, \quad 50.2 \ \text{in}^4
\end{aligned} \tag{11.37}$$

and from Eq. (11.35):

$$\tan \theta_{p1} = \frac{(I_{xx} - I_{p1})}{I_{xy}} = \frac{(288 - 309.8)}{-72} = 0.3028 \tag{11.38}$$

$$\rightarrow \quad \theta_{p1} = 16.85°$$

and, consequently, $\theta_{p2} = 106.85°$. These results are shown in Figure 11.13(b).

In the last example, we could use the results from tables directly. In some
cases, however, we may have to use the parallel axis theorem before we
can determine the principal area moments, as the next example shows.

Example 11.4

Determine the principal area moments of the circular area in Figure 11.14(a)
with respect to a set of axes through the origin O and the corresponding
principal directions.

From the table of areas in Figure 11.20, we can find the area moments
with respect to the centroidal axes

$$I_{x'x'} = I_{y'y'} = \pi a^4/4, \quad I_{x'y'} = 0 \tag{11.39}$$

Then using the parallel axis theorem to transfer these values to the (x, y)
axes, we find

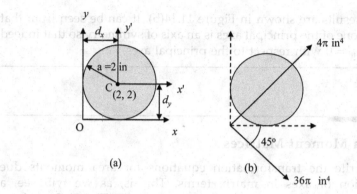

FIGURE 11.14
(a) A circular area whose principal area moments and directions are to be found, and (b) those principal area moments and their directions.

$$I_{yy} = I_{y'y'} + Ax_c^2$$

$$= \frac{1}{4}\pi(2)^4 + \pi(2)^2(2)^2 = 20\pi \text{ in}^4$$

$$I_{xx} = I_{x'x'} + Ay_c^2$$

$$= \frac{1}{4}\pi(2)^4 + \pi(2)^2(2)^2 = 20\pi \text{ in}^4$$
(11.40)

$$I_{xy} = I_{x'y'} + Ax_c y_c$$

$$= 0 + \pi(2)^2(2)(2) = 16\pi \text{ in}^4$$

The principal area moments are then given by

$$I_{p1,p2} = \frac{I_{xx} + I_{yy}}{2} \pm \sqrt{\left(\frac{I_{xx} - I_{yy}}{2}\right)^2 + I_{xy}^2}$$

$$= \frac{20\pi + 20\pi}{2} \pm \sqrt{\left(\frac{20\pi - 20\pi}{2}\right)^2 + (16\pi)^2}$$
(11.41)

$$= 20\pi \pm 16\pi = 36\pi, \ 4\pi \text{ in}^4$$

and the principal angles are (using both formulae):

$$\tan\theta_{p1} = \frac{(I_{xx} - I_{p1})}{I_{xy}} = \frac{(20\pi - 36\pi)}{16\pi} = -1$$

$$\rightarrow \ \theta_{p1} = -45°$$
(11.42)

$$\tan\theta_{p2} = \frac{(I_{xx} - I_{p2})}{I_{xy}} = \frac{(20\pi - 4\pi)}{16\pi} = 1$$

$$\rightarrow \ \theta_{p1} = 45°$$

and these results are shown in Figure 11.14(b). It can be seen from that figure that one of the principal axes is an axis of symmetry so that indeed we have $I_{x'y'} = 0$ with respect to the principal axes.

11.6 Area Moment Matrices

We can write the transformation equations for area moments due to a rotation of axes in matrix terms. This is, as we will see, a very useful way to consider these equations in MATLAB. Consider a rotation of axes again, as shown in Figure 11.15(a). The transformation equations, Eq. (11.25), written in terms of the angle θ, we repeat here:

$$I_{x'x'} = I_{xx} \cos^2 \theta + I_{yy} \sin^2 \theta - 2I_{xy} \sin \theta \cos \theta$$
$$I_{y'y'} = I_{xx} \sin^2 \theta + I_{yy} \cos^2 \theta + 2I_{xy} \sin \theta \cos \theta \qquad (11.43)$$
$$I_{x'y'} = I_{xx} \sin \theta \cos \theta - I_{yy} \sin \theta \cos \theta + I_{xy} (\cos^2 \theta - \sin^2 \theta)$$

We can write these relations in terms of components of the unit vectors $(\mathbf{n}, \mathbf{t})$ acting along the (x', y')axes, respectively, where (see Figure 11.15(b)):

$$\mathbf{n} = \cos \theta \, \mathbf{e}_x + \sin \theta \, \mathbf{e}_y = n_x \mathbf{e}_x + n_y \mathbf{e}_y$$
$$\mathbf{t} = -\sin \theta \, \mathbf{e}_x + \cos \theta \, \mathbf{e}_y = t_x \mathbf{e}_x + t_y \mathbf{e}_y \qquad (11.44)$$

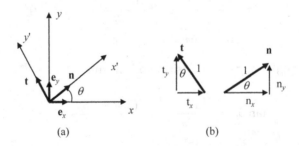

(a)   (b)

FIGURE 11.15
(a) A rotation of axes and unit vectors along the original and rotated axes. (b) Components of the unit vectors along the rotated axes.

and where $(\mathbf{e}_x, \mathbf{e}_y)$ are unit vectors along the (x, y) axes. We find

$$I_{x'x'} = I_{xx}n_x^2 + I_{yy}n_y^2 - 2I_{xy}n_x n_y$$
$$I_{y'y'} = I_{xx}t_x^2 + I_{yy}t_y^2 - 2I_{xy}t_x t_y \qquad (11.45)$$
$$I_{x'y'} = -I_{xx}n_x t_x - I_{yy}n_y t_y + I_{xy}(n_x t_y + n_y t_x)$$

which can be written in matrix notation as

$$\begin{bmatrix} I_{x'x'} & -I_{x'y'} \\ -I_{x'y'} & I_{y'y'} \end{bmatrix} = \begin{bmatrix} n_x & n_y \\ t_x & t_y \end{bmatrix} \begin{bmatrix} I_{xx} & -I_{xy} \\ -I_{xy} & I_{yy} \end{bmatrix} \begin{bmatrix} n_x & t_x \\ n_y & t_y \end{bmatrix} \qquad (11.46)$$

and which can also be written as

$$[I'] = [Q]^T [I][Q] \qquad (11.47)$$

where

$$[Q] = \begin{bmatrix} n_x & t_x \\ n_y & t_y \end{bmatrix} = \begin{bmatrix} \cos(x, x') & \cos(x, y') \\ \cos(y, x') & \cos(y, y') \end{bmatrix} \qquad (11.48)$$

is called the *direction cosine matrix*. This name comes from the fact that terms such as $t_x = \cos(x, y')$, for example, represent the cosine of the angle between the x and y' axes with similar expressions for the other terms in the matrix, as seen in Eq. (11.48).

Example 11.5

Consider the triangular area previously considered (Figure 11.13), where we found the area moments with respect to the x- and y-axes. Using MATLAB, determine the area moments with respect to a set of (x', y') axes that make an angle of 30° with the x-axis (Figure 11.16).

The area moments were given in Eq. (11.36). We will let the (x', y') axes be the (n, t) axes where $\mathbf{n}$ and $\mathbf{t}$ are unit vectors along those axes (Figure 11.16). In MATLAB, we find

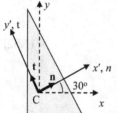

FIGURE 11.16
Determining area moments about a rotated set of axes.

M = [288 72; 72 72];   % area moment matrix. Note: this matrix uses – Ixy
n = [cosd(30) sind(30)];   % unit vectors along rotated axes
t = [-sind(30) cosd(30)];
Q = [n.' t.'];   % direction cosine matrix. Unit vectors in columns
Mr = Q.'*M*Q   % area moment matrix in rotated coordinates
Mr = 296.3538 -57.5307
      -57.5307   63.6462

Thus, we find, approximately,

$$I_{nn} = 296.4 \ \text{in}^4$$
$$I_{tt} = 63.6 \ \text{in}^4 \tag{11.49}$$
$$I_{nt} = 57.5 \ \text{in}^4$$

where again it is the negative of the mixed area moment that appears in the matrix.

Using area moment matrices, it is also very easy to find the principal area moments and principal directions. Suppose we can find a set of principal coordinates (N, T) where the unit vectors along the principal axes have the components (N_x, N_y) and (T_x, T_y), respectively. Then by Eq. (11.47), we have

$$\begin{bmatrix} I_1 & 0 \\ 0 & I_2 \end{bmatrix} = \begin{bmatrix} N_x & N_y \\ T_x & T_y \end{bmatrix} \begin{bmatrix} I_{xx} & -I_{xy} \\ -I_{xy} & I_{yy} \end{bmatrix} \begin{bmatrix} N_x & T_x \\ N_y & T_y \end{bmatrix} \tag{11.50}$$

which we could also write in matrix form as

$$[I_p] = [Q]^T [I][Q] \tag{11.51}$$

where (I_1, I_2) are the principal area moments and the mixed area moment is zero in the principal coordinates. If we multiply both sides of Eq. (11.50) by $[Q]$ and use the fact that the direction cosine matrix multiplied by its transpose gives the unit matrix:

$$[Q][Q]^T = [I] = \begin{bmatrix} 1 & 0 \\ 0 & 1 \end{bmatrix} \tag{11.52}$$

We find

$$\begin{bmatrix} N_x & T_x \\ N_y & T_y \end{bmatrix}\begin{bmatrix} I_1 & 0 \\ 0 & I_2 \end{bmatrix} = \begin{bmatrix} I_{xx} & -I_{xy} \\ -I_{xy} & I_{yy} \end{bmatrix}\begin{bmatrix} N_x & T_x \\ N_y & T_y \end{bmatrix} \tag{11.53}$$

which gives

$$\begin{bmatrix} N_x I_1 & T_x I_2 \\ N_y I_1 & T_y I_2 \end{bmatrix} = \begin{bmatrix} I_{xx} & -I_{xy} \\ -I_{xy} & I_{yy} \end{bmatrix}\begin{bmatrix} N_x & T_x \\ N_y & T_y \end{bmatrix} \tag{11.54}$$

Equation (11.54) is equivalent to the two sets of equations:

$$\begin{Bmatrix} N_x I_1 \\ N_y I_1 \end{Bmatrix} = \begin{bmatrix} I_{xx} & -I_{xy} \\ -I_{xy} & I_{yy} \end{bmatrix}\begin{Bmatrix} N_x \\ N_y \end{Bmatrix}$$
$$\begin{Bmatrix} T_x I_2 \\ T_y I_2 \end{Bmatrix} = \begin{bmatrix} I_{xx} & -I_{xy} \\ -I_{xy} & I_{yy} \end{bmatrix}\begin{Bmatrix} T_x \\ T_y \end{Bmatrix} \tag{11.55}$$

Thus, to find either the principal area moment I_1 and its principal direction defined by the unit vector **N**, or the principal area moment I_2 and its principal direction defined by the unit vector **T**, we need to solve the system of equations:

$$\begin{bmatrix} I_{xx} & -I_{xy} \\ -I_{xy} & I_{yy} \end{bmatrix}\begin{Bmatrix} U_x \\ U_y \end{Bmatrix} = \begin{Bmatrix} U_x I \\ U_y I \end{Bmatrix} \tag{11.56}$$

where the unit vector **U** can be either **N** or **T** and I can be either I_1 or I_2. The system of equations in Eq. (11.56) can be written in matrix-vector form as

$$[I]\{U\} = I\{U\} \tag{11.57}$$

The solution of Eq. (11.57) is called the solution to a matrix *eigenvalue problem*, where the solution is a scalar *eigenvalue*, I, which here corresponds to a principal area moment, and an *eigenvector*, $\{U\}$, which is a unit vector along the corresponding principal direction. Eigenvalue problems of this type appear frequently in engineering problems in other contexts, so their solution is important. This eigenvalue problem can also be written as

$$\begin{bmatrix} I_{xx} - I & -I_{xy} \\ -I_{xy} & I_{yy} - I \end{bmatrix}\begin{Bmatrix} U_x \\ U_y \end{Bmatrix} = \begin{Bmatrix} 0 \\ 0 \end{Bmatrix} \tag{11.58}$$

so to find a solution we must solve the system of two equations given by

$$
\begin{aligned}
(I_{xx} - I)U_x - I_{xy}U_y &= 0 \\
-I_{xy}U_x + (I_{yy} - I)U_y &= 0
\end{aligned}
\tag{11.59}
$$

But this is a homogeneous set of two equations in two unknowns which only has the solution $\{U\} = 0$, unless the determinant of the matrix of coefficients is zero, i.e.

$$
\begin{vmatrix} I_{xx} - I & -I_{xy} \\ -I_{xy} & I_{yy} - I \end{vmatrix} = (I_{xx} - I)(I_{yy} - I) - I_{xy}^2 = 0
\tag{11.60}
$$

Expanding Eq. (11.60), we obtain a quadratic equation for I:

$$
I^2 - (I_{xx} + I_{yy})I + (I_{xx}I_{yy} - I_{xy}^2) = 0
\tag{11.61}
$$

which has two roots

$$
I_1, I_2 = \frac{(I_{xx} + I_{yy})}{2} \pm \sqrt{\left(\frac{I_{xx} - I_{yy}}{2}\right)^2 + I_{xy}^2}
\tag{11.62}
$$

and which we recognize as just the principal area moments. If we place one of these principal values back into Eq. (11.59), then we can solve for the corresponding principal direction. However, since we have set the determinant of the matrix of coefficients equal to zero, the two equations in Eq. (11.59) are not independent. Thus, we can solve only one of them for a ratio of unit vector components. Thus, for example, from the first equation in Eq. (11.59) and using $I = I_1$, we have

$$
\frac{U_x}{U_y} = \frac{I_{xy}}{(I_{xx} - I_1)}
\tag{11.63}
$$

But since $\{U\}$ is a unit vector, we have

$$
\sqrt{U_x^2 + U_y^2} = 1
\tag{11.64}
$$

which we can use to solve for U_y in terms of the ratio given in Eq. (11.63) as

$$
U_y = \frac{\pm 1}{\sqrt{(U_x/U_y)^2 + 1}}
\tag{11.65}
$$

We can then find U_x from Eq. (11.63) since the ratio U_x/U_y is known. Note that we only get a solution for the unit vector $\{U\}$ to within a plus or minus sign since both $\{U\}$ and $-\{U\}$ are along principal directions. If we repeat this same process for I_2, then we can get the second principal direction. Of course, the second principal direction is at right angles to the first principal direction so we could also use that fact to determine the second principal direction.

Note that if we let $U_y/U_x = \tan \theta_{p1}$ and $I_1 = I_{p1}$ in Eq. (11.63), where θ_{p1} is the angle that the first principal direction makes with respect to the x-axis, then Eq. (11.63) gives

$$\tan \theta_{p1} = \frac{(I_{xx} - I_{p1})}{I_{xy}} \qquad (11.66)$$

which is just Eq. (11.35). We used Eq. (11.35) previously, without proof, to associate a principal area moment with a particular angle. A similar expression for the second principal direction is true as well, as given in Eq. (11.35).

Solving eigenvalue problems in MATLAB is very easy as there is a built-in function called eig that can obtain the solution. We first must place the area moments and mixed area moments in a matrix:

$$M = \begin{bmatrix} I_{xx} & -I_{xy} \\ -I_{xy} & I_{yy} \end{bmatrix}$$

and then use that matrix as an argument to the function eig, which solves the eigenvalue problem. In MATLAB, the function is called as

$$[\text{pdirs, pvals}] = \text{eig}(M)$$

The matrix pdirs will then have the principal direction unit vector components (in columns) as

$$\text{pdirs} = \begin{bmatrix} (U_x)_1 & (U_x)_2 \\ (U_y)_1 & (U_y)_2 \end{bmatrix}$$

and the matrix pvals will contain the corresponding principal values:

$$\text{pvals} = \begin{bmatrix} I_1 & 0 \\ 0 & I_2 \end{bmatrix}$$

To illustrate the use of eig, let's consider the triangle used in Example 11.5.

Example 11.6

Using the MATLAB function eig, determine the principal area moments and the principal directions for the triangle considered in Example 11.5, where the area moments and mixed area moment were

$$I_{xx} = 288 \ in^4$$
$$I_{yy} = 72 \ in^4$$
$$I_{xy} = -72 \ in^4$$

The results in MATLAB are

```
M = [288 72; 72 72];        % area moments matrix
[pdirs, pvals] = eig(M)
pdirs = 0.2898  -0.9571     % principal directions (in columns)
       -0.9571  -0.2898
pvals = 50.2002    0        % principal area moments
          0    309.7998
atand(pdirs(2,2)/pdirs(1,2))  % angle principal axis associated with the 309.8
ans = 16.8450               % principal value makes with respect to the x-axis
```

We see that the eig function produces two matrices. The pvals matrix gives the principal area moments along its diagonal. Since the off-diagonal terms are zero, this is actually the full area moment matrix for the principal directions. The eig function also generates a pdirs matrix with the components of the unit vectors along the principal directions in its columns. As shown, if we want to find the angles of these axes, we can do so easily by taking the ratio of their components. *Note, however, that eig does not always choose principal directions that form a right-handed system. Thus, you may have to change a sign on one of the unit vectors to ensure this is the case. An easy test to see if the unit vectors given do form a right-handed system is to evaluate the determinant of the pdirs matrix. If it is +1, the system is right-handed. It is important to always have right-handed coordinates, so we can use, for example, our previous definitions of the cross product.*
    If we evaluate the determinant in this triangle case, we find

```
det(pdirs)
ans = -1
```

so the unit vectors *do not* form a right-handed system. If we change the sign on $\{U_2\}$, then we have the column unit vectors:

$$\{U_1\} = [0.2898; \quad -0.9571]$$
$$\{U_2\} = [0.9571; \quad 0.2898]$$

which do form a right-handed system. Note that the angle that $\{U_2\}$ makes with respect to the x-axis does not change but its direction does.

In solving eigenvalue problems $[I]\{U\} = I\{U\}$, it can be shown that the eigenvalues (I_1, I_2) are always real and the eigenvectors $\{U_1\},\{U_2\}$ are always real and orthogonal to each other if the matrix $[I]$ is a real, symmetric matrix (i.e., the off-diagonal terms are the same). This is always the case for area moments.

Using MATLAB makes determining principal area moments and principal directions very easy. If you do not have MATLAB available (as on a test), then you can always use our first approach where we found explicit expressions for the principal moments as

$$I_1, I_2 = \frac{(I_{xx} + I_{yy})}{2} \pm \sqrt{\left(\frac{I_{xx} - I_{yy}}{2}\right)^2 + I_{xy}^2}$$

and if we examine, say, I_1, then the angle from the positive x-axis to the I_1 principal direction is found from

$$\tan \theta_1 = \frac{(I_{xx} - I_1)}{I_{xy}}$$

and the other principal direction is at $\theta_2 = \theta_1 \pm 90°$.

11.7 Mass Moments of Inertia

One of the reasons for treating the transformation of area moments by a matrix approach is that it easily generalizes to more complex problems. For example, in dynamics the three-dimensional angular motion of a body (such as a spinning satellite, for example) is controlled by the mass moments of inertia defined as

$$I_{xx}^m = \int \rho(y^2 + z^2)dV$$

$$I_{yy}^m = \int \rho(x^2 + z^2)dV$$

$$I_{zz}^m = \int \rho(x^2 + y^2)dV$$

$$I_{xy}^m = I_{yx}^m = \int \rho(xy)dV$$

$$I_{xz}^m = I_{zx}^m = \int \rho(xz)dV$$

$$I_{yz}^m = I_{zy}^m = \int \rho(yz)dV$$

$$(11.67)$$

where ρ is the mass density (mass/unit volume) and dV is a volume element. We see that there can be as many as six mass moments involved. If we let unit vectors **n**, **t**, **v** be along the x', y', z' axes (which are assumed to be orthogonal to each other and right-handed), (Figure 11.17), then the transformation of mass moments follows the same pattern we saw for area moments. Change of mass moments of inertia with a 3-D rotation of axes, for example, can again be expressed as a matrix multiplication of a mass moment of inertia matrix with direction cosine matrices:

$$\begin{bmatrix} I_{x'x'}^m & -I_{x'y'}^m & -I_{x'z'}^m \\ -I_{y'x'}^m & I_{y'y'}^m & -I_{y'z'}^m \\ -I_{z'x'}^m & -I_{z'y'}^m & I_{z'z'}^m \end{bmatrix} = \begin{bmatrix} n_x & n_y & n_z \\ t_x & t_y & t_z \\ v_x & v_y & v_z \end{bmatrix} \begin{bmatrix} I_{xx}^m & -I_{xy}^m & -I_{xz}^m \\ -I_{yx}^m & I_{yy}^m & -I_{yz}^m \\ -I_{zx}^m & -I_{zy}^m & I_{zz}^m \end{bmatrix} \begin{bmatrix} n_x & t_x & v_x \\ n_y & t_y & v_y \\ n_z & t_z & v_z \end{bmatrix}$$

$$(11.68)$$

which again can be written in matrix form as

$$[I^{m'}] = [Q]^T [I^m][Q] \qquad (11.69)$$

In this case, there are three principal mass moments of inertia and three corresponding principal directions that are orthogonal to each other. The

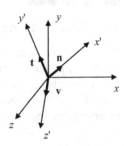

FIGURE 11.17
A three-dimensional rotation of coordinate system axes.

principal mass moments of inertia and principal directions are again determined by a solution of the eigenvalue problem:

$$[I^m]\{U\} = I^m\{U\} \tag{11.70}$$

Example 11.7

Consider a body whose mass moment of inertia matrix is given as (in units of mass×length^2)

$$[I^m] = \begin{bmatrix} 100 & 20 & 30 \\ 20 & 300 & 50 \\ 30 & 50 & 200 \end{bmatrix}$$

Determine the principal mass moments of inertia and the principal directions.

In MATLAB, we have

```
Im = [100 20 30; 20 300 50; 30 50 200]   % mass moment of inertia matrix
Im = 100   20    30
      20  300    50
      30   50   200
[pdirs, pvals] = eig(Im)              % solve eigenvalue problem
pdirs = 0.9670  -0.2166  0.1337       % the principal directions
       -0.0322   0.4169  0.9084
       -0.2525  -0.8827  0.3962
pvals = 91.4995       0        0      % the principal mass moments of inertia
              0  183.7468        0
              0        0  324.7537
det(pdirs)                            % the principal directions
ans = 1                               % do form a right-handed system
```

In this case, the principal directions do form a right-handed system but if that was not the case, then we can easily change a sign on a unit vector to make the system right-handed.

11.7.1 Parallel Axis Theorem

There is a parallel axis theorem for masses that is very similar to that for areas. Consider, for example, a set of parallel axes, as shown in Figure 11.18(a), where the origin of the (x', y', z') axes is at the center of mass. Then, the mass moment of inertia, I_{xx}, is given by

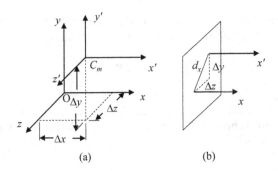

FIGURE 11.18
(a) Two parallel sets of axes. (b) The distance, d_x, between the x and x' axes in a plane perpendicular to those axes.

$$I_{xx}^m = \int (y^2 + z^2)\,dm = \int [(y' + \Delta y)^2 + (z' + \Delta z)^2]\,dm$$
$$= \int [(y')^2 + (z')^2]\,dm + \int (\Delta y^2 + \Delta z^2)\,dm + 2\Delta y \int (y')\,dm + 2\Delta z \int z'\,dm$$
$$= I_{x'x'}^m + m(\Delta y^2 + \Delta z^2)$$
$$= I_{x'x'}^m + md_x^2$$

$$(11.71)$$

where m is the mass of the body and the first moment integrals vanish because y' and z' are measured from the center of mass. The distance d_x is the distance between the x and x' axes in a plane perpendicular to those axes, as shown in Figure 11.18(b). In an entirely similar fashion, we find

$$I_{yy}^m = I_{y'y'}^m + m(\Delta x^2 + \Delta z^2) = I_{y'y'}^m + md_y^2$$
$$I_{zz}^m = I_{z'z'}^m + m(\Delta x^2 + \Delta y^2) = I_{z'z'}^m + md_z^2$$

$$(11.72)$$

For the mixed mass moments of inertia, we have instead

$$I_{xy}^m = I_{x'y'}^m + m\Delta x\Delta y$$
$$I_{xz}^m = I_{x'z'}^m + m\Delta x\Delta z$$
$$I_{yz}^m = I_{y'z'}^m + m\Delta y\Delta z$$

$$(11.73)$$

These results are often very useful when dealing with composite bodies made of simple shapes such as the examples shown in Figure 11.19.

11.7.2 Radius of Gyration

Just as for areas, we can define a length parameter called the radius of gyration for mass moments of inertia. Since a mass moment of inertia has the dimensions of mass × length^2, a radius of gyration for a body of mass m and mass moment of inertia I^m is given by

$$k = \sqrt{\frac{I^m}{m}} \tag{11.74}$$

If we have a particle of mass m that is located at a distance l from a point P, then the mass moment of inertia of that particle about P is just $I^m = ml^2$. Thus, we can view the radius of gyration as the distance at which we can concentrate all the mass of a body which will give the same mass moment of inertia as the actual body. The radius of gyration can be related physically to the dynamics of a body. For example, in swinging a baseball bat, there is a "sweet spot" where the bat does not produce a "sting" on the hands when striking a baseball. The distance from the batter's hands to this sweet spot is called the center of percussion, which is a point along the bat that can be related to the distance to the center of mass of the bat and its radius of gyration.

11.8 Problems

P11.1 Determine the area moment I_{xx} for the shaded area in Fig. P11.1. Let $b = 4$ in and $h = 3$ in.

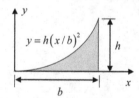

Figs. P11.1, P11.2, P11.3

Choices (in^4)

1. 3.62
2. 3.95
3. 4.33
4. 5.02
5. 5.14

P11.2 Determine the area moment I_{yy} for the shaded area in Fig. P11.2. Let $b = 4$ in and $h = 3$ in.

Choices (in^4)

1. 38.4
2. 39.5
3. 45.9
4. 48.2
5. 49.1

P11.3 Determine the mixed area moment I_{xy} for the shaded area in Fig. P11.3. Let $b = 4$ in and $h = 3$ in.

Choices (in^4)

1. 8.0
2. 10.0
3. 12.0
4. 14.0
5. 16.0

P11.4 Determine the area moment I_{xx} for the shaded area in Fig. P11.4. Let $b = 25$ mm and $h = 20$ mm.

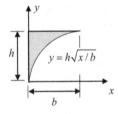

Figs. P11.4, P11.5, P11.6

P11.5 Determine the area moment I_{yy} for the shaded area in Fig. P11.5. Let $b = 25$ mm and $h = 20$ mm.

P11.6 Determine the mixed area moment I_{xy} for the shaded area in Fig. P11.6. Let $b = 25$ mm and $h = 20$ mm.

P11.7 Determine the area moment about the x-axis and the mixed area moment about the x- and y-axes for a rectangular plate with a circular hole at its center, as shown in Fig. P11.7, if the radius of the hole $r = 2$ in. The distances $a = 6$ in and $b = 4$ in.

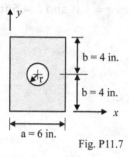

Fig. P11.7

P11.8 The origin of the x- and y-axes of the Z-section shown in Fig. P11.8 is at the centroid of the cross section. Determine the area moment about the x-axis. The thickness $t = 20$ mm and the distances $a = 100$ mm and $b = 140$ mm.

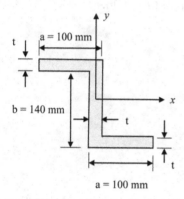

Figs. P11.8, P11.9, P11.10

P11.9 The origin of the x- and y-axes of the Z-section shown in Fig. P11.9 is at the centroid of the cross section. Determine the area moment about the y-axis. The thickness $t = 20$ mm and the distances $a = 100$ mm and $b = 140$ mm.

P11.10 The origin of the x- and y-axes of the Z-section shown in Fig. P11.10 is at the centroid of the cross section. Determine the mixed area moment about the x- and y-axes. The thickness $t = 20$ mm and the distances $a = 100$ mm and $b = 140$ mm.

P11.11 Determine the area moments and mixed area moment of the rectangular area in Fig. P11.11 with respect to the rotated *u*- and *v*- Cartesian axes if the angle $\theta = 30°$. The distances $a = 4$ ft and $b = 5$ ft.

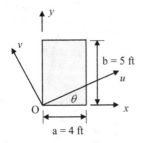

Figs. P11.11, P11.13

P11.12 Determine the area moments and the mixed area moment of the triangular area in Fig. P11.12 with respect to the rotated *u*- and *v*-axes if the angle $\theta = 30°$. The distances $a = 12$ in and $b = 6$ in.

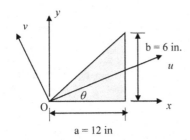

Figs. P11.12, P11.14

P11.13. Determine the principal area moments and principal directions for a set of axes with origin at O for the rectangular area in Fig. P11.13. The distances $a = 4$ ft and $b = 5$ ft.

P11.14 Determine the principal area moments and principal directions for a set of axes with origin at O for the triangular area in Fig. P11.14. The distances $a = 12$ in and $b = 6$ in.

P11.15 A pendulum consists of two homogeneous thin rods that each weigh 3 lb, as shown in Fig. P11.15. Determine the pendulum mass moment of inertia about an axis passing through the *z*-axis at O. The distances $a = 2$ in and $b = 4$ in.

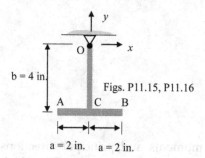

Figs. P11.15, P11.16

P11.16 A pendulum consists of two homogeneous thin rods that each weigh 3 lb, as shown in Fig. P11.16. Determine the pendulum mass moment of inertia about an axis passing through a z-axis at the center of mass of the pendulum. The distances $a = 2$ in and $b = 4$ in.

11.8.1 Review Problems

These problems typically have the level of difficulty found on exams. They should be done by hand (i.e., with a calculator).

R11.1 Determine area moment about the x-axis for the composite area of Fig. R11.1. The distances $a = 100$ mm and $b = 20$ mm.

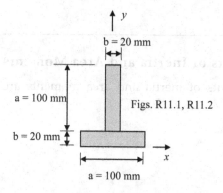

Figs. R11.1, R11.2

R11.2 Determine the area moment about a horizontal axis through the centroid of the composite area shown in Fig. R11.2. The distances $a = 100$ mm and $b = 20$ mm.

R11.3 Determine the principal area moments and principal directions with respect to the origin O for the triangular area in Fig. R11.3. The distances $a = 3$ in and $b = 6$ in.

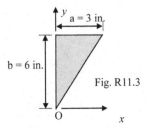

Fig. R11.3

R11.4 Determine the principal area moments and principal directions with respect to the origin O for the L-shaped area shown in Fig. R11.4. The distances $a = 25$ mm and $b = 100$ mm.

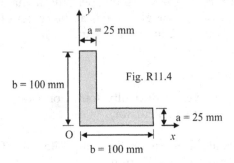

Fig. R11.4

11.9 Tables of Mass Moments of Inertia and Area Moments

Tabulated values of mass moments of inertia and area moments are given in Figures 11.19 and 11.20.

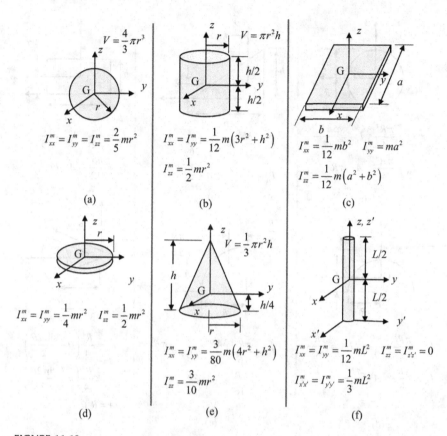

FIGURE 11.19

Mass moments of inertia and the location of the center of mass (also the center of gravity, G, and the centroid) for some simple homogeneous bodies. (a) A sphere. (b) A cylinder. (c) A thin rectangular plate. (d) A thin circular disk (see the cylinder for $h \ll r$). (d) A cone. (f) A slender rod (see the cylinder for $r \ll h$).

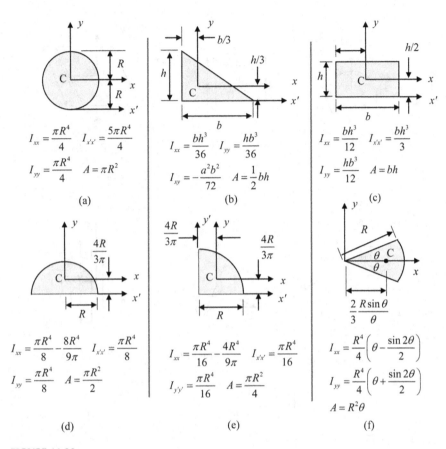

FIGURE 11.20
Area moments and areas for some simple shapes. (a) Circular area. (b) Triangular area.
(c) Rectangular area. (d) Semi-circular area. (e) Circular quadrant. (f) Circular sector.

Appendices

A MATLAB® Overview

OBJECTIVES

- To examine the basic constants, functions, and operations available in MATLAB® that are relevant to statics.
- To describe vectors and matrices and their use in the solution of the system of linear equations often found in statics.
- To introduce symbolic algebra and symbolic solutions.

This book uses MATLAB frequently as a computation tool for solving statics problems. MATLAB is very sophisticated with many capabilities beyond what is used in statics so we will provide in this appendix a summary of only the most relevant items. There are many books and resources on the web that give you a much fuller understanding of the capabilities present in MATLAB. For a tutorial on how to use MATLAB for statics problems, go to the companion website www.eng-statics.org.

A.1 Constants, Operations, and Built-in Functions

MATLAB contains built-in constants and variables. Some important ones are:

ans   most recent answer
eps   small constant ~ 10^{-16}
inf   infinity
pi   3.14159 …

Some of the standard mathematical operations are:

± addition, subtraction
* multiplication
/ division
^ exponentiation e.g., y^n = y^n

In statics, we often use trigonometric functions. A few of the many that are built into MATLAB are listed here:

sin(x) sine	x is in radians
cos(x) cosine	x is in radians
sind(x) inverse sine	x is in degrees
cosd(x) inverse cosine	x is in degrees
x = asin (u) inverse sine	x is in radians
x = acos(u) inverse cosine	x is in radians
x = asind(u) inverse sine	x is in degrees
x = acosd(u) inverse cosine	x is in degrees

Other functions are:

exp(x)	exponential
sqrt(x)	square root
log(x)	natural log (base e)
log10(x)	log to base 10

In MATLAB, when we enter or compute items, we may want to output the results to the MATLAB command window to see them directly, but often we want to store the items but not display them. We can do a suppression of any displayed output in MATLAB with a semicolon at the end of an input line. For example, if we type in the MATLAB command window:

x = 3

and enter return, MATLAB stores and echoes the result:

x = 3

but if we enter:

x = 3;

the variable x is stored with the value of 3 but no result is echoed. Generally, we want to use the semicolon at the end of a line unless we need to see a particular result and the result is not too complicated. In this text, we often omit the semicolon when doing MATLAB calculations so you can see intermediate as well as final results.

One may want to add comments to a MATLAB line to explain the calculations. This is done by placing a percent sign in front of the comment as in:

% This is a comment.

As we solve a problem with MATLAB, the MATLAB workspace will be populated with the variables and other items we create. When we move to a new problem, those quantities may conflict with the ones we want to create for the new problem, so it is good practice to clear the workspace and start anew. This can be done by typing clear in the command window:

clear

The command window itself may become cluttered with various variables and commands we generate, so we can also clear that window by typing:

clc

Note that this does not erase the previous commands, but simply removes them from display in the command window. We can access previous commands by using the up and down arrow keys on the keyboard.

In Section A.5, we will show how MATLAB can be used to generate plots, which can be presented in multiple figure windows. The clear command does not affect those windows so one can close (remove) all old figure windows and start anew by typing:

close all

A.2 Vectors

Vectors are frequently used in statics. In MATLAB, a vector can be generated either as a row vector or a column vector. For example, suppose a force vector has Cartesian components $F_x = 3$ lb, $F_y = 4$ lb, $F_z = 12$ lb. Then, we could enter this force in MATLAB as a row vector by placing the values of the components in square brackets [ ], separated by spaces:

```
F = [3 4 12];  % we can also use commas between values as
               % F = [3, 4, 12];
```

We can instead input this vector as a column vector by using semicolons between values. The semicolons tell MATLAB to start a new row:

```
F = [3; 4; 12]     % column vector. Output is displayed here
F = 3
    4
    12
```

We can compute the *magnitude of a vector* with the norm function. For our force F, we find

```
norm(F)
ans = 13
```

The magnitude of the force is 13 pounds. Do not confuse the magnitude of a vector with its length. The length of a vector in MATLAB is simply the number of its components. For example, if we have a two-dimensional vector F2 = [3 4], then the MATLAB length and norm functions give:

```
F2 = [3 4];
length(F2)
ans = 2

norm(F2)
ans = 5
```

because there are two components, and the magnitude of the force is 5. The norm function is used to generate unit vectors since we get a unit vector by dividing a vector by its magnitude. Consider a force vector F (written as a row vector), for example:

```
F = [3 4 12];
eF = F/norm(F)
eF = 0.2308   0.3077   0.9231
```

so eF is a unit vector in the direction of F. We can check that indeed eF is a unit vector:

```
norm(eF)
ans = 1
```

We can recover the original vector force by writing it as its magnitude times this unit vector:

```
Fv = norm(F)*eF
Fv = 3   4   12
```

Vectors in MATLAB can be easily added. Suppose we have two force vectors Force1 and Force2 and we want to add them. We simply write

```
Force1 = [100 200 300];
Force2 = [50 100 100];
Force1 + Force2
ans = 150   300   400
```

There are several ways that we can multiply vectors. One way is with the dot product. MATLAB has a built-in function dot that performs the dot product. One application in statics of the dot product is to find the Cartesian scalar component of a force in a given direction. For example, suppose we want to find the Cartesian scalar component of the vector Force1 that acts in the direction of Force2 for the two forces we have just defined above. That scalar component is the dot product of Force1 with a unit vector along Force2. In MATLAB, we can easily compute the unit vector along Force2, eF2, and then find the component of Force1 along eF2 with the dot product. We will call that component F1c:

```
eF2 = Force2/norm(Force2);
F1c = dot(Force1, eF2)
F1c = 366.6667
```

MATLAB also has a way to multiply vectors element-by-element with a special multiplication operator .* (dot star) If, for example, we have

```
v1 = [3 2 5];
v2 = [1 2 3];
v1.*v2
ans = 3   4   15
```

The dot product of two vectors is just the sum of the products of its Cartesian components. Thus, if we use element-by-element multiplication of two vectors followed by the MATLAB function sum which sums up the products of the elements, we will also get the dot product.

Consider our previous example of computing the component of Force1 in the direction of Force2. We could also obtain that component through

```
sum(Force1.*eF2)
ans = 366.6667
```

Element-by-element multiplication and other element-by-element operations have many other uses in MATLAB that we will discuss later.

Another way we can multiply vectors is with the vector cross product. In statics, for example, if we have a force vector given by Force = [100 200 50] and a position vector from some point P to that force, position = [2 3 4], then the moment of the force about P is the cross product of the position vector with the force vector, which we can get with the built-in MATLAB function cross:

```
Force = [100  200  50];
position = [2  3  4];
moment = cross(position, Force)
moment = -650   300   100
```

The MATLAB function cross requires that the vectors in the cross product both be three-dimensional vectors. Thus, if we have a two-dimensional problem where the forces and position vectors only have components in, say, the x- and y-directions, then we must include zero z-components for those vectors before we can compute the moment with cross product. Here is an example:

```
Force2D = [20  30  0];
position2D = [2  4  0];
moment2D = cross(position2D, Force2D)
moment2D = 0   0  -20
```

The moment in such a two-dimensional problem will always be plus or minus z-direction, as found here.

In some cases, we may want to find one or more components of a given vector. Suppose, for example, we have a vector v = [5 6 3 8 7 9]. We can find, say, the fourth component by entering v(4):

```
v = [5  6  3  8  7  9];
v(4)
ans = 8
```

If we want a range of components, we can indicate the range with a colon between the start and end values of the range we want:

```
v(1:3)
ans = 5 6 3
```

If we want all the components from a given component to the last (end) component, we can enter:

```
v(3:end)
ans = 3 8 7 9
```

In MATLAB, if we have the symbolic math toolbox, we can also deal with symbolic vectors. For example, suppose we want to write the components of force vectors and position vectors symbolically. We must first declare those components as symbolic variables and then we can manipulate them symbolically. For example:

```
syms Ax Ay Az Bx By Bz rx ry rz
A = [Ax Ay Az];
B = [Bx By Bz];
r = [rx ry rz];
```

sets up components of two symbolic vectors A and B (which might be force vectors, for example) and components of a vector r that might be a position vector. We can add vectors symbolically:

```
A + B
  ans = [Ax + Bx, Ay + By, Az + Bz]
```

and we can implement the dot product symbolically as

```
sum(A.*B)
ans = Ax*Bx + Ay*By + Az*Bz
```

Similarly, for the symbolic cross product of r and A:

```
cross(r, A)
ans = [Az*ry - Ay*rz, Ax*rz - Az*rx, Ay*rx - Ax*ry]
```

which you can verify gives the correct components for the moment vector. If we try to calculate the dot product of two symbolic vectors with the dot function, we find

```
dot(A, B)
  ans = Bx*conj(Ax) + By*conj(Ay) + Bz*conj(Az)
```

which looks unusual, where conj here indicates "complex conjugate". If we have a complex numerical or symbolic variable a + ib, its complex conjugate is a − ib, where $i = \sqrt{-1}$. For real numerical and symbolic variables, there is no difference between a number and its complex conjugate, so our above result from the dot product is correct, but it is more complex looking than necessary since in statics we do not deal with complex quantities. When using the dot product with symbolic variables, we can declare those variables to be real when they are first created so that the dot product appears in its normal form. For example:

```
syms Ax Ay Az Bx By Bz real
A = [Ax Ay Az];
B = [Bx By Bz];
dot(A, B)
ans = Ax*Bx + Ay*By + Az*Bz
```

A.3 Matrices

Matrices are rectangular or square arrays of components. For a numerical array, there are M rows of numbers and N columns of numbers in the array. The dimension of a matrix is said to be $M \times N$ (number of rows by number of columns). Vectors, therefore, are either a $1 \times N$ array (a row vector with N components) or a $M \times 1$ array (a column vector with M components). We can enter matrices easily in MATLAB in the same fashion as vectors:

```
matrix = [4 0 3; 0 3 5; 3 −5 7]
matrix = 4    0    3
         0    3    5
         3   −5    7
```

where the semicolon is again used to start a new row. We can access individual components by giving the row number and column. For the element in the second row and third column, for example:

```
matrix(2, 3)
ans = 5
```

If we want all the elements in a row or column, we can use a colon to indicate that request. Here, for example, are all the elements in the second column of the array:

```
matrix(:,2)
ans = 0
      3
     -5
```

and here are all the elements in the third row:

```
matrix(3, :)
ans = 3 -5 7
```

There are some useful matrix functions. Here are a few functions for the matrix M:

size(M)	returns number of rows, nr, and number of columns, % nc, as a vector [nr, nc]
trace(M)	trace of M (sum of diagonal terms)
det(M)	determinant of M
M.'	transpose of M (interchange rows and columns).

If M is real, then we can use M' instead to obtain the transpose. However, if M is complex or symbolic, then M' will interchange rows and columns and perform a complex conjugation. Thus, it is advisable to always use .' for the transpose or, in the case of symbolic variables, to declare them to be real when they are created, as mentioned previously. The transpose is often written in texts as M^T.

Let's look at these functions with the matrix of our previous example:

```
size(matrix)
ans = 3  3
trace(matrix)
ans = 14
det(matrix)
ans = 157
matrix.'    % can also use M' since M is real
ans = 4  0   3
      0  3  -5
      3  5   7
```

The transpose can also be used to turn row vectors into column vectors (and vice versa):

```
u = [1  2  3  4];
u .'   % can also use u' since u is real
ans = 1
      2
      3
      4
```

The determinant function can be used in several ways in statics. As indicated in Chapter 2, we can write the cross product symbolically by placing the symbolic unit vectors in the first row of a determinant matrix and the position vector components in the second row and force components in the third row. Here is an example:

```
F = [50  20  10];          % force vector
r = [2  3  1];             % position vector
syms ex ey ez              % symbolic unit vectors
mat = [ex ey ez; r; F]     % symbolic matrix for determinant
mat = [ex, ey, ez]
      [2, 3, 1]
      [50, 20, 10]
M = det(mat)               % moment of the force is the determinant
M = 10*ex + 30*ey - 110*ez % the moment in symbolic form
```

We can also compute the scalar component of a moment in a given direction, called the moment of a force about a line, as done in Chapter 2, by using a determinant. Suppose, for example, we want to find the scalar component of the moment of a force about a line AB, where u = [3/13 4/13 12/13] is a *unit vector* along that line. We can find this component by placing the components of the unit vector, the position vector (from a point on the line to the force), and the force in the first, second, and third rows of the determinant matrix, respectively. Here is an example with the force and position vector just given:

```
u = [3  4  12]/13;          % unit vector along line AB
mat2 = [u; r; F]
mat2 = 0.2308   0.3077   0.9231
       2.0000   3.0000   1.0000
      50.0000  20.0000  10.0000
Mab = det(mat2)
Mab = -90                   % scalar moment component along line AB
```

There are also some special matrices that can easily be constructed:

zeros(m,n)   matrix of all zeros with m rows and n columns
ones(m,n)   matrix of all ones with m rows and n columns
eye(m,n)   identity matrix with m rows and n columns

Here are some examples:

zeros(3,3)
ans = 0 0 0
      0 0 0
      0 0 0
ones(3,3)
ans = 1 1 1
      1 1 1
      1 1 1
eye(3,3)
ans = 1 0 0
      0 1 0
      0 0 1

Matrices can be added or subtracted by adding or subtracting each corresponding element in the arrays:

A = [1 2; 3 4]
A = 1 2
    3 4
B = [3 3; 5 5]
B = 3 3
    5 5
A + B
ans = 4 5
      8 9

A matrix can also be multiplied by a scalar, which multiplies all the elements by that scalar. Consider, for example, the matrix B we just defined:

2*B
ans = 6   6
      10 10

Matrices, like vectors, can be multiplied element by element with the ".*" multiplication symbol. For our A and B matrices, for example, we have

```
A.*B
ans = 3   6
     15  20
```

"Ordinary" matrix multiplication (which is done in MATLAB with the * multiplication symbol), however, is used more frequently in statics than element-by-element multiplication and follows a more detailed process. Let A_{mn} be the element in an array, A, of dimensions M×N at the mth row position and nth column position. Similarly, for a matrix B of dimensions N×K, let B_{nk} be the element at the nth row and kth column position for that matrix. Then the matrix product of A and B, C = A*B, is a M×K matrix defined for each element of C as

$$C_{mk} = \sum_{n=1}^{N} A_{mn} B_{nk}$$

For this product to be meaningful, we must have the number of columns of A be equal to the number of rows of B but the other dimensions can be different. Here is an example:

```
A = [2 3; 1 5];
B = [2 2; 6 3];
C = A*B
C = 22  13
    32  17
```

This result comes from our multiplication rule:

$$C_{11} = A_{11}B_{11} + A_{12}B_{21} = (2)(2) + (3)(6) = 22$$
$$C_{12} = A_{11}B_{12} + A_{12}B_{22} = (2)(2) + (3)(3) = 13$$
$$C_{21} = A_{21}B_{11} + A_{22}B_{21} = (1)(2) + (5)(6) = 32$$
$$C_{22} = A_{21}B_{12} + A_{22}B_{22} = (1)(2) + (5)(3) = 17$$

Since vectors are one-dimensional row or column matrices, they also follow the matrix multiplication rule. Thus, consider a row vector vr and the same vector written as a column vector vc:

```
vr = [1 2 3 4];
vc = [1;2 ; 3 ; 4];
```

If we form the product d = vr*vc then we are multiplying a (1×4) vector by a (4×1) vector, giving a 1×1 vector d, which is just a scalar:

d = vr*vc
d = 30

This result is also the dot product, and since the vector vr and vc are just the same vector written as row and column vectors, the result is also in this case the square of the magnitude of the vector. Now, considering reversing the order, p = vc*vr so we are multiplying a (4×1) vector by a (1×4) vector, which is possible since the sizes follow the rule of matrix multiplication. In this case, we get a (4×4) matrix for p where the elements of p are $p(m, n) = vc(m)*vr(n)$:

p = vc*vr
p = 1 2 3 4
    2 4 6 8
    3 6 9 12
    4 8 12 16

We can also multiply matrices by vectors if we follow the matrix rule for multiplication:

C = [1 2; 3 4]
C = 1 2
    3 4
x = [5; 6]
x = 5
    6
b = C*x
b = 17
    39

Here C is a (2×2) matrix, x is a (2×1) vector, so b is a (2×1) vector. The matrix-vector equation C*x = b is equivalent to the linear system of equations:

$$C_{11}x_1 + C_{12}x_2 = b_1$$
$$C_{21}x_1 + C_{22}x_2 = b_2$$

which is the type of equations we get frequently in statics from the equations of equilibrium, where the x-vector may be unknown forces or moments and the b-vector may be known forces or moments. In that case, if we define the components of the matrix C and the vector b, we

can solve for the unknowns x in MATLAB using the "backslash"
operator "\". For example:

```
C = [1 2; 3 7];
b = [1; 1];
x = C\b
x = 5.0000
   -2.0000
```

which you can verify satisfies these equations.

In some cases, we may want to eliminate rows or columns from a
matrix. We can do this with the use of the colon notation and an empty
set of brackets, i.e., [ ]. For example, suppose we have the matrix A:

```
A = [1 2 3; 4 5 6; 7 8 9]
      1   2   3
A = 4   5   6
      7   8   9
```

Suppose we want to eliminate the second row. We can do that as follows:

```
A(2,:) = [ ]
A = 1 2 3
    7 8 9
```

Now, suppose we want to eliminate the first column. We can do that in a
similar fashion:

```
A(: ,1) = [ ]
A = 2 3
    8 9
```

We will see uses of such eliminations, for example, when we discuss
modifying the equilibrium matrices found in statically indeterminate
problems.

A.4 Solutions of Simultaneous Linear Equations

In statics, we solve equilibrium equations for unknown forces or
moments. These equations are usually a set of simultaneous linear
equations. In this section, we will examine the various ways we can set
up and solve the equilibrium equations, using numerical and symbolic

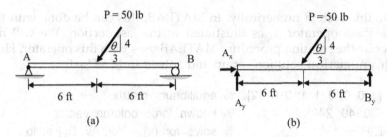

FIGURE A.1
(a) A simply supported beam. (b) The free body diagram.

methods. It is useful to consider a specific example, so we will use the beam problem shown in Figure A.1 as representative of an equilibrium problem.

A.4.1 Numerical Solution

First, we will obtain a completely numerical solution where we give the length of the beam as 12 ft, and let $P = 50$ lb. The angle θ is also known, where $\tan\theta = 4/3$, as shown in Figure A.1. The equations of equilibrium are then

$$\begin{aligned}
\Sigma F_x &= 0 \quad A_x - (3/5)(50) = 0 \\
\Sigma F_y &= 0 \quad A_y + B_y - (4/5)(50) = 0 \\
\Sigma M_{Az} &= 0 \quad 12B_y - (6)(4/5)(50) = 0
\end{aligned} \tag{A.1}$$

We can solve these equations easily by hand but let's instead treat them in a general manner. We can write them in matrix-vector terms as an equilibrium matrix, [E], and an applied force column vector $\{P\}$, where $[E]\{x\} = \{P\}$, with $\{x\}$ being a column vectors of unknowns $\{x\} = [A_x; A_y; B_y]$. We need to fill in the equilibrium matrix and applied force vector, giving

$$\begin{bmatrix} 1 & 0 & 0 \\ 0 & 1 & 1 \\ 0 & 0 & 12 \end{bmatrix} \begin{Bmatrix} A_x \\ A_y \\ B_y \end{Bmatrix} = \begin{Bmatrix} 30 \\ 40 \\ 240 \end{Bmatrix} \tag{A.2}$$

which are three equations in three unknowns since this is a statically determinate problem, i.e., we can solve it by use of equilibrium only. The equilibrium matrix is a 3×3 square matrix and if the equilibrium equations are also independent equations, as they are here, we can solve this system numerically. In MATLAB, this can be done with the back slash operator \, as illustrated in the last section. We will not describe the solution procedure MATLAB uses with this operator. Here is this numerical solution set up and solved in MATLAB:

```
E = [1 0 0; 0 1 1; 0 0 12];  % equilibrium matrix
P = [30; 40; 240];           % known force column vector
x = E\P                      % solve for {x} = [Ax; Ay; By] in lb
x = 30
    20
    20
```

A.4.2 Symbolic Solution in Terms of Numerical Variables

We can also treat this problem in a symbolic fashion where we write the equations of equilibrium, Eq. (A.1), in the same form in MATLAB as we place on paper, but in terms of symbolic variables:

```
syms Ax Ay By                    % symbolic unknowns
Eq(1) = Ax − (3/5)*(50) == 0;    % Equilibrium equations in a symbolic vector Eq
Eq(2) = Ay + By − (4/5)*(50) == 0;
Eq(3) = 12*By − (6)*(4/5)*(50) == 0;
S = solve(Eq)                    % solve symbolic system
S = struct with fields: Ax: 30
                        Ay: 20
                        By: 20
```

We used the logical symbol == to symbolically represent the equals sign and we placed the equilibrium equations in a symbolic vector that we called Eq. The MATLAB function solve then uses symbolic algebra to obtain a symbolic solution of these equations.

[Note: It is not necessary to include the == 0 in the equations. The solve function still works the same. We will see an example later in this appendix when we convert algebraic equilibrium equations to an equilibrium matrix.]

We have placed the result in a variable S which is a MATLAB structure S containing fields S.Ax, S.Ay, S.By. The values in these fields are just the forces we solved for. We can access any of these field values by typing, for example, S.Ax

```
S.Ax
ans = 30
```

If you want to know more about MATLAB structures, you can examine the MATLAB documentation. Here, we only need to know how to access the content of the structure, as just illustrated. You should realize, however, that the values obtained are still symbolic values. We can easily change them to double precision numerical values through the MATLAB function double:

Forces = double([S.Ax S.Ay S.By])
Forces = 30   20   20

where the symbolic values of the structure fields are placed in a vector which is then used as the argument of the double function.

This symbolic approach is very attractive for statics problems because (1) we simply write on paper the equilibrium equations from our free body diagrams, (2) place them in MATLAB in a form almost identical to our written equations, and then (3) solve them directly in that form with the solve function. We do not have to construct an equilibrium matrix or a known force vector and the symbolic names of the unknowns are attached to the symbolic values of the solution obtained. It is usually good to convert the symbolic answers to numerical values, as in some problems the symbolic answers are exact answers written in terms of fractions of very large integers, whose values are difficult to understand.

A.4.3 Symbolic Solution in Terms of Symbolic Variables

We can also obtain a solution of the equilibrium equations for the symbolic unknowns in terms of other symbolic variables. This is especially useful if we want to examine the solution behavior as those symbolic variables change their values. For example, suppose we change the angle of the applied force (see Figure A.1) to be the symbolic angle t (for θ, in degrees). Then, the equilibrium equations can be written as

$$\sum F_x = 0 \quad A_x - \cos(t)(50) = 0$$
$$\sum F_y = 0 \quad A_y + B_y - \sin(t)(50) = 0 \quad\quad\quad (A.3)$$
$$\sum M_{Az} = 0 \quad 12B_y - (6)\sin(t)(50) = 0$$

and in MATLAB we can find a symbolic solution, but we must specify the symbolic variables we are seeking when we call the solve function since there are more symbolic variables than those we want to solve for:

```
syms Ax Ay By t
Eq(1) = Ax - 50*cosd(t) == 0;        % enter equations with angle t in degrees
Eq(2) = Ay + By - 50*sind(t) == 0;
Eq(3) = 12*By - 6*sind(t)*50 == 0;
S = solve(Eq, [Ax Ay By])           % solve for Ax, Ay, By
S = struct with fields:
    Ax: 50* cos((pi*t)/180)
    Ay: 25* sin((pi*t)/180)
    By: 25* sin((pi*t)/180
```

Note that if we want to have the angle *t* to be in degrees, then when we enter the equilibrium equations in MATLAB we must use functions such as sind and cosd. When solving these equations symbolically, we see the symbolic solution uses the MATLAB sin and cos functions but enters the additional factor of pi/180 which converts degrees to radians, since the argument of the MATLAB sin and cos functions must be in radians.

Now, we can substitute values for the angle, *t*, in degrees, or a range of values that we can then plot. Let's do the latter to obtain the reaction force, By, as a function of the angle:

```
ang = linspace(0, 90, 500); % 500 angles ranging from 0 to 90 degrees
Byv = subs(S.By, t, ang);   % substitute angles into symbolic By variable t
plot(ang, Byv)              % plot the variable Byv versus the angles
xlabel('angle, degrees')
ylabel('By')
```

The plot of Byv versus θ is then given in Figure A.2

In statics, the problems usually solved are statically determinate problems where the number of equations equals the number of unknown forces/moments and the equilibrium matrix is a square matrix. In problems like the beam shown in Figure A.3 (a), however, there are more unknowns than there are equilibrium equations and thus the problem is statically indeterminate, i.e., equilibrium alone is not able to solve the problem. The equilibrium matrix is now a rectangular matrix. From Figure A.3 (b), the equilibrium equations are

$$\sum F_x = 0 \quad A_x - (3/5)(50) = 0$$
$$\sum F_y = 0 \quad A_y + B_y + C_y - (4/5)(50) = 0 \qquad \text{(A.4)}$$
$$\sum M_{Az} = 0 \quad 6C_y + 12B_y - (6)(4/5)(50) = 0$$

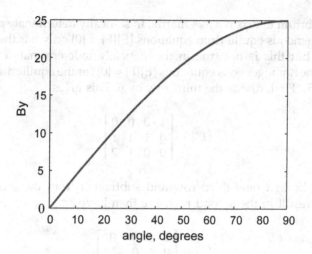

FIGURE A.2
The beam reaction force versus the angle θ.

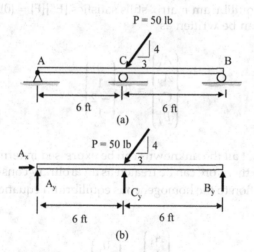

FIGURE A.3
(a) A statically indeterminate beam, and (b) its free body diagram.

which we can write as $[E]\{F\} = \{P\}$, where $\{F\} = [A_x; A_y; C_y; B_y]$ is a column vector and

$$[E] = \begin{bmatrix} 1 & 0 & 0 & 0 \\ 0 & 1 & 1 & 1 \\ 0 & 0 & 6 & 12 \end{bmatrix}, \quad \{P\} = \begin{Bmatrix} 30 \\ 40 \\ 240 \end{Bmatrix} \qquad (A.5)$$

The equilibrium matrix is a 3×4 matrix. In statically determinate problems, the homogeneous equilibrium equations [E]{F} = {0} only has the solution {F} = {0}, but this is not true in the statically indeterminate case. Let's examine the homogeneous equations [E]{F} = {0} for the equilibrium matrix of Eq. (A.5). First, divide the third row by 6. This gives

$$[E'] = \begin{bmatrix} 1 & 0 & 0 & 0 \\ 0 & 1 & 1 & 1 \\ 0 & 0 & 1 & 2 \end{bmatrix} \tag{A.6}$$

Second, take that new third row and subtract it from the second row. Place the result in the second row. We then have

$$[E''] = \begin{bmatrix} 1 & 0 & 0 & 0 \\ 0 & 1 & 0 & -1 \\ 0 & 0 & 1 & 2 \end{bmatrix} \tag{A.7}$$

Note that this equilibrium matrix stills satisfies [E'']{F} = {0}. In this form, the equations can be written as

$$\begin{Bmatrix} A_x \\ A_y \\ C_y \\ B_y \end{Bmatrix} = B_y \begin{Bmatrix} 0 \\ 1 \\ -2 \\ 1 \end{Bmatrix} \tag{A.8a}$$

which shows that all the unknowns can be expressed in terms an arbitrary B_y value, which therefore can be treated as an arbitrary constant, c. Thus, a non-trivial solution to the homogeneous equilibrium equations is the force vector

$$\begin{Bmatrix} A_x \\ A_y \\ C_y \\ B_y \end{Bmatrix} = \begin{Bmatrix} 0 \\ 1 \\ -2 \\ 1 \end{Bmatrix} \tag{A.8b}$$

or any constant times it. You can verify that fact by multiplying the original equilibrium matrix by the column vector of Eq. (A.8b). In MATLAB, there is a function null that can obtain the homogeneous solutions to the underdetermined equilibrium equations of statically indeterminate problems. Let's define this equilibrium matrix in MATLAB and call the null function:

E = [1 0 0 0; 0 1 1 1; 0 0 6 12];
null(E, 'r')
ans = 0
       1
     - 2
       1

which yields the same homogeneous solution. The 'r' option placed in the null function gives a result in terms of integers or fractions when the elements of the equilibrium matrix are whole numbers or ratios of whole numbers, as is the case here. If we use the null function without this argument, we find

null(E)
ans = 0
      0.4082
    - 0.8165
      0.4082

which is just our previous result times a constant, so it also is a solution to the homogeneous equations. The magnitude of this vector is one, as you can easily verify since the null function without the 'r' option generates a normalized homogeneous solution. We will not describe the details of how these solutions are found with the null function. You can find more information in the MATLAB documentation. However, as shown here, it is not difficult to find homogeneous solutions. See Chapter 10 on how homogeneous solutions can be used to solve statically indeterminate problems.

There may be more than one homogeneous solution in statically indeterminate problems. Consider the beam in Figure A.4 (a), for example. The equilibrium equations are

$$\sum F_x = 0 \quad A_x - (3/5)(50) = 0$$
$$\sum F_y = 0 \quad A_y + C_y + D_y + B_y - (4/5)(50) = 0 \qquad (A.9)$$
$$\sum M_{Az} = 0 \quad 4C_y + 8D_y + 12B_y - (6)(4/5)(50) = 0$$

which produces the equilibrium matrix for the column vector force $\{F\}= [A_x; A_y; C_y; D_y; B_y]$

$$[E] = \begin{bmatrix} 1 & 0 & 0 & 0 & 0 \\ 0 & 1 & 1 & 1 & 1 \\ 0 & 0 & 4 & 8 & 12 \end{bmatrix} \qquad (A.10)$$

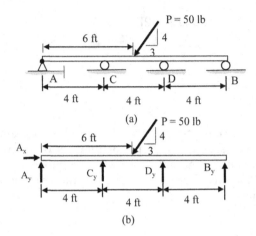

FIGURE A.4
(a) A statically indeterminate beam and (b) its free body diagram.

If we use the null function, we find

E = [1 0 0 0 0; 0 1 1 1 1; 0 0 4 8 12];
S = null(E, 'r')

$$
S = \begin{matrix} 0 & 0 \\ 1 & 2 \\ -2 & -3 \\ 1 & 0 \\ 0 & 1 \end{matrix}
$$

so there are two column vectors, each of which is a homogeneous solution.

   Homogeneous solutions to the equilibrium equations found in statically indeterminate problems play an important role in solving those problems. As mentioned previously, the details can be found in Chapter 10. Note that to get the homogeneous solutions to the equilibrium equations, we need the equilibrium matrix. If we write our equations out in symbolic form, then we can get that matrix with a MATLAB function equationsToMatrix. The function takes a set of equilibrium equations written symbolically (where the == 0 need not be included) and generates the equilibrium matrix [E] and the "force" vector {P} which satisfy the equilibrium equations [E]{U} = {P} The vector U contains the symbolic unknowns being solved for in its components. The calling sequence for the function is

[E, P] = equationsToMatrix(Eq, U);

As an example, consider the equilibrium equations in Eq. (A.9), which are written below without including the == 0. Here are the MATLAB statements for extracting the equilibrium matrix and the vector {P} (see Eq. (A.10)):

```
syms Ax Ay By Cy Dy
Eq(1) = Ax - (3/5)*50;
Eq(2) = Ay + Cy + Dy + By - (4/5)*(50);
Eq(3) = 4*Cy + 8*Dy + 12*By - 6*(4/5)*50;
U = [Ax Ay Cy Dy By];
[E, P] = equationsToMatrix(Eq, U);
E
E = [1,  0,  0,  0,  0]
    [0,  1,  1,  1,  1]
    [0,  0,  4,  8,  12]
P
P = 30
    40
    240
```

This equilibrium matrix can then be used to solve this statically indeterminate problem. The details of how this is done are in Chapter 10.

A.5 Element-by-Element Operations, Plotting, and Logical Vectors

A.5.1 Element-by-Element Operations

We discussed the element-by-element multiplication operator .* before. There are other operations that can be done element by element:

±	addition, subtraction
.*	multiplication
./	division
.^	exponentiation

Here are some examples:

```
x = [1 2 3 4];
y = x + 2
y = 3  4  5  6
z = x .^2
z = 1  4  9  16
f = x .*x .^2
f = 1  8  27  64
```

These are very useful in generating multiple values for functions that can then be directly plotted.

A.5.2 Plotting

MATLAB has a variety of useful plotting functions. We can, for example, perform simple plotting. We first generate a series of values and then plot a function of those values:

```
x = linspace(0, 2*pi, 100);
y = cos(x);
plot(x, y)
```

The plot output is shown in Figure A.5(a). The plot function chooses what it deems are "good" limits on the x- and y-axes to display the results. The range of x- and y-values in the plot can be changed with the axis function which has the form:

```
axis([xmin xmax ymin ymax]);
```

For example, if we want to limit the x-axis to only those values plotted and change the maximum and minimum y-values to be 2.0 and −2.0, we can enter:

```
axis([0 2*pi − 2 2])
```

which generates the plot shown in Figure A.5(b).

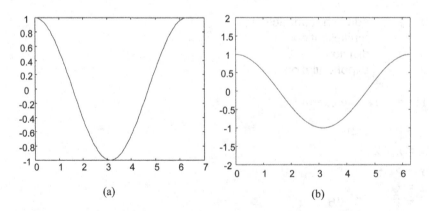

(a)   (b)

FIGURE A.5
(a) A simple plot. (b) The same plot with different limits on the x- and y-axes.

We can place multiple plots in the same plot window in multiple ways. One way is by successively giving the *x*- and *y*-values in the same plot call:

```
x = linspace(0, 2*pi, 100);
y1 = cos(x);
y2 = sin(x);
plot(x, y1, x, y2)
```

Another way is to make a plot with one call to the plot function, keep the same plot window with the hold command, and then make a second plot call. The hold off command then releases that window so further plots will occur in new windows:

```
x = linspace(0, 2*pi, 100);
y1 = cos(x);
plot(x, y1)
hold on
y2 = sin(x);
plot(x, y2)
hold off
```

Either method gives the same plots shown in Figure A.6.

If one wants to place the second plot in a separate window one can also do that by typing:

```
figure(2)
```

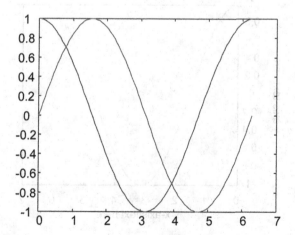

FIGURE A.6
Multiple plots in the same plot window.

before executing the second call to plot (the first plot by default is in the window figure(1)).

We can also add labels to the *x*- and *y*-axes and give a title to a plot. Here is an example:

```
x = linspace(0, 2*pi, 100);
y = cos(x);
plot(x, y)
xlabel('x - axis text here')
ylabel('y axis text here')
title('title text here')
```

A label or title text is, as shown, contained within single quotes. These represent MATLAB strings. The result is shown in Figure A.7. We can also use different line styles to distinguish different plots:

```
x = linspace(0, 2*pi, 100);
y1 = cos(x);
y2 = cos(x + 1);
y3 = cos(x + 2);
plot(x, y1, '- -', x, y2, ':', x, y3 ,'. -')
```

A few different line styles are shown in Figure A.8.

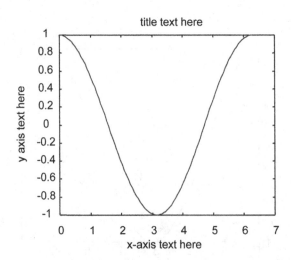

FIGURE A.7
Adding labels to the *x*- and *y*-axes and a plot title.

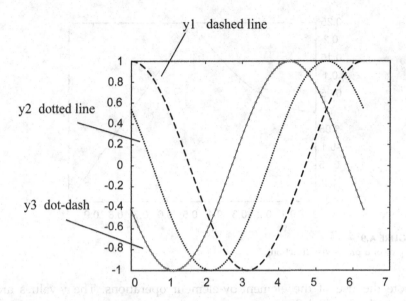

FIGURE A.8
Plotting with different line styles.

A.5.3 Logical Vectors

Logical (0–1) vectors are vectors having values of either one or zero depending on whether a logical statement is true or false, respectively. For example, let's define an x vector and then type either $x > 3$ or $x <= 4$. MATLAB interprets these to be logical statements and returns the appropriate vector of ones and zeros:

```
x = [1 2 3 4 5];
x > 3
ans = 0  0  0  1  1
x < = 4
ans = 1  1  1  1  0
```

Logical vectors are very useful for plotting "piece-wise" functions whose behavior changes over different ranges of values. In the beam problems of Chapter 8, for example, the internal shear force and bending moments exhibit such piece-wise behavior. Here is a simple example where a function is a quadratic function over a given range and linear over another:

```
x = linspace(0, 1, 500);
y = (x.^2).*(x < 0.5) + (0.75 − x).*(x > =0.5);
plot(x, y)
```

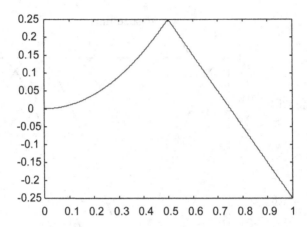

FIGURE A.9
A plot of a piece-wise function.

Note the use of the element-by-element operations. The *y*-values are shown in Figure A.9.

A.5.4 Constants eps and inf

At the beginning of this appendix, we mentioned that there are built-in constants eps and inf in MATLAB. These can be useful for evaluating and plotting functions where the functions are needed for values of zero or infinity. For example, the sin*x*/*x* function is a well-behaved function (see Figure A.10) but at *x* = 0 a formal evaluation gives 0/0, which is

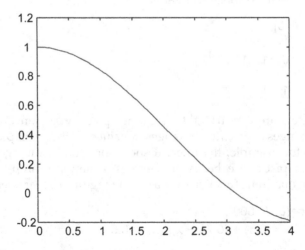

FIGURE A.10
The sin*x*/*x* function.

undefined. However, if at $x = 0$ we let $x =$ eps, which is a very small number (approximately 10^{-16}), then the limit of the function can be calculated as 1, which is correct since near $x = 0$ the sine function behaves like x so $\sin(x)/x = x/x = 1$ at $x = 0$. Here is this example in MATLAB:

```
x = linspace(0, 4, 100);
y = sin(x)./x;
Warning: Divide by zero.   % direct evaluation give 0/0 so we get a warning
x = x + eps*(x = = 0);
y = sin(x)./x;             % now we get the correct limit
plot(x, y)
```

We may also need to evaluate a function $f(x)$ when x goes to infinity. We cannot represent infinity numerically but we can get the proper limit with the use of inf. Here is an example:

```
x = inf;
y = 3/(1 + 4/x)
y = 3
```

A.6 Functions and Scripts

MATLAB functions are defined in the MATLAB editor and saved as m-files. The name of the m-file is the name of the function with a .m extension, e.g., myfunction.m. As shown in the example below, functions can have multiple outputs (as well as multiple inputs).

Suppose the following function is written in the editor and saved as test.m:

```
function [y, z] = test(x)
y = x.^2;
z = 10*exp(-x)
```

Then, in the MATLAB command window, we can define a vector of x-values and call the function. We then plot the two outputs of the function. The plot is shown in Figure A.11.

```
x = linspace(0, 2, 100);
[s, t] = test(x);
plot(x, s, x, t)
```

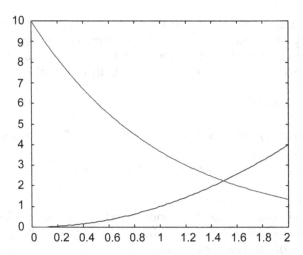

FIGURE A.11
Plots of the functions x^2 and $10e^{-x}$.

All the variables appearing within a function are local to that function, i.e., they do not change any similarly named variables in the MATLAB workspace. For example, the y- and z-vectors defined in the test function do not change any values of y and z, if they were defined previously in the MATLAB workspace. There are many more details about functions that could be covered but we will leave those details to other documentation sources.

A MATLAB script is a sequence of ordinary MATLAB statements, defined in the MATLAB editor and then saved as a file. The file has the name of the script followed by a .m extension, e.g., myscript.m. Typing the script name at the MATLAB prompt in the MATLAB command window causes the script to be executed. Variables in the script change any values of the variables of the same name that exist in the MATLAB workspace in the current MATLAB session. Here is a script that is written and saved as testscript.m. The script evaluates the same y and z variables found in the function test:

```
% testscript
x = linspace(0, 2, 100);
y = x.^2;
z = 10*exp(-x);
plot(x, y, x, z)
```

In the command window, entering:

```
testscript
```

produces the same plot shown in Figure A.11.

In this text, we will use MATLAB as a sophisticated "calculator" by entering commands directly into the MATLAB command window. Thus, we will not do MATLAB programming where it is necessary to write MATLAB functions or scripts. The one exception will be the use of anonymous (in-line) functions that are described in the next section.

A.7 Anonymous Functions

You can easily create functions directly in the MATLAB command window without saving them as m-files. This is done by generating what is called a function handle with the @ symbol. For example, we can write an anonymous function that cubes an x-variable:

```
cub = @(x) x.^3;
```

Here, cub is the function handle and x is the variable in the function. We can use cub to evaluate the function as done with an ordinary function

```
cub(2)
ans = 8
```

The main advantage of using anonymous functions is that you do not have to generate a separate m-file for a function that is relatively simple. An anonymous function can, like an ordinary function, contain other known variables:

```
a = 2;
b = 3;
func = @(x) a*x + b;
func(1)
ans = 5
```

However, if those known variables are changed, then the anonymous function must be re-evaluated.

A.8 Symbolic Calculations

We have given a few examples of doing vector calculations symbolically and symbolically obtaining the solutions of a set of linear equations. In this section, we want to outline some of the other symbolic functions and operations.

A.8.1 Symbolic Integration

If one wants to compute single integrals symbolically such as

$$I = \int_a^b f(x, c_1, c_2, ...)dx$$

where $(x, a, b, c_1, c_2, ...)$ are all symbolic variables, in MATLAB we can write

I = int(f, x, [a b])

where f is the integrand (a symbolic function), x is the integration variable (also a symbolic function), and the integration interval is from a to b. For example, let $f = 3x + c$. Then in MATLAB

```
syms x a b c
f(x, c) = 3*x + c;                     % symbolic function
I = int(f, x, [a b])                   % do integral of f from a to b
I(c) = -((a - b)*(3*a + 3*b + 2*c))/2
```

If we want to substitute specific values for the symbolic variables, we use the subs function and call the numeric output variable, In,:

```
In = subs(I, [a b c], [1 2 3]) % substitute a = 1,  b = 2, c = 3
In(c) = 15/2
```

For double integrations such as

$$I = \iint_R f(x, y, c_1, c_2)dxdy$$

over some two-dimensional region, R, we need to use nested int functions. The integration limits may be dependent on x or y and they generally depend on the order we perform the integration.

Here is an example taken from Example 11.1 in Chapter 11 where we are computing an area moment, I_{xx}, given by the integral

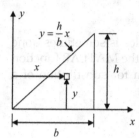

FIGURE A.12
A triangular area for demonstrating two-dimensional symbolic integration.

$$I_{xx} = \iint_A y^2 dx dy$$

for the triangular area A shown in Figure A.12 (which is the same as Fig. 11.6). The integrand here is $f = y^2$. Doing the y-integration first, we must integrate from $y = 0$ to the $y(x)$ at the straight line and then from $x = 0$ to $x = b$:

```
syms b h x y
Ixx = int(int(y^2, y,  [0 h*x/b]), x, [0 b])
Ixx = (b*h^3)/12
```

If instead we do the x-integration first, we must integrate from the $x(y)$ of the straight line to $x = b$ and then from $y = 0$ to $y = h$:

```
Ixx2 = int(int(y^2, x, [b*y/h b]), y, [0 h])
Ixx2 = (b*h^3)/12
```

A.8.2 Symbolic Differentiation

We can also do symbolic differentiation with the diff function. Here is a simple example:

```
syms a b c x
f = a*x^3 + b*x^2 + c*x;
diff(f, x)
   ans = 3*a*x^2 + 2*b*x + c
```

One can also perform two (or more) derivatives:

```
diff(f, x, 2)
   ans = 2*b + 6*a*x
diff(f, x, 3)
   ans = 6*a
```

A.8.3 Substitution

Substituting numerical values for symbolic variables is sometimes done in this text. Simple substitutions can be made with the MATLAB function subs in several different forms. The simplest form for substitution of a single symbolic variable is shown here:

```
syms a b c x
y = a*x^2 + b*x + c;
y1 = subs(y, a, 2)
y1 = 2*x^2 + b*x + c
```

If one wants to do multiple substitutions, this can be done by placing the variables and their values in vectors:

```
y2 = subs(y, [a  b], [2 1])
y2 = 2*x^2 + x + c
```

In some cases, one may want to place multiple values in one variable but single values in others. This often occurs, for example, when we want to plot a function versus one of the variables. In this case, we must place the values in a cell vector, which allows the values to be of different types (scalars, vectors, matrices). A cell vector is indicated by the use of curly brackets { }:

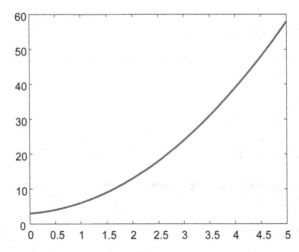

FIGURE A.13
A plot generated by substitution into a symbolic expression.

```
xv = linspace(0, 5, 200);
y3 = subs(y,[a  b  c  x],{2  1  3  xv});
plot(xv, y3)
```

The plot generated after substitution is shown in Figure A.13.

B MATLAB® Files

OBJECTIVES

- To define the MATLAB® singularity functions for use in beam problems.
- To describe where additional MATLAB® files can be found.

The functions built into MATLAB and its toolkits generally provide all the support needed for solving statics problems, but we have defined a few new functions. We added, for example, the singularity functions described in Chapter 8, which are extremely useful for plotting shear force and bending moment diagrams for beams. We described these basic functions in Fig. 8.19. Here, we give more complete commented listings of these singularity functions. They are available for download on the statics website for the text at www.eng-statics.org. In Chapter 3, we also mentioned a special function R_loc that can be used to locate resultants of force systems. It is also available for download at the website and is described in more detail there in the lecture notes.

step singularity function

```
function y = step_sf(x,a)
% y = step_sf(x, a) generates a singularity function that is zero for x < a
% and is one for x > a. For a concentrated load P the shear
% force is V = P*step_sf(x,a) while for a concentrated couple C
% we have the bending moment M = C*step_sf(x,a).

if x(1) == a        % if the step is located at the first point then let the
    y = (x>a);      % value at the first point be zero, otherwise
```

```
else                % let that value be unity. This allows
                    % MATLAB to plot
    y = (x>= a);    % the jump in the function properly.
end
```

linear singularity function

```
function y = lin_sf(x,a)
% y = lin_sf(x, a) generates a singularity function that is zero for x < a
% and varies like (x - a) for x > a. For a concentrated load P the bending
% moment is M = P*lin_sf(x,a) while for a uniform load w that starts at
% x = a we have the shear force V = w*lin_sf(x,a).

y = (x -a).*(x > a);
```

quadratic singularity function

```
function y = quad_sf(x,a)
% y = quad_sf(x, a) generates a singularity function that is zero for
x < a
% and varies like (x - a).^2 for x > a. For a uniform distributed load
% of intensity w force/unit length that starts at x = a, the bending
moment
% is M =(w/2)*quad_sf(x,a) while for a linearly increasing distributed
load
% whose slope is s force/unit length^2 the shear force is
% V = (s/2)*quad_sf(x,a).

y = ((x-a).^2).*(x >a);
```

cubic singularity function

```
function y = cubic_sf(x,a)
% y = cubic_sf(x, a) generates a singularity function that is zero for x < a
% and varies like (x - a).^3 for x > a. for a linearly increasing
% distributed load that starts at x = a and whose slope is
% s force/unit length^2 the bending moment is M = (s/6)*cubic_sf(x,a).

y = ((x-a).^3).*(x > a);
```

Problem Answers

The answers to all the problems at the end of the chapters are given except for those problems that are multiple choice problems or where the answers are given graphically such as in free body diagrams or in shear force and bending moment diagrams.

Chapter 1

P1.7 $\theta = 49.9°$, $R = 1077$ lb

P1.8 (a) $F_{OA} = -496.4$ lb, (b) $F_{OB} = -59.8$ lb, (c) $\mathbf{F} = 433\mathbf{i} + 250\,\mathbf{j}$ lb

P1.10 $F_{AB} = -77.2$ lb, $F_{AC} = 130$ lb

P1.11 $F_1 = 125$ N, $F_2 = -60.9$ N, $F_3 = 30$ N

P1.12 $\theta_x = 26.2°$, $\theta_y = 101°$, $\theta_z = 66.4°$

P1.13 $F_{OA} = -28.0$ N, $\mathbf{F}_{OA} = -24.3\mathbf{i} -14.0\mathbf{j}$ N

P1.14 (a) $\mathbf{R} = -473.5\mathbf{i} +380.6\mathbf{j} +408.5\mathbf{k}$ lb, (b) $\theta = 25.6°$

P1.15 $T_{CD} = 46$ N

R1.1 (a) $\mathbf{F} = 212\mathbf{i} -170\mathbf{j} +127\mathbf{k}$ lb, (b) $\theta_y =124°$, (c) $F_{AB} = -212$ lb

R1.2 (a) $\mathbf{R} = 113\mathbf{i} + 113\mathbf{j} + 229\mathbf{k}$ lb, (b) $\theta_x = 66.1°$, $\theta_y = 66.1°$, $\theta_z = 34.9°$

R1.3 (a) $\mathbf{T} = -36.7\mathbf{i} + 36.7\mathbf{j} +18.3\mathbf{k}$ lb, (b) $\theta_x = 132.8°$, $\theta_y = 48.19°$, $\theta_z = 70.53°$, (c) 0

R1.4 (1) $R = 509$ N, $\theta = 36.2°$, (2) $F_u = 474$ N, $F_v = 63.3$ N

Chapter 2

P2.6 $\mathbf{M}_o = -2\mathbf{k}$ kN-m

P2.8 (a) $\mathbf{M}_O = -725\mathbf{i} -649.5\mathbf{j}$ in-lb, (b) $|\mathbf{M}_{OA}| = 649.5$ in-lb

P2.9 $\mathbf{M}_O = -766\mathbf{i} +963\mathbf{j} +405\mathbf{k}$ in-lb, $M_{OA} = 451$ in-lb

P2.10 $F_x = F_y = 0$, $M_o = 314$ in-lb

R2.1 $\mathbf{C}_R = -1500\mathbf{i} + 2000\mathbf{j}$ in-lb, (b) $\mathbf{M}_{AE} = -1828\mathbf{i} +1632\mathbf{j}$ in-lb

R2.2 $\mathbf{M}_{AE} = -56,517\mathbf{i} + 23,684\mathbf{j}$ N-mm

R2.3 $\mathbf{M}_A = -520\mathbf{i} + 1350\mathbf{j} + 1400\mathbf{k}$ ft-lb

R2.5 (a) $\mathbf{C} = 11,400\mathbf{j} + 9,200\mathbf{k}$ in-lb, (b) $\mathbf{M}_{AD} = 2,880\mathbf{j} - 2,160\mathbf{k}$ in-lb

Chapter 3

P3.1 $R = 410$ lb, $\theta = 20.9°$, $d = 14.8$ in. to the right of A

P3.3 $R = 12.8$ kN, $\theta = -38.7°$, $d = 2.19$ m to the right of A

P3.4 $\mathbf{F} = -145.5\mathbf{i} + 218.3\mathbf{j} + 145.5\mathbf{k}$ lb, $\mathbf{C} = -437\mathbf{i} +582\mathbf{j} -1310\mathbf{k}$ ft-lb

P3.5 $\mathbf{R} = -354\mathbf{i}$ lb at $y = 0.893$ ft

P3.6 $\mathbf{R} = -5\mathbf{k}$ kN at $(3.2, 10.8, 0)$ m

P3.7 $\mathbf{R} = 75\mathbf{j} -150\mathbf{k}$ lb, $\mathbf{M}_o = -3000\mathbf{i} +1530\mathbf{j}$ in-lb

P3.8 $\mathbf{R} = 125\mathbf{i} -80\mathbf{k}$ lb, $z = 8.64$ in.

P3.9 $\mathbf{R} = 75\mathbf{j} + 150\mathbf{k}$ lb, $\mathbf{M}_{par} = 180\mathbf{j} +360\mathbf{k}$ in-lb at $x = 1.2$ in., $z = -9.6$ in.

P3.10 $\mathbf{R} = -47.8\mathbf{i} -59.2\mathbf{j} +129.6\mathbf{k}$ lb, $\mathbf{M}_{par} = 30.6\mathbf{i} +37.9\mathbf{j} - 83.0\mathbf{k}$ ft-lb at
  $x = 3.29$ ft, $y = 2.05$ ft

P3.11 $R = 2771$ lb at $x = 7.2$ ft

P3.12 $R = 125$ lb at $x = 4.33$ ft

P3.13 $z_c = 2.5$ in.

P3.14 $x_c = y_c = r/\pi$, $z_c = h/4$

R3.1 $\mathbf{R} = 166.6\mathbf{i} -20\mathbf{j}$ N at $x = 10.6$ m

R3.2 $\mathbf{R} = 20\mathbf{k}$ lb at $x = -11$ ft, $y = 10.5$ ft

R3.3 $\mathbf{R} = 10\mathbf{j}$ lb, $\mathbf{M}_{par} = 30\mathbf{j}$ ft-lb at $x = 1$ ft

R3.4 $R = 225$ lb at $x = 7.07$ ft

Chapter 4

P4.15 $A_x = 225$ N, $A_y = 450$ N, $B = 225$ N

P4.17 $\theta = 20.1°$

P4.18 $A = B = C = 116.7$ lb

P4.19 $T = 150$ lb, $A_y = -25$ lb, $A_z = 89.4$ lb, $B_y = 154.9$ lb, $B_z = -14.4$ lb

P4.20 $P = 840$ N, $A_x = 200$ N, $A_z = 480$ N, $B_x = 500$ N, $B_y = 0$, $B_z = 360$ N

R4.1 $T = 999$ N, $A_x = 905$ N, $A_y = 328$ N

R4.2 $A = 9.94$ kN, $M_A = 50.0$ kN-m

R4.3 $P = 588.4$ N

R4.4 $P = 150$ lb

R4.5 $T_B = 70.7$ lb

Chapter 5

P5.1 $T_{AB} = 393$ lb T, $T_{AC} = 552$ lb T, $T_{BC} = 920$ lb C, $A_x = -866$ lb, $A_y = -236$ lb, $C_y = 736$ lb

P5.2 $T_{AB} = 732$ lb C, $T_{AC} = 634$ lb T, $T_{BC} = 1360$ lb T, $A_x = 366$ lb, $B_x = 994$ lb, $B_y = 634$ lb

P5.3 $T_{AB} = 866$ lb C, $T_{AD} = 1000$ lb T, $T_{BD} = 1732$ lb C, $A_x = 0$, $A_y = -500$ lb, $B_y = 1500$ lb

P5.4 $T_{AB} = 208$ lb C, $T_{AD} = 6125$ lb T, $T_{BC} = 3542$ lb C, $T_{BD} = 3000$ lb T, $T_{CD} = 2125$ lb T, $A_x = -6000$ lb, $A_y = 167$ lb, $C_y = 2833$ lb

P5.5 $T_{AB} = 750$ lb C, $T_{AD} = 1000$ lb T, $T_{BC} = 3750$ lb C, $T_{BD} = 3606$ lb T, $T_{CD} = 3000$ lb T, $A_x = -1000$ lb, $A_y = 750$ lb, $C_y = 2250$ lb

P5.6 $T_{AB} = 3000$ lb T, $T_{AD} = 7071$ lb T, $T_{AE} = 0$, $T_{BC} = 5831$ lb T, $T_{BD} = 5000$ lb C,

$T_{CD} = 3000$ lb C, $T_{DE} = 8000$ lb C, $A_x = -5000$ lb, $A_y = -8000$ lb, $E_y = 8000$ lb

P5.7 $T_{AC} = 455$ lb T, $T_{AD} = 273$ lb T, $T_{AE} = 0$, $T_{CD} = 420$ lb C, $T_{DE} = 630$ lb C, $A_x = -630$ lb, $A_y = 350$ lb, $E_x = 630$ lb

P5.8 $T_{AB} = 0$, $T_{AD} = 200$ N T, $T_{BC} = 250$ N C, $T_{BD} = 250$ N C, $T_{CD} = 150$ N T, $A_x = 0$, $A_y = 200$ N, $C_y = 200$ N

P5.9 BH, DG, DF

P5.10 FH, DI, CI, BI

P5.11 $T_{AB} = 500$ lb C, $T_{BC} = 500$ lb C, $T_{BD} = 1166$ lb T, $T_{AD} = 640$ lb C, $T_{CD} = 640$ lb C, $A_x = -800$ lb, $A_y = 500$ lb, $A_z = 300$ lb, $B_y = -1000$ lb, $C_x = 800$ lb, $C_y = 500$ lb, $C_z = 300$ lb

P5.12 $T_{AB} = 0$, $T_{AC} = 0$, $T_{AD} = 424$ lb C, $T_{BC} = 189$ lb C, $T_{BD} = 267$ lb T, $T_{CD} = 18$ lb C, $A_x = -300$ lb, $A_y = 0$, $A_z = 300$ lb, $B_z = -314$ lb, $C_y = -200$ lb, $C_z = 14$ lb

P5.13 $T_{AB} = 300$ lb C, $T_{AE} = 721$ lb C, $T_{BC} = 400$ lb C, $T_{BE} = 781$ lb T, $T_{CD} = 300$ lb C, $T_{CE} = 781$ lb T, $T_{DE} = 721$ lb C, $A_x = -400$ lb, $A_y = 600$ lb, $A_z = 300$ lb, $D_x = 400$ lb, $D_y = 600$ lb, $D_z = 300$ lb, $B_y = -600$ lb, $C_y = -600$ lb

P5.14 $T_{CD} = 317$ lb C

P5.15 $F_{BG} = 165$ lb T

P5.16 $F_{GH} = 225$ lb T

P5.17 $F_{CD} = 733$ lb C

P5.18 $F_{CG} = 577$ lb T

P5.19 $F_{BC} = 867$ lb C

R5.4 $F_{BC} = 100$ kN C

R5.5 $F_{DG} = 60$ kN C

R5.6 $F_{CD} = 120$ kN C

Chapter 6

P6.2 $A_x = 16.9$ lb, $A_y = 67.5$ lb, $B_x = -16.9$ lb, $B_y = 22.5$ lb

P6.4 $A_x = -350$ N, $A_y = -394$ N, $B_x = -525$ N, $B_y = -394$ N, $D_x = 875$ N, $D_y = 788$ N

P6.6 $A_x = 375$ lb, $A_y = 125$ lb, $B_x = -375$ lb, $B_y = 125$ lb, $C_x = -500$ lb, $C_y = 0$

P6.7 $P = 3880$ lb, $A_x = 3880$ lb, $A_y = 2800$ lb, $F_{EI} = 847$ lb, $G_x = 847$ lb, $G_y = 2200$ lb, $F_{BF} = -2450$ lb, $D_x = -2248$ lb, $D_y = 1427$ lb

P6.9 $E = 44.4$ lb, $G = 356$ lb, $F_{BF} = 562$ lb C, $C_x = 178$ lb, $C_y = -133$ lb, $F_{DF} = 754$ lb C

P6.10 $B_x = -200$ lb, $B_y = 0$

P6.11 $E = 608$ lb, $A_x = -308$ lb, $A_y = 125$ lb, $D_y = 0$, $B_y = 0$, $B_x = -200$ lb, $D_x = 500$ lb, $C_x = -108$ lb, $C_y = 0$

P6.12 $T = 1374$ lb

R6.1 $C_y = 6.96$ kN

R6.2 $B_x = 890$ lb, $B_y = 200$ lb

R6.3 $A_y = 142$ lb, $B_x = 0$, $B_y = 250$ lb, $D_y = 8.33$ lb

R6.4 $C_x = -520$ lb, $C_y = 325$ lb

R6.5 $D_x = -37.5$ lb, $D_y = 130$ lb

Chapter 7

P7.7 $x_c = 25$ in., $y_c = 10$ in.

P7.8 $x_c = 26.55$ in., $y_c = 9.15$ in.

P7.9 $z_c = 6.25$ in.

P7.10 $x_c = z_c = 0$, $y_c = 2\,h/3$, $V = \pi r^2 h/2$

P7.11 $V = \pi r^2 h/3$, $x_c = y_c = 0$, $z_c = h/4$

P7.12 $x_c = 4$ in., $y_c = 3$ in.

P7.13 $x_c = 18.89$ mm, $y_c = 23.89$ mm

P7.14 $x_c = 0$, $y_c = 123.8$ mm

P7.15 $x_c = 56.1$ mm, $y_c = 56.1$ mm

P7.16 $R = 367.5$ lb, $x = 7.2$ ft

P7.17 $R = 800$ lb, $x = 7.2$ ft

P7.18 $x_c = 5$ in. $y_c = 4.875$ in., $z_c = 6.375$ in., $x_g = 5$ in., $y_g = 5.16$ in., $z_g = 5.97$ in.

R7.1 $x_c = 3.91$ mm, $y_c = 6.91$ mm

R7.2 $x_c = 2.4$ in., $y_c = 0.571$ in.

R7.4 $R = 4.57$ lb

Chapter 8

R8.1 $M = -5$ kN-m

R8.2 $M = 40 - 15x$ kN-m

R8.4 M = 33 kN-m

R8.5 $M = -5x^2 +33.5x -21$ kN-m

Chapter 9

P9.2 d > 500 mm

P9.3 P = 118.2 lb

P9.4 P =32.6 lb, tipping impends

P9.5 (a) P = 1 lb, tipping impends, (b) P = 1.2 lb, slipping impends

P9.7 P = 41.1 lb, slipping impends

P9.8 d = 3.18 ft

P9.9 θ = 10.4°

P9.10 θ = 35.5°

R9.1 P = 117.8 lb

R9.2 P = 5.45 lb

R9.3 θ = 31°

R9.4 P = 240 lb

R9.5 μ > 0.27

Chapter 10

P10.1 F_A = 9.1 lb, F_B = 109.1 lb, F_C = 81.8 lb, u_A = 0.0091 in., θ = 0.0007 rad

P10.2 F_{AC} = 902 lb, F_{BC} = −7.95 lb, F_{DC} = 525 lb

P10.3 F_{AB} = 208 lb C, F_{AD} = 2000 lb T, F_{BC} = 3542 lb C, F_{BD} = 3000 lb T, F_{CD} = 2000 lb C, A_x = −1875 lb, A_y = 166.7 lb, C_x = −4125 lb, C_y = 2833 lb

P10.4 F_{AB} = 750 lb C, F_{AD} = 1000 lb C, F_{BC} = 3750 lb C, F_{BD} = 3606 lb T, F_{CD} = 1000 lb T, A_x = 1000 lb, A_y = 750 lb, C_x = −2000 lb, C_y = 2250 lb

P10.5 F_{AB} = 5500 lb T, F_{AD} = 3536 lb T, F_{AE} = 2500 lb T, F_{BC} = 5831 lb T, F_{BD} = 2500 lb C, F_{BE} = 3536 lb C, F_{CD} = 3000 lb C, F_{DE} = 5500 lb C, A_x = −5000 lb, A_y = −8000 lb, E_y = 8000 lb

P10.6 F_{AB} = 2528 lb T, F_{AE} = 172 lb T, F_{AD} = 711 lb T, F_{BC} = 2080 lb T, F_{BD} = 345 lb T, F_{BE} = 448 lb C, F_{CD} = 1920 lb C, F_{DE} = 2467 lb C, A_x = −2880 lb, A_y = 1600 lb, E_x = 2880 lb, $u_{Ax} = u_{Ay}$ = 0 in., u_{Bx} = 0.0171 in, u_{By} = −0.0614 in., u_{Cx} = −0.0632 in., u_{Cy} = −0.3385 in., u_{Dx} = −0.0355 in., u_{Dy} = −0.0635 in., u_{Ex} = 0 in., u_{Ey} = −0.0021 in.

P10.7 F_{AB} = 74.4 N C, F_{AD} = 150.4 N T, F_{BC} = 312 N C, F_{BD} = 188 N C, F_{CD} = 112.8 N T, A_x = −74.4 N, A_y = 150.4 N, C_x = 74.4 N, C_y = 249.6 N

P10.8 F_{AB} = 41.6 N C, F_{AC} = 50 N T, F_{AD} = 172.3 N T, F_{BC} = 284.7 N C, F_{BD} = 215.3 N C, F_{CD} = 129.2 N T, A_x = 0 N, A_y = 200 N, C_y = 200 N

10.9 (a) T_{CA} = 1170 lb T, T_{CB} = 205 lb T, T_{CE} = 1215 lb C, T_{CD} = 22 lb C,
  A_x = −1021 lb, A_y = −255 lb, A_z = 511 lb, B_x = −179 lb, B_y = 45 lb, B_z =
  89 lb, D_x = 21 lb, D_y = 5 lb, D_z = 0, E_x = 1179 lb, E_y = −295 lb, E_z = 0 lb

(b) T_{CA} = 687 lb T, T_{CB} = 687 lb T, T_{CE} = 618 lb C, T_{CD} = 618 lb C, A_x =
  −600 lb, A_y = −150 lb, A_z = 300 lb, B_x = −600 lb, B_y = 150 lb, B_z = 300 lb,
  D_x = 600 lb, D_y = 150 lb, D_z = 0, E_x = 600 lb, E_y = −150 lb, E_z = 0 lb

R10.1 F_A = 18.75 T, F_B = 18.75 lb C, F_C = 75.0 lb

R10.2 F_A = 38.4 lb T, F_C = 28.8 lb C, B_x = −81.6 lb, B_y = −28.8 lb

R10.3 F_{CA} = 25 lb T, F_{CB} = 50 lb C, R_A = 25 lb, R_B = 50 lb

R10.4 F_B = 90 lb T, F_C = 90 lb T, A_x = 0, A_y = −30 lb

R10.5 F_1 = 100 lb T, F_2 = 50 lb C, B_x = 150 lb, B_y = 200 lb

Chapter 11

P11.4 I_{xx} = 40,000 mm^4

P11.5 I_{yy} = 14,881 mm^4

P11.6 I_{xy} = 20,833 mm^4

P11.7 I_{xx} = 810 in^4, I_{xy} = 425 in^4

P11.8 I_{xx} = 22.6×10^6 mm^4

P11.9 I_{yy} = 9.81×10^6 mm^4

P11.10 I_{xy} = −11.2×10^6 mm^4

P11.11 I_{xr} = 65.1 ft^4, I_{yr} = 208 ft^4, I_{xyr} = 76 ft^4

P11.12 I_{xr} = 249 in^4, I_{yr} = 2559 in^4, I_{xyr} = −705 in^4

P11.13 I_{p1} = 241 ft^4, I_{p2} = 32.3 ft^4, θ_{p1} = −36.6°, θ_{p2} = 53.4°

P11.14 I_{p1} = 2757 ft^4, I_{p2} = 51 ft^4, θ_{p1} = −75.7°, θ_{p2} = 14.3°

P11.15 I_O = 2.11 slug-in^2

R11.1 I_{xx} = 11.73×10^6 mm^4

R11.2 I_{xx} = 5.33×10^6 mm^4

R11.3 I_{p1} = 172.3 in^4, I_{p2} = 3.17 in^4, θ_{p1} = −14.3°, θ_{p2} = 75.7°

R11.4 I_{p1} = 11.75×10^6 mm^4, I_{p2} = 5.69×10^6 mm^4, θ_{p1} = −45°, θ_{p2} = 45°

Index

Printed in the United States
by Baker & Taylor Publisher Services